AF815038

METHODS IN MOLECULAR BIOLOGY

Series Editor
John M. Walker
School of Life and Medical Sciences
University of Hertfordshire
Hatfield, Hertfordshire, UK

For further volumes:
http://www.springer.com/series/7651

For over 35 years, biological scientists have come to rely on the research protocols and methodologies in the critically acclaimed *Methods in Molecular Biology* series. The series was the first to introduce the step-by-step protocols approach that has become the standard in all biomedical protocol publishing. Each protocol is provided in readily-reproducible step-by-step fashion, opening with an introductory overview, a list of the materials and reagents needed to complete the experiment, and followed by a detailed procedure that is supported with a helpful notes section offering tips and tricks of the trade as well as troubleshooting advice. These hallmark features were introduced by series editor Dr. John Walker and constitute the key ingredient in each and every volume of the *Methods in Molecular Biology* series. Tested and trusted, comprehensive and reliable, all protocols from the series are indexed in PubMed.

Structural Bioinformatics

Methods and Protocols

Edited by

Zoltán Gáspári

Information Technology and Bionics, Pázmány Péter Catholic University, Budapest, Hungary

Humana Press

Editor
Zoltán Gáspári
Information Technology and Bionics
Pázmány Péter Catholic University
Budapest, Hungary

ISSN 1064-3745 ISSN 1940-6029 (electronic)
Methods in Molecular Biology
ISBN 978-1-0716-0269-0 ISBN 978-1-0716-0270-6 (eBook)
https://doi.org/10.1007/978-1-0716-0270-6

© Springer Science+Business Media, LLC, part of Springer Nature 2020
Chapter 13 is licensed under the terms of the Creative Commons Attribution 4.0 International License (http://creativecommons.org/licenses/by/4.0/). For further details see license information in the chapter.
This work is subject to copyright. All rights are reserved by the Publisher, whether the whole or part of the material is concerned, specifically the rights of translation, reprinting, reuse of illustrations, recitation, broadcasting, reproduction on microfilms or in any other physical way, and transmission or information storage and retrieval, electronic adaptation, computer software, or by similar or dissimilar methodology now known or hereafter developed.
The use of general descriptive names, registered names, trademarks, service marks, etc. in this publication does not imply, even in the absence of a specific statement, that such names are exempt from the relevant protective laws and regulations and therefore free for general use.
The publisher, the authors, and the editors are safe to assume that the advice and information in this book are believed to be true and accurate at the date of publication. Neither the publisher nor the authors or the editors give a warranty, express or implied, with respect to the material contained herein or for any errors or omissions that may have been made. The publisher remains neutral with regard to jurisdictional claims in published maps and institutional affiliations.

Cover Caption: Cover image prepared using UCSF Chimera, developed by the Resource for Biocomputing, Visualization, and Informatics at the University of California, San Francisco, with support from NIH P41-GM103311.

This Humana imprint is published by the registered company Springer Science+Business Media, LLC part of Springer Nature.
The registered company address is: 233 Spring Street, New York, NY 10013, U.S.A.

Preface

Structural biology is built on the axiom that understanding biomolecular processes in detail requires explicit knowledge of the atomic-level structure of the macromolecules involved. However, the level of this knowledge can vary greatly and may substantially affect our thinking and ability to interfere with the biological actions of cells. The past decade has seen an unprecedented increase in the complexity of our description of biomolecules: on the one hand, larger and larger supramolecular assemblies can be studied in increasing detail; on the other hand, the quantitative description of the dynamical aspect of proteins and their complexes is now within our reach. The Nobel Prize awarded for cryo-electron microscopy in 2017 marks an important milestone in the former aspect, whereas the growing number of methods and applications of ensemble-based descriptions of both folded and unfolded proteins to account for their dynamic nature is proof of the latter.

The ever-increasing amount of structural information—as of writing these lines, the PDB holds nearly 150,000 entries—gives rise to many kinds of comparative analyses of structures as well as approaches to predict and evaluate protein-ligand interactions. In this book, Chapter 1 describes LiteMol, an interactive visualization tool with a number of practical features to select and analyze the details of interest in a given protein. Chapter 2 is about Bio3D-Web, designed for straightforward comparative analysis of related protein structures. The classic protein structure comparison method Dali is detailed in Chapter 3. CATH, one of the most fundamental resources in protein structure classification and functional annotation, is presented in Chapter 4. One aspect of understanding the residue-residue interaction networks within protein structures is analysis of thermal stability. The method HoTMuSiC, discussed in Chapter 5, offers a way to explore and design point mutations with respect to changes in melting temperature. A contact-based protein structure analysis tool, CAD-score, also applicable to protein-ligand complexes, is described in Chapter 6, whereas Chapter 7 is about a suite of graph-based approaches for the analysis of protein-ligand interactions. The method Wrap'n'Shake, explained in Chapter 8, is suitable for a comprehensive enumeration and analysis of possible ligand binding sites on protein surfaces. The growing number of high-resolution cryo-EM structures made it possible to determine the membrane interactions of transmembrane proteins based on experimental data as demonstrated in Chapter 9 describing the recently developed method MemBlob.

Protein structure determination has always relied on extensive computations and the search for synergy between a priori known structural features—such as bond lengths, angles, and other preferences—but today the power of combining measurements and calculations is more evident than ever when the structures of large protein complexes are determined by docking methods making use of experimental data. This is well exemplified by the methods PyDockSaxs and HADDOCK, capable of incorporating SAXS and cryoEM-based data, presented in Chapters 10 and 11, respectively. Other selected approaches of protein complex modeling making use of pairwise interactions or interaction graphs, CombDock and DockStar, are presented in Chapter 12. Chapter 13 describes the use of VAST+, a tool for comparative analysis of protein complexes.

NMR spectroscopy is clearly the best tool for experimental characterization of protein internal dynamics in solution. The comprehensive resource BioMagResBank is an invaluable complex resource of parameters determined by NMR spectroscopy. Its organization and usage is detailed in Chapter 14. Chapters 15 and 16 describe BME and CoNSEnsX$^+$, tools for the generation and analysis of ensemble-based descriptions of proteins that reflect dynamical features determined primarily by NMR spectroscopy.

By providing practical guidance in the usage of the tools listed above, we hope to provide the reader a state-of-the-art practical reference in the continuously changing and growing field of structural bioinformatics.

Budapest, Hungary ***Zoltán Gáspári***

Contents

Contributors

DAVID B. ASCHER • *Structural Biology and Bioinformatics, Department of Biochemistry and Molecular Biology, Bio21 Institute, University of Melbourne, Melbourne, VIC, Australia; Computational Biology and Clinical Informatics, Baker Heart and Diabetes Institute, Melbourne, VIC, Australia; Instituto René Rachou, Fundação Oswaldo Cruz, Belo Horizonte, Brazil; Department of Biochemistry, University of Cambridge, Cambridge, UK*

MÓNIKA BÁLINT • *Department of Pharmacology and Pharmacotherapy, Medical School, University of Pécs, Pécs, Hungary*

KUMARAN BASKARAN • *BMRB, Biochemistry Department, University of Wisconsin-Madison, Madison, WI, USA*

TONE BENGTSEN • *Structural Biology and NMR Laboratory, Department of Biology, Linderstrøm-Lang Centre for Protein Science, University of Copenhagen, Copenhagen N, Denmark*

KAREL BERKA • *RCPTM, Department of Physical Chemistry, Faculty of Science, Palacký University, Olomouc, Czech Republic*

PAU BERNADÓ • *Centre de Biochimie Structurale, CNRS, INSERM, Université de Montpellier, Montpellier, France*

ALEXANDRE M. J. J. BONVIN • *Computational Structural Biology Group, Bijvoet Centre for Biomolecular Research, Faculty of Science—Chemistry, Utrecht University, Utrecht, The Netherlands*

SANDRO BOTTARO • *Structural Biology and NMR Laboratory, Department of Biology, Linderstrøm-Lang Centre for Protein Science, University of Copenhagen, Copenhagen N, Denmark*

STEPHEN H. BRYANT • *National Center for Biotechnology Information, National Library of Medicine, National Institutes of Health, Bethesda, MD, USA*

GEORGINA CSIZMADIA • *Department of Biophysics and Radiation Biology, Semmelweis University, Budapest, Hungary; MTA-SE Molecular Biophysics Research Group, Hungarian Academy of Sciences, Budapest, Hungary*

NATALIE L. DAWSON • *Institute of Structural and Molecular Biology, University College London, London, UK*

DÁNIEL DUDOLA • *Information Technology and Bionics, Pázmány Péter Catholic University, Budapest, Hungary*

BIANKA FARKAS • *Department of Biophysics and Radiation Biology, Semmelweis University, Budapest, Hungary; MTA-SE Molecular Biophysics Research Group, Hungarian Academy of Sciences, Budapest, Hungary; Faculty of Information Technology and Bionics, Pázmány Péter Catholic University, Budapest, Hungary*

JUAN FERNÁNDEZ-RECIO • *Barcelona Supercomputing Center (BSC), Barcelona, Spain; Institut de Biologia Molecular de Barcelona (IBMB), Consejo Superior de Investigaciones Científicas (CSIC), Barcelona, Spain; Instituto de Ciencias de la Vid y del Vino (ICVV), Consejo Superior de Investigaciones Científicas (CSIC), Logroño, Spain*

TOSHIMICHI FUJIWARA • *PDBj-BMRB, Institute for Protein Research, Osaka University, Suita, Osaka, Japan*

ZOLTÁN GÁSPÁRI • *Information Technology and Bionics, Pázmány Péter Catholic University, Budapest, Hungary*

BARRY J. GRANT • *Section of Molecular Biology, Division of Biological Sciences, University of California, San Diego, La Jolla, CA, USA*
TAMÁS HEGEDŰS • *Department of Biophysics and Radiation Biology, Semmelweis University, Budapest, Hungary; MTA-SE Molecular Biophysics Research Group, Hungarian Academy of Sciences, Budapest, Hungary*
CSABA HETÉNYI • *Department of Pharmacology and Pharmacotherapy, Medical School, University of Pécs, Pécs, Hungary*
JEFFREY C. HOCH • *BMRB, Department of Molecular Biology and Biophysics, UConn Health, Farmington, CT, USA*
LIISA HOLM • *Faculty of Biological and Environmental Sciences and Institute of Biotechnology, University of Helsinki, Helsinki, Finland*
TAKESHI IWATA • *PDBj-BMRB, Institute for Protein Research, Osaka University, Suita, Osaka, Japan*
BRIAN JIMÉNEZ-GARCÍA • *Barcelona Supercomputing Center (BSC), Barcelona, Spain; Bijvoet Center for Biomolecular Research, Faculty of Science—Chemistry, Utrecht University, Utrecht, The Netherlands*
MALANCHA KARMAKAR • *Structural Biology and Bioinformatics, Department of Biochemistry and Molecular Biology, Bio21 Institute, University of Melbourne, Melbourne, VIC, Australia; Computational Biology and Clinical Informatics, Baker Heart and Diabetes Institute, Melbourne, VIC, Australia*
ESZTER KATONA • *Department of Biophysics and Radiation Biology, Semmelweis University, Budapest, Hungary; University College London, London, UK*
NAOHIRO KOBAYASHI • *PDBj-BMRB, Institute for Protein Research, Osaka University, Suita, Osaka, Japan*
JAROSLAV KOČA • *CEITEC—Central European Institute of Technology, Masaryk University Brno, Brno-Bohunice, Czech Republic; National Centre for Biomolecular Research, Faculty of Science, Brno-Bohunice, Czech Republic*
BERTALAN KOVÁCS • *Information Technology and Bionics, Pázmány Péter Catholic University, Budapest, Hungary*
GENJI KURUSU • *PDBj-BMRB, Institute for Protein Research, Osaka University, Suita, Osaka, Japan*
JEAN MARC KWASIGROCH • *Computational Biology and Bioinformatics, Université Libre de Bruxelles, Brussels, Belgium*
CHRISTOPHER LANCZYCKI • *National Center for Biotechnology Information, National Library of Medicine, National Institutes of Health, Bethesda, MD, USA*
KRESTEN LINDORFF-LARSEN • *Structural Biology and NMR Laboratory, Linderstrøm-Lang Centre for Protein Science, Department of Biology, University of Copenhagen, Copenhagen N, Denmark*
JOÃO P. V. LINHARES • *Instituto René Rachou, Fundação Oswaldo Cruz, Belo Horizonte, Brazil; Bioinformatics Program, Universidade Federal de Minas Gerais, Belo Horizonte, Brazil*
THOMAS MADEJ • *National Center for Biotechnology Information, National Library of Medicine, National Institutes of Health, Bethesda, MD, USA*
ARON MARCHLER-BAUER • *National Center for Biotechnology Information, National Library of Medicine, National Institutes of Health, Bethesda, MD, USA*
JOHN L. MARKLEY • *BMRB, Biochemistry Department, University of Wisconsin-Madison, Madison, WI, USA*

DIMITRI MAZIUK • *BMRB, Biochemistry Department, University of Wisconsin-Madison, Madison, WI, USA*
ADAM MIDLIK • *CEITEC—Central European Institute of Technology, Masaryk University Brno, Brno-Bohunice, Czech Republic; National Centre for Biomolecular Research, Faculty of Science, Brno-Bohunice, Czech Republic*
YOOCHAN MYUNG • *Structural Biology and Bioinformatics, Department of Biochemistry and Molecular Biology, Bio21 Institute, University of Melbourne, Melbourne, VIC, Australia; Computational Biology and Clinical Informatics, Baker Heart and Diabetes Institute, Melbourne, VIC, Australia*
KLIMENT OLECHNOVIČ • *Institute of Biotechnology, Life Sciences Center, Vilnius University, Vilnius, Lithuania*
CHRISTINE ORENGO • *Institute of Structural and Molecular Biology, University College London, London, UK*
DOUGLAS E. V. PIRES • *Structural Biology and Bioinformatics, Department of Biochemistry and Molecular Biology, Bio21 Institute, University of Melbourne, Melbourne, VIC, Australia; Computational Biology and Clinical Informatics, Baker Heart and Diabetes Institute, Melbourne, VIC, Australia; Instituto René Rachou, Fundação Oswaldo Cruz, Belo Horizonte, Brazil*
STEPHANIE PORTELLI • *Structural Biology and Bioinformatics, Department of Biochemistry and Molecular Biology, Bio21 Institute, University of Melbourne, Melbourne, VIC, Australia; Computational Biology and Clinical Informatics, Baker Heart and Diabetes Institute, Melbourne, VIC, Australia*
LUKÁŠ PRAVDA • *CEITEC—Central European Institute of Technology, Masaryk University Brno, Brno-Bohunice, Czech Republic; National Centre for Biomolecular Research, Faculty of Science, Brno-Bohunice, Czech Republic*
FABRIZIO PUCCI • *Computational Biology and Bioinformatics, Université Libre de Bruxelles, Brussels, Belgium*
PÂMELA M. REZENDE • *Instituto René Rachou, Fundação Oswaldo Cruz, Belo Horizonte, Brazil; Bioinformatics Program, Universidade Federal de Minas Gerais, Belo Horizonte, Brazil*
CARLOS H. M. RODRIGUES • *Structural Biology and Bioinformatics, Department of Biochemistry and Molecular Biology, Bio21 Institute, University of Melbourne, Melbourne, VIC, Australia; Computational Biology and Clinical Informatics, Baker Heart and Diabetes Institute, Melbourne, VIC, Australia*
PEDRO R. ROMERO • *BMRB, Biochemistry Department, University of Wisconsin-Madison, Madison, WI, USA*
MARIANNE ROOMAN • *Computational Biology and Bioinformatics, Université Libre de Bruxelles, Brussels, Belgium*
DINA SCHNEIDMAN-DUHOVNY • *School of Computer Science and Engineering and the Institute of Life Sciences, The Hebrew University of Jerusalem, Jerusalem, Israel*
DAVID SEHNAL • *CEITEC—Central European Institute of Technology, Masaryk University Brno, Brno-Bohunice, Czech Republic; National Centre for Biomolecular Research, Faculty of Science, Brno-Bohunice, Czech Republic*
MICHAEL SILK • *Structural Biology and Bioinformatics, Department of Biochemistry and Molecular Biology, Bio21 Institute, University of Melbourne, Melbourne, VIC, Australia; Computational Biology and Clinical Informatics, Baker Heart and Diabetes Institute, Melbourne, VIC, Australia*
LARS SKJÆRVEN • *Department of Biomedicine, University of Bergen, Bergen, Norway*

RADKA SVOBODOVÁ • *CEITEC—Central European Institute of Technology, Masaryk University Brno, Brno-Bohunice, Czech Republic; National Centre for Biomolecular Research, Faculty of Science, Brno-Bohunice, Czech Republic*
MIKAEL TRELLET • *Computational Structural Biology Group, Bijvoet Centre for Biomolecular Research, Faculty of Science—Chemistry, Utrecht University, Utrecht, The Netherlands*
GÁBOR E. TUSNÁDY • *"Momentum" Membrane Protein Bioinformatics Research Group, Institute of Enzymology, RCNS, Hungarian Academy of Sciences, Budapest, Hungary*
ELDON L. ULRICH • *BMRB, Biochemistry Department, University of Wisconsin-Madison, Madison, WI, USA*
GYDO VAN ZUNDERT • *Computational Structural Biology Group, Bijvoet Centre for Biomolecular Research, Faculty of Science—Chemistry, Utrecht University, Utrecht, The Netherlands*
WANDRÉ N. P. VELOSO • *Bioinformatics Program, Universidade Federal de Minas Gerais, Belo Horizonte, Brazil; Institute of Technological Sciences, Universidade Federal de Itajubá, Itabira, Brazil*
ČESLOVAS VENCLOVAS • *Institute of Biotechnology, Life Sciences Center, Vilnius University, Vilnius, Lithuania*
JONATHAN R. WEDELL • *BMRB, Biochemistry Department, University of Wisconsin-Madison, Madison, WI, USA*
HAIM J. WOLFSON • *Blavatnik School of Computer Science, Tel Aviv University, Tel Aviv, Israel*
JOICYMARA S. XAVIER • *Instituto René Rachou, Fundação Oswaldo Cruz, Belo Horizonte, Brazil; Bioinformatics Program, Universidade Federal de Minas Gerais, Belo Horizonte, Brazil; Institute of Agricultural Sciences, Universidade Federal dos Vales do Jequitinhonha e Mucuri, Unaí, Brazil*
HONGYANG YAO • *BMRB, Biochemistry Department, University of Wisconsin-Madison, Madison, WI, USA*
XIN-QIU YAO • *Department of Chemistry, Georgia State University, Atlanta, GA, USA*
MASASHI YOKOCHI • *PDBj-BMRB, Institute for Protein Research, Osaka University, Suita, Osaka, Japan*
DACHUAN ZHANG • *National Center for Biotechnology Information, National Library of Medicine, National Institutes of Health, Bethesda, MD, USA*

Chapter 1

Visualization and Analysis of Protein Structures with LiteMol Suite

David Sehnal, Radka Svobodová, Karel Berka, Lukáš Pravda, Adam Midlik, and Jaroslav Koča

Abstract

LiteMol suite is an innovative solution that enables near-instant delivery of model and experimental biomacromolecular structural data, providing users with an interactive and responsive experience in all modern web browsers and mobile devices. LiteMol suite is a combination of data delivery services (CoordinateServer and DensityServer), compression format (BinaryCIF), and a molecular viewer (LiteMol Viewer). The LiteMol suite is integrated into Protein Data Bank in Europe (PDBe) and other life science web applications (e.g., UniProt, Ensemble, SIB, and CNRS services), it is freely available at https://litemol.org, and its source code is available via GitHub. LiteMol suite provides advanced functionality (annotations and their visualization, powerful selection features), and this chapter will describe their use for visual inspection of protein structures.

Key words Protein visualization, Atom selection, Validation report, Ligand representation, Electron density

1 Introduction

Visualization is a critical step in understanding and making effective use of macromolecular structure data. The review by O'Donoghue et al. [1] describes a range of use cases requiring interactive visualization to help answer biological questions, from the basic display of secondary structure to the determination of complex structure-sequence relationships or analysis of ligand binding sites. Moreover, visual inspection of the data obtained from X-ray diffraction experiments (i.e., electron densities) or electron microscopy imaging (i.e., electric potential maps) allows users to assess the quality of the models derived from data.

For these reasons, we have developed LiteMol suite [2], an innovative open-source solution consisting of a 3D molecular visualizer (LiteMol Viewer), data delivery services (CoordinateServer and DensityServer), and a data compression format (BinaryCIF).

Zoltán Gáspári (ed.), *Structural Bioinformatics: Methods and Protocols*, Methods in Molecular Biology, vol. 2112, https://doi.org/10.1007/978-1-0716-0270-6_1, © Springer Science+Business Media, LLC, part of Springer Nature 2020

LiteMol Viewer provides an interactive, web browser-based visualization of 3D structures together with information on experimental evidence and annotation of the biological context. CoordinateServer and DensityServer offer a data delivery approach that enables a dramatic reduction in data transfer size by providing on-demand access to relevant data, thus eliminating the need to send complete data files. Finally, the BinaryCIF format provides very high compression ratios to decrease the amount of transferred data. Thanks to the combination of all these components, the LiteMol suite enables near-instant data delivery and visualization of large macromolecular structures and associated experimental data. The LiteMol suite is integrated into the Protein Data Bank in Europe (PDBe) and other life science web applications (e.g., UniProt, Ensemble, SIB, and CNRS services). Its components are freely available for integration into other online services. The LiteMol suite can be accessed at https://litemol.org.

This book chapter describes in detail how to use full power of LiteMol suite and perform advanced tasks focused on visual inspection of protein structures (*see* **Note 1**).

2 Materials

2.1 Implementation of LiteMol Suite

All components of the LiteMol suite are implemented using the TypeScript language, which is a typed version of JavaScript. This means that most of the code (apart from file system access, which is only available on the server) can be run in both the web browser and on the server (using Node.js; https://nodejs.org). LiteMol suite runs in all modern browsers without the need of additional plugins. The LiteMol suite is an open-source collection of software, and its source codes are available on GitHub (https://github.com/dsehnal/LiteMol). An example of LiteMol suite integration into a web page is also available here: https://www.ebi.ac.uk/pdbe/pdb-component-library/doc.html#a_LiteMol.

3 Methods

3.1 Visualization of Annotations

LiteMol Viewer is able to display multiple annotations together with structure itself—it can give annotation about the structure quality, which is otherwise buried within validation reports, it can visualize sequence annotations from multiple sequence databases, and it can annotate carbohydrate structures using Symbol Nomenclature for Graphical Representation of Glycans.

3.1.1 Annotation of Structure Quality

LiteMol Viewer is able to display structure quality information extracted from wwPDB Validation Reports (VR) [3] and Ligand Validation Reports (LVR) [4] on a model of biomacromolecule (Fig. 1).

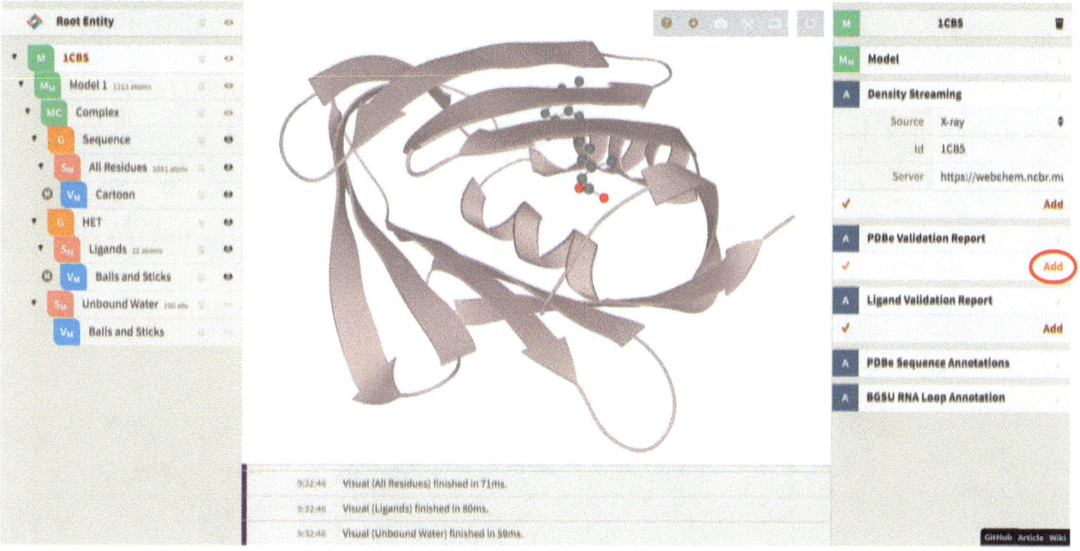

Fig. 1 How to map validation report (VR) annotation on macromolecular model (PDB ID 1cbs)

Visualization of Annotations from VR

VR include not only information about geometric quality of the molecule but also assessment of the experimental data. The VR for a molecule can be downloaded from the PDBe site (find molecule of your interest; click Details in Experiments and Validation section; and then find Full Validation Report in the right panel). LiteMol suite can automatically fetch VR and color cartoon visual of biomacromolecule accordingly. The color of each residue depends on the number of validation issues it exhibits: green = 0, yellow = 1, orange = 2, red = 3, or more. Note that LiteMol uses the same coloring scheme as PDBe. Moreover, the information from the VR is also available as a text description, when you hover a mouse on individual residues. Some of the quality criteria are displayed (whenever they apply), including clashes, side chain outliers, Ramachandran outliers, and more. Figure 2a shows an example of VR annotation visualization over 3D structure of drug-metabolizing enzyme cytochrome P450 3A4. Note that most of the issues are located on the surface of the protein and not in its binding site core.

How to visualize the information from VR on a structure? In the left menu, click on the line with the PDB ID. Then click on Add in the section PDBe Validation Report in the right menu (*see* Fig. 1). Finally, click on Add in the section Apply Coloring.

Visualization of Annotations from LVR

LVR includes validation data for the ligands in this structure. These data come from ValidatorDB [4], a database comparing the real ligand structure with its model structure, obtained from wwPDB Chemical Component Dictionary (wwPDB CCD) [5]. Specifically, these validation data include information about missing rings, missing atoms, and chirality mismatches. The residues with the

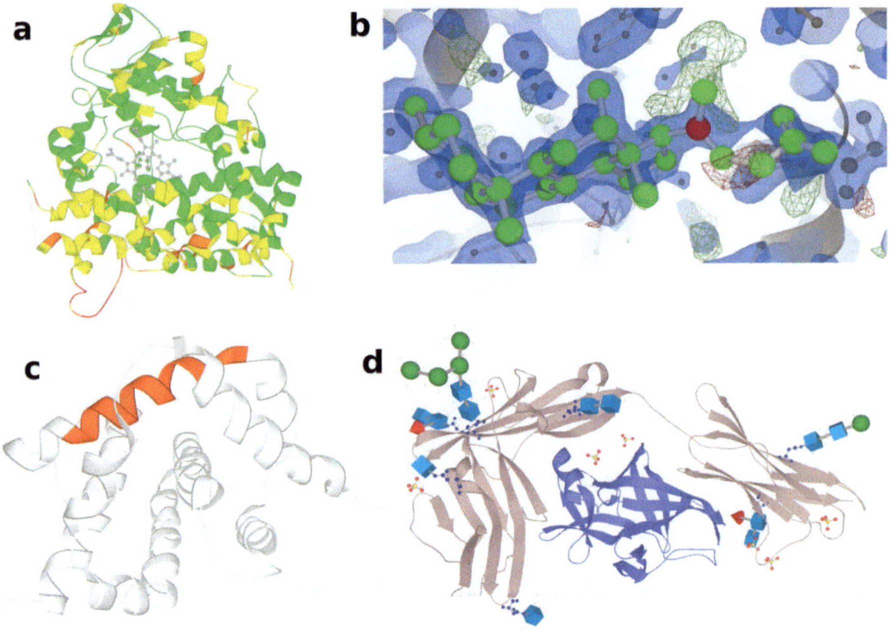

Fig. 2 Visualization of annotations in LiteMol Viewer: (**a**) cytochrome P450 3A4 (1tqn), coloring based on Validation Report; (**b**) ferrochelatase (3aqi), coloring based on Ligand Validation Report—red atom in cholic acid has wrong chirality (balls-and-sticks representation with electron density isosurfaces); (**c**) BAX proapoptotic protein (1f16) with highlighted BH3 domain (InterPro annotations); (**d**) IL-18 in complex (3wo3) containing bound glycans (protein in cartoon, glycans in balls-and-sticks, and 3D-SNFG representation)

missing atoms or rings are depicted in red. Also, the atoms with mismatched chirality are shown in red. Figure 2b shows an example of a quality annotation of the ligand from LVR within ferrochelatase. Notice the surplus (green wireframe) and deficient (red wireframe) electron densities indicating wrong model fit. Note: There are currently not present any types of annotations specific for NMR.

How to visualize the information from LVR on a structure?: In the left menu, click on the line with PDB ID. Then click on Add in the section Ligand Validation Report in the right menu. Then click on Add in the section Apply Coloring.

3.1.2 Sequence Annotation

The databases such as InterPro, Pfam, CATH, SCOP, or UniProt provide many highly useful information about various biologically important regions located in the sequence of the biomacromolecule (e.g., binding sites, activation or inhibition regions, mutations). LiteMol suite is able to display some of the information aggregated by SIFTS service and accessible over the PDBe API. These sequence-based annotations are in turn mapped on a protein structure. For example, Fig. 2c shows visualization of the BH3 peptide domain region from InterPro in proapoptotic protein BAX—this domain is important for initiation of apoptosis.

How to visualize the sequence annotation on the structure? The same way as for VRs, but use the section Ligand Validation Report instead of PDBe Validation Report.

3.1.3 Annotations of Carbohydrates

The schematic representation of carbohydrates in 3D space facilitates their interpretation; however, none of the state-of-the-art visualization software supports this notation. For this reason, LiteMol Viewer integrates annotation and visualization of monosaccharides in glycans using the 3D Symbol Nomenclature for Graphical Representation of Glycans (3D-SNFG) [6] nomenclature (derived from original SNFG annotation, e.g., Man, GalNAc, ManN [7]). These symbols allow user to readily identify which monosaccharide residues are present in the structure and where they are located. Additionally, a name given by SNFG nomenclature is displayed on mouse hover. Figure 2d shows visualization of protein containing glycans.

How to annotate and visualize glycans? During reading of the input structure, LiteMol Viewer automatically identifies glycans and annotates them SNFG conventions. Afterward, the glycans in the structure are by default depicted via 3D-SNFG symbols. In the left menu, the carbohydrate part is mentioned as Std. Carbohydrates. Its default visualization methods are 3D-SNFG and Balls and Sticks.

3.2 Selection Functionality in the User Interface

LiteMol suite has a selection language available, similar to other structure visualization tools such as VMD or PyMOL. The LiteMol selection algebra is inspired by PatternQuery [8], which was developed by our team. The selection functionality is integrated directly into LiteMol suite user interface. Its commands are described in the Table 1 (*see* **Note 1**, part Selection Language). More details can be found in the LiteMol Wiki pages (https://webchem.ncbr.muni.cz/Wiki/LiteMol:UserManual).

How to perform selections?: In the left menu, click on the line with Model. Then, in the right menu, click on Selection and type your query. Then, click on Add.

3.3 Visualization of Large Structures by Distance-Based Coloring

One of the ways to color macromolecular structure is a rainbow-coloring based on the distance from the particle center (*see* Fig. 3). This is a standard way to display, for example, viral capsids in the structural virology. LiteMol Viewer can rainbow-color any structure representation by their distance to the center of mass in just a few clicks. This functionality combined with low memory footprint of CoordinateServer-delivered data enables to display viral capsids even on mobile devices.

How to color structure representation by the distance from the structure center? First, we need to select a molecule visual of a deemed molecule model selection (e.g., surface representation of polymer) from the left-hand entity tree menu. Next, click on the line Particle coloring in the right menu and click Add.

Table 1
Queries in LiteMol suite—selection functionality in the user interface and in CoordinateServer

Type of object	Selection queries in the UI (including examples)	CoordinateServer queries (including examples)	Description of query (and corresponding mmCIF fields)
Basic query			
Atoms[a]	atomsByElement('C','O')	–	Atoms based on their element symbol (field *_atom_site.type_symbol*)
	atomsByName('N','CA')	–	Atoms based on their atom name (field *_atom_site.label_atom_id*)
	atomsById(1,2,3)	–	Atoms based on their integer identifier (field *_atom_site.id*)
Residues[b]	residuesByName('ALA', 'PO4','HEM')	/residues?authName=HEM	Residues based on their residue name (field *_atom_site.label_comp_id*)
	residuesById(42,157)	/residues?authSeqNumber=42	Residues based on integer identified (field *_atom_site.auth_seq_id*)
	–	/residueRange?authAsymId=A&range=1-5:20-30	Residues with asym id (field *_atom_site.auth_asym_id*) in the range
Chain parts	backbone()	/backbone	Extracts a backbone of a protein or nucleic acid
	–	/trace	Atoms named CA and P from polymer entities + optionally HET and/or water atoms
	sidechain()	/sidechain	Complement to the backbone query, i.e., all *polymer* atoms without such name are reported
	hetGroups()	/het	*HETATM* atoms defined by the field *_atom_site.group_PDB. CoordinateServer does not include waters in /het command*
	nonHetPolymer()	–	*ATOM* atoms defined by the field *_atom_site.group_PDB*
	cartoon()	/cartoon	Extracts atoms vital for *polymer* cartoon visualization based on their names: *CA, O, O5′, C3′, N3*
Chains[c]	chainsById('A', 'B')	/chains?authAsymId=A	Polymer chains based on their id (field *_atom_site.auth_asym_id*)
Larger objects	–	/assembly?id=1	Constructs assembly with the given id
	–	/entities?type=polymer	Entities that satisfy the given parameters
	everything()	/full	All atoms in the active context

(continued)

Table 1 (continued)

Type of object	Selection queries in the UI (including examples)	CoordinateServer queries (including examples)	Description of query (and corresponding mmCIF fields)
Advanced query			
Inside	residuesByName('GLY'). inside(chainsById('A'))	–	Finds selection within another selection
Surrounding residues in a sphere	residuesByName ('HEM'). ambientResidues(5)	/ambientResidues? authName=HEM& radius=5	Surrounds the inner selection by residues that have at least one atom within the given radius
Whole surrounding residues	atomsByElement('Pt'). wholeResidues()	–	Surrounds the inner selection by all atoms of residue's origin
Symmetry mates	–	/symmetryMates? radius=5	Identifies symmetry mates within the given radius
Interacting ligands	–	/ligandInteraction? &authName=HET &radius=5	Identifies ligands, which interact with defined residue and are in the defined radius
Logical query			
Or	or(residuesByName ('HEM'). ambientResidues(5), chainsById('A'))	–	Merges several selections
Intersection	residuesByName ('HEM'). ambientResidues(5). intersectWith (chainsById('A'))	–	Finds intersection between two selections
Complement	chainsById('A'). complement()	–	Finds the complement of the inner selection to the active context

[a]CoordinateServer does not support selection of individual atoms

[b]Via the selection query, the user can define more residue names or IDs. CoordinateServer provides many other possibilities, representing corresponding mmCIF fields, such as entityId, asymId, and insCode

[c]Via the selection query, the user can define more Chain IDs

3.4 CoordinateServer and Its Selection Functionality

CoordinateServer is a web service for delivering a subset of mmCIF coordinate data for a PDB entry held in the archive. The server is able to return the specific portions of the structure, as specified in a user's query (*see* **Note 2**), for example, the coordinates of the atoms within a 5 Å radius around the ligand binding site, including symmetry mates. As a result, it greatly reduces the time needed to transmit and manipulate the data. The outputs of the

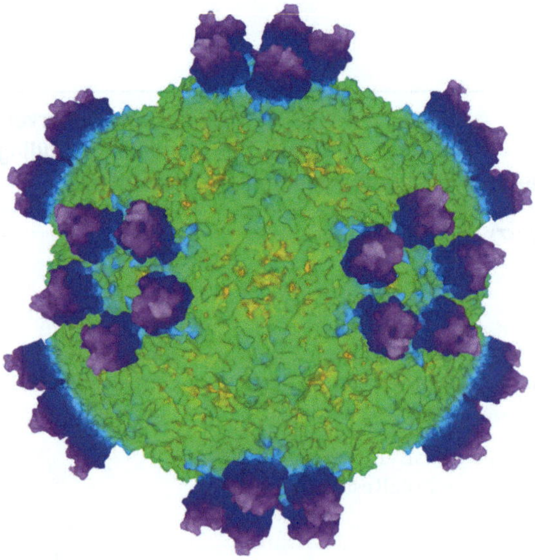

Fig. 3 Surface representation of the slow bee paralysis virus (5j96) using rainbow-color coding based on the distance from the center of the assembly

CoordinateServer is used as data delivery method for LiteMol Viewer. Moreover, the usage of the CoordinateServer is not limited to LiteMol Viewer and can be used by any software that supports the mmCIF format [9]. The CoordinateServer queries are described in Table 1. More details can be found on the web of CoordinateServer (https://webchem.ncbr.muni.cz/CoordinateServer). The data can be encoded in mmCIF or BinaryCIF (*see* **Note 3**).

How to use CoordinateServer queries? First, prepare a web link with the query. It should contain the following parts:

<URL of the CoordinateServer> / <PDB ID> / <CoordinateServer query>

For example:

https://www.ebi.ac.uk/pdbe/coordinates/1gzt/ligandInteraction?name=FUC

The prepared query then serves as an input for LiteMol Viewer. It should be submitted to LiteMol Viewer in the following way: In the right menu, set Source: URL in the Molecule section. Then paste the query into the field URL. Finally, click on Add.

3.5 Using DensityServer to Explore Electron Density Data

DensityServer is a web service for streaming slices of 3D volumetric data—the electron density data available in the Electron Density Server (EDS) and electron microscopy imaging data from the Electron Microscopy Data Bank (EMDB). DensityServer provides near-instant access to user-defined slices of detailed density data (e.g., 5 Å box around a ligand) in full resolution or a downsampled surface of the entire structure for quick visualization.

DensityServer includes three commands: Box, Cell, and Data Header.

Box returns density data inside the specified box for the given entry. For X-ray data, it returns 2Fo-Fc and Fo-Fc volumes (*see* **Note 4**) in a single response. The box is described by a position of its bottom left corner and top right corner. The data can be encoded in mmCIF or BinaryCIF (*see* **Note 5**).

Cell returns (downsampled) volume data for the entire "data cell." For X-ray data, it returns unit cell of 2Fo-Fc and Fo-Fc volumes and for EM data returns everything. The user can define the level of detail of the volumetric data. Possible values are in the range from 0 (0.52 M voxels) to 6 (25.17 M voxels). Again, the data can be encoded in mmCIF or BinaryCIF.

Data Header returns a JSON response specifying if the requested data are available and the maximum region that can be queried.

The functionality of DensityServer is integrated into LiteMol suite and allows quick visualization of local density (e.g., in binding site), as well as a quick overview of global density (e.g., for whole virus). Similar to CoordinateServer, DensityServer can be used by third-party solutions.

4 Example Application: Step-by-Step Visual Analysis of Carbohydrate-Binding Protein

For detailed demonstration of the visual analysis available in LiteMol, we will use the structure of Nipah virus G glycoprotein (3d12). There are 30 instances of 11 different carbohydrates found in this protein model. Many of them exhibit validation issues. In this analysis, we will show how to inspect them and their quality visually.

Step 1: Visualization of glycoprotein in LiteMol suite:

- Method: Load the structure with PDB ID 3d12 into LiteMol Viewer (use the section Molecule in the right menu).
- Results: *See* Fig. 4a. The structure is visualized in default representation—cartoon for protein, balls-and-sticks and 3D-SNFG for carbohydrates, and balls-and-sticks for ligands.

Step 2: Rotation of glycoprotein to see glycans:

- Method: Use the left mouse button for rotation, the right button for zoom, and the middle button for movement.
- Results: *See* Fig. 4b; it shows many different glycans. The most common carbohydrate residue is NAG (*N*-acetyl-D-glucosamine) depicted as a blue cube occurring ten times. We will focus on the quality issues of this carbohydrate.

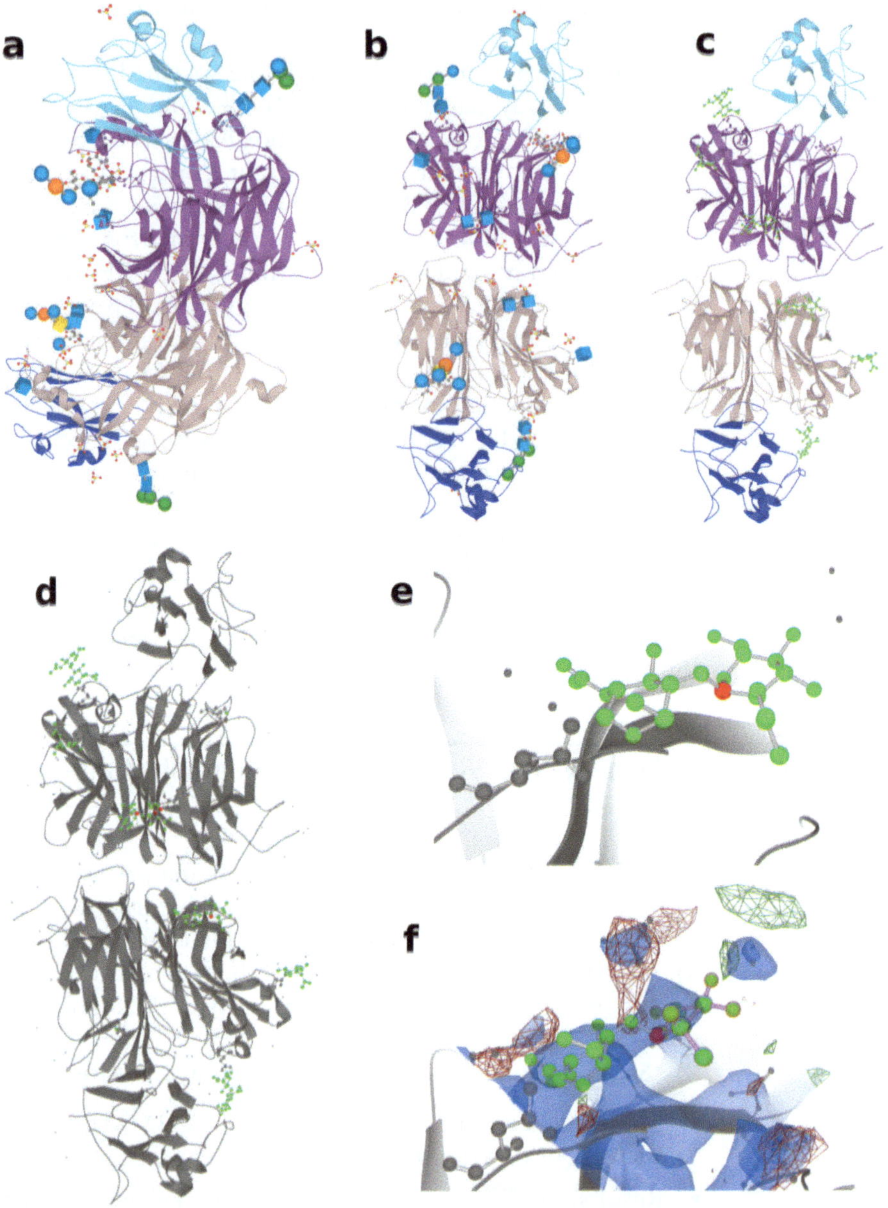

Fig. 4 Step-by-step analysis of a carbohydrate-binding protein (Nipah virus G glycoprotein, PDB ID 3d12): (**a**) visualization of the glycoprotein; (**b**) rotation of the glycoprotein; (**c**) selection and marking of NAGs; (**d**) display of ligand quality; (**e**) detailed look at one of the problematic NAGs; (**f**) NAG electron density visualization

Step 3: Selection and marking of NAGs:

- Method: Create a new selection containing only the NAG residues (use query residuesByName('NAG'), details in Subheading 3.2). Color the selection in green (in the section Visual in the right menu, set Type: Balls and Sticks, Coloring: Uniform Color; click the plus icon to show more coloring options; set

Uniform to green; confirm by Add). Hide all other heteroatoms (in the left menu, click on the eye icon in the line HET).

- Results: *See* Fig. 4c. NAG residues are visible in balls-and-sticks representation in green.

Step 4: Inspection of the ligand's quality:

- Method: Add Ligand validation report (details in Subheading 3.1.1).
- Results: *See* Fig. 4d. Two out of ten NAG ligands exhibit errors (i.e., at least one green atom turned red).

Step 5: Detailed look at one of the erroneous NAG ligands:

- Method: Click on one of the NAG molecules containing a red atom.
- Results: *See* Fig. 4e. The atom C1 exhibits wrong chirality.

Step 6: Visualization of the electron density:

- Method: Add density (select the line 3D12 in the left menu, then use section Density Streaming in the right menu). Rotate the scene to see the density around the C1 atom.
- Results: *See* Fig. 4f. It is obvious that the C1 atom is placed on the edge of the electron density cloud. In parallel, its neighbor atom O4 is placed in a red cloud (marking regions where atoms should not occur). It seems that the positions of both these atoms are incorrect.

5 Notes

1. For the first release of the LiteMol suite, development was focused mainly on designing and implementing methods for fast data delivery and visualization of large molecular data sets in a web browser environment. As a result, some "standard features" are not present in the current version of LiteMol Viewer.
   - Multiple models/trajectories: The viewer can parse and display multiple models inside a single scene. But there is no "play" support similar to the functionality provided by desktop viewers such as PyMOL or VMD. We plan to include this functionality in a future version of the viewer.
   - Selection language: The selection language in LiteMol is currently not very well suited for nonexpert users of the application, as it is targeted to expert developers who integrate LiteMol Viewer as a plugin into their services. We plan to include support for PyMOL/VMD/etc. selection expressions in a future version of the viewer.

- State saving: There is currently no built-in way to save the state of the application. However, the state can be saved and reconstructed manually depending on the particular use case of the plugin. In a future iteration of LiteMol Viewer, we plan to provide built-in support for this functionality.

2. The selection query specifies the subset of atoms for which the data will be sent. However, it is not possible to specify which types (columns) of data will be sent—the server response always includes all columns (residue name, element symbol, coordinates, etc.).
3. CoordinateServer currently supports only PDBx/mmCIF, PDB, and MOL/SDF data formats. We plan to support additional formats, such as Gromacs/Amber trajectory files, in future releases.
4. A brief introduction to different types of electron density maps is available at https://www.rcsb.org/pages/help/edmaps.
5. DensityServer outputs the density data in mmCIF or BinaryCIF format. However, the input data for DensityServer are typically in CCP4/MAP format (mode 0 and 2), which is provided by PDBe for both PDB and EMDB structures. Other input formats are currently not supported—we will extend this as necessary as additional data formats become available.

Acknowledgments

This work has been financially supported by the ELIXIR-EXCELERATE project, grant agreement no. 676559; ELIXIR CZ research infrastructure project (MEYS grant no. LM2015047); European Regional Development Fund-Project ELIXIR-CZ (no. CZ.02.1.01/0.0/0.0/16_013/0001777); and RIAT-CZ (ATCZ40). K.B. acknowledges European Regional Development Fund project no. CZ.02.1.01/0.0/0.0/16_019/0000754 of the Ministry of Education, Youth and Sports of the Czech Republic. A.M. was also financed by Brno City Municipality (Ph.D. Talent Scholarship).

References

1. O'Donoghue SI, Goodsell DS, Frangakis AS, Jossinet F, Laskowski RA, Nilges M, Saibil HR, Schafferhans A, Wade RC, Westhof E, Olson AJ (2010) Visualization of macromolecular structures. Nat Methods 7 (3 Suppl):S42–S55. https://doi.org/10.1038/nmeth.1427
2. Sehnal D, Deshpande M, Svobodová Vařeková R, Mir S, Berka K, Midlik A, Pravda L, Velankar S, Koča J (2017) LiteMol suite: interactive web-based visualization of large-scale macromolecular structure data. Nat Methods 14(12):1121–1122. https://doi.org/10.1038/nmeth.4499

3. Gore S, Sanz Garcia E, Hendrickx PMS, Gutmanas A, Westbrook JD, Yang H et al (2017) Validation of structures in the Protein Data Bank. Structure 25(12):1916–1927. https://doi.org/10.1016/j.str.2017.10.009
4. Sehnal D, Svobodová Vařeková R, Pravda L, Ionescu CM, Geidl S, Horský V, Jaiswal D, Wimmerová M, Koča J (2015) ValidatorDB: database of up-to-date validation results for ligands and non-standard residues from the Protein Data Bank. Nucleic Acids Res 43(Database issue):D369–D375. https://doi.org/10.1093/nar/gku1118
5. Westbrook JD, Shao C, Feng Z, Zhuravleva M, Velankar S, Young J (2014) The chemical component dictionary: complete descriptions of constituent molecules in experimentally determined 3D macromolecules in the Protein Data Bank. Bioinformatics 31(8):1274–1278. https://doi.org/10.1093/bioinformatics/btu789
6. Thieker DF, Hadden JA, Schulten K, Woods RJ (2016) 3D implementation of the symbol nomenclature for graphical representation of glycans. Glycobiology 26(8):786–787. https://doi.org/10.1093/glycob/cww076
7. Varki A, Cummings RD, Aebi M, Packer NH, Seeberger PH, Esko JD, Stanley P, Hart G, Darvill A, Kinoshita T, Prestegard JJ, Schnaar RL, Freeze HH, Marth JD, Bertozzi CR, Etzler ME, Frank M, Vliegenthart JF, Lütteke T, Perez S, Bolton E, Rudd P, Paulson J, Kanehisa M, Toukach P, Aoki-Kinoshita KF, Dell A, Narimatsu H, York W, Taniguchi N, Kornfeld S (2015) Symbol nomenclature for graphical representations of glycans. Glycobiology 25(12):1323–1324. https://doi.org/10.1093/glycob/cwv091
8. Sehnal D, Pravda L, Ionescu CM, Svobodová Vařeková R, Koča J (2015) PatternQuery: web application for fast detection of biomacromolecular structural patterns in the entire Protein Data Bank. Nucleic Acids Res 43(W1):W383–W388. https://doi.org/10.1093/nar/gkv561
9. Bourne PE, Berman HM, McMahon B, Watenpaugh KD, Westbrook J, Fitzgerald PMD (1997) The macromolecular crystallographic information file (mmCIF). Methods Enzymol 277:571–590

Chapter 2

Comparative Protein Structure Analysis with Bio3D-Web

Barry J. Grant, Lars Skjærven, and Xin-Qiu Yao

Abstract

Bio3D-web is an online application for the interactive analysis of sequence-structure-dynamics relationships in user-defined protein structure sets. Major functionality includes structure database searching, sequence and structure conservation assessment, inter-conformer relationship mapping and clustering with principal component analysis (PCA), and flexibility prediction and comparison with ensemble normal mode analysis (eNMA). Collectively these methods allow users to start with a single sequence or structure and characterize the structural, conformational, and internal dynamic properties of homologous proteins for which there are high-resolution structures available. Functionality is also provided for the generation of custom PDF, Word, and HTML analysis reports detailing all user-specified analysis settings and corresponding results. Bio3D-web is available at http://thegrantlab.org/bio3d/webapps, as a Docker image https://hub.docker.com/r/bio3d/bio3d-web/, or downloadable source code https://bitbucket.org/Grantlab/bio3d-web.

Key words Protein structure, Protein dynamics, Protein flexibility, Sequence-structure-function relationships, Structural bioinformatics

1 Introduction

Bio3D-web is an online webserver for the user-friendly analysis of protein structures [1]. Bio3D-web runs on all modern web browsers and provides functionality for the following: (1) the identification of related protein structure sets to user-specified thresholds of similarity, (2) the multiple alignment and structure superposition, (3) sequence and structure conservation analysis, (4) inter-conformer relationship mapping with principal component analysis (PCA), and (5) comparison of predicted internal dynamics via ensemble normal mode analysis (eNMA). This integrated functionality provides a complete workflow for the investigation of sequence-structure-dynamics relationships within large protein structure sets. In addition to a convenient easy-to-use interface for exploring the effects of parameter and method choices, Bio3D-web also records the complete user input and subsequent graphical results of a user's session. This allows users to easily share and reproduce the sequence of analysis steps that created their

Zoltán Gáspári (ed.), *Structural Bioinformatics: Methods and Protocols*, Methods in Molecular Biology, vol. 2112,
https://doi.org/10.1007/978-1-0716-0270-6_2, © Springer Science+Business Media, LLC, part of Springer Nature 2020

results. In particular, custom summary reports can be created in multiple formats that capture all user-defined analysis choices and optionally enable collaborators to visit previous analysis sessions.

Bio3D-web is powered by a subset of the well-established Bio3D R package for structural bioinformatics [2, 3]. In contrast to the conventional Bio3D package, Bio3D-web does not require any installation or programming skills. Rather, you explore through an interactive online interface (instead of programing your analysis workflow with the R-Bio3D language at the Unix like command line). The design of Bio3D-web emphasizes simplicity over exhaustive inclusion of the many additional analysis methods available in the full Bio3D package (*see* **Note 1**). This effectively reduces the required technical expertise and thus facilitates advanced structural bioinformatics analysis for a broader range of students and researchers. For example, Bio3D-web is used in undergraduate- and graduate-level bioinformatics and structural biology courses at UC San Diego and elsewhere. In research settings, Bio3D-web is most often used to quickly explore protein structure datasets; map their structural, conformational, and internal dynamic properties, and thus understand general trends that can inform more specialized analyses.

2 Materials

The main Bio3D-web server is available without restriction at http://thegrantlab.org/bio3d/webapps. Full source code is made available under a GPL2 license from https://bitbucket.org/Grantlab/bio3d/. The most convenient way to install and run your own Bio3D-web instance is via the Dockerized version hosted at https://hub.docker.com/r/bio3d/bio3d-web/. This image includes all necessary dependencies and can be run on any computing infrastructure or cloud platform supporting Docker (*see* **Note 2**).

3 Methods

3.1 Overview

Bio3D-web analysis typically proceeds through five consecutive and dependent steps, (namely **SEARCH**, **ALIGN**, **FIT**, **PCA**, and **eNMA**). Each step is implemented as a consecutive navigation tab of the Bio3D-web interface (*see* Fig. 1 navigation tabs) and described further below.

1. *Structure search and selection (SEARCH)*. This tab enables the identification and selection of PDB structures related to a user input PDB code or protein sequence. Identified structures are presented in rank order of decreasing sequence similarity to the query. Selected structures from this set will be subject to ensemble analysis of their sequence structure and conformational relationships in additional tabs.

Fig. 1 (**a**) Main navigation tabs and analysis steps of Bio3D-web. Bio3D-web is divided into five major steps, each represented by consecutive navigation tabs. Each tab is divided into multiple panels representing a task or analysis. (**b**) Hit selection for further analysis. The plot shows a summary of the search results where each dot represents one particular PDB ID. Dots above the dashed red line are above the specified threshold and can be selected for further analysis. Further structure annotation and optional filtering options not shown

2. *Multiple sequence alignment analysis (ALIGN)*. In this tab, all previously selected structures are subject to multiple sequence alignment and initial sequence similarity and sequence conservation analysis.
3. *Structure fitting and analysis (FIT)*. In this tab, aligned structures are superimposed on their invariant structural core. Analysis of pairwise structural deviations (RMSD), fluctuations (RMSF), and multiple structure visualization is also provided along with RMSD clustering heatmaps, dendrograms, and histograms.
4. *Principal component analysis (PCA)*. In this tab, PCA is performed on the coordinates of all superimposed structures to characterize inter-conformer relationships. This analysis effectively captures and summarizes the main conformational features and structural displacements of the ensemble. This tab also provides clustering of the structures based on the calculated principal components.

5. *Ensemble normal mode analysis (eNMA).* In this tab, normal mode analysis (NMA) of all structures is performed to predict large-scale motions. Here, NMA is performed on all structures in the ensemble in a way that facilitates the interpretation of structural similarity and dissimilarity trends. This tab also provides clustering of the structures based on the calculated normal modes and fluctuation profiles.

In the next section, we present a detailed protocol for the investigation of protein sequence-structure-dynamics relationship using Bio3D-web. The procedure will be identical whether you are using the public webserver or your own local version.

3.2 Example Application to Ribose-Binding Protein

Ribose-binding proteins (RBPs) function in bacterial chemotaxis and transport [4]. RBPs scavenge for ribose in the cell's environment by coupling ligation to interaction with chemotaxis receptors and transporter proteins in the inner membrane. Like many other proteins of this class, RBPs undergo conformational changes upon ligation to increase protein-ligand interactions and expose new surface residues that are recognized by RBP binding partners. These conformational changes have also been adapted to construct engineered biosensors that transduce ligand binding to a variety of physical signals [4]. Understanding the mechanistic details of RBP conformational changes therefore is important not only for understanding their biological function but also for furthering protein engineering applications. In subsequent sections, we demonstrate the use of Bio3D-web for the investigation and visualization of the sequence-structure-dynamics relationships of RBPs using the full ensemble of experimental structures available from PDB. We note that analogous workflows using structure ensembles from other sources, such as molecular dynamics simulations, is a major feature of the full Bio3D package (*see* **Note 1**).

3.2.1 SEARCH: Structure Search and Selection

To start the analysis, open a web browser and go to the Bio3D-web application (http://thegrantlab.org/bio3d/webapps). This will bring you to the first part of the application—the **SEARCH** tab (*see* **Note 3**). This tab contains a total of three sub-steps (labeled **A–C**).

Input structure(s) or sequence: Here we will use a single PDB structure code as input and type the RBP PDB code **2DRI** into the input text box (*see* **Note 4** and Fig. 1 user input). When the four characters of a PDB entry have been entered, the search will automatically start with a progress bar appearing at the very top of the screen to indicate that the server is working. When the search is completed, you will find a short summary of your query protein in the middle panel of the first row. This includes the protein name and species, as well as PFAM annotation data. You can use the link to PFAM in this section to learn more about the protein family undergoing analysis (*see* **Note 5**). The third and final panel of the first row (right-hand side) provides a simple interactive **Input Structure Visualization.** Click and drag the mouse pointer over the protein to rotate and scroll to zoom. Different display and coloring options are also available.

Hit selection for further analysis: To proceed, click the blue **Next (Hit selection)** button in the first panel or simply scroll down to panel **(b) Hit selection for further analysis.** As the title indicates, this panel controls the selection of hits to be analyzed in subsequent steps. It includes setting a similarity threshold cutoff value (**Adjust inclusion BitScore cutoff** slider), in which structures above this cutoff can be chosen for further analysis, and the **Limit total number of included structures** slider to set the maximum number of structures to be used. Next to this panel, a plot provides a schematic representation of the search results. In this plot, each dot represents a particular hit (i.e., structure with similar sequence) in the PDB. Dots above the red dashed line are hits above the cutoff, while blue circles indicate selected hits (*see* Fig. 1). Note that a minimum of three structures are required for the analysis to proceed.

Optional filtering of related structures for further analysis: This panel (not shown in Fig. 1) allows for optional finer-grained selection of structures and provides annotation data to help in this process. This table also links directly to the PDB if you want to explore the individual PDB entries, as well as their bound ligands (*see* **Note 6**). Proceed to the next main step by clicking on the **ALIGN** tab on the top of the page.

3.2.2 ALIGN: Sequence Alignment and Analysis

The next step in the application includes alignment of all PDB structures selected in the previous (**SEARCH**) tab. This is automatically performed upon entering the **ALIGN** tab. When the sequence alignment has been completed, panels providing (a) a summary of the sequence alignment and (b) basic analyses of the sequence alignment become available (*see* Fig. 2).

Alignment summary: This panel shows a short summary of the alignment providing details on the number of sequence rows (equivalent to the number of PDB structures), as well as the number of position columns including a specification of the number of gap and non-gap containing columns. This panel also shows which PDB structures (if any) contain missing in-structure residues (e.g., amino acid residues which have not been resolved in the X-ray crystallography experiment).

The figure on the right-hand side in the first row provides a schematic representation of the sequence alignment. Here, the gray areas represent non-gap positions, while white areas in the alignment correspond to gaps. A representation of the sequence conservation is shown above the alignment with red areas indicating

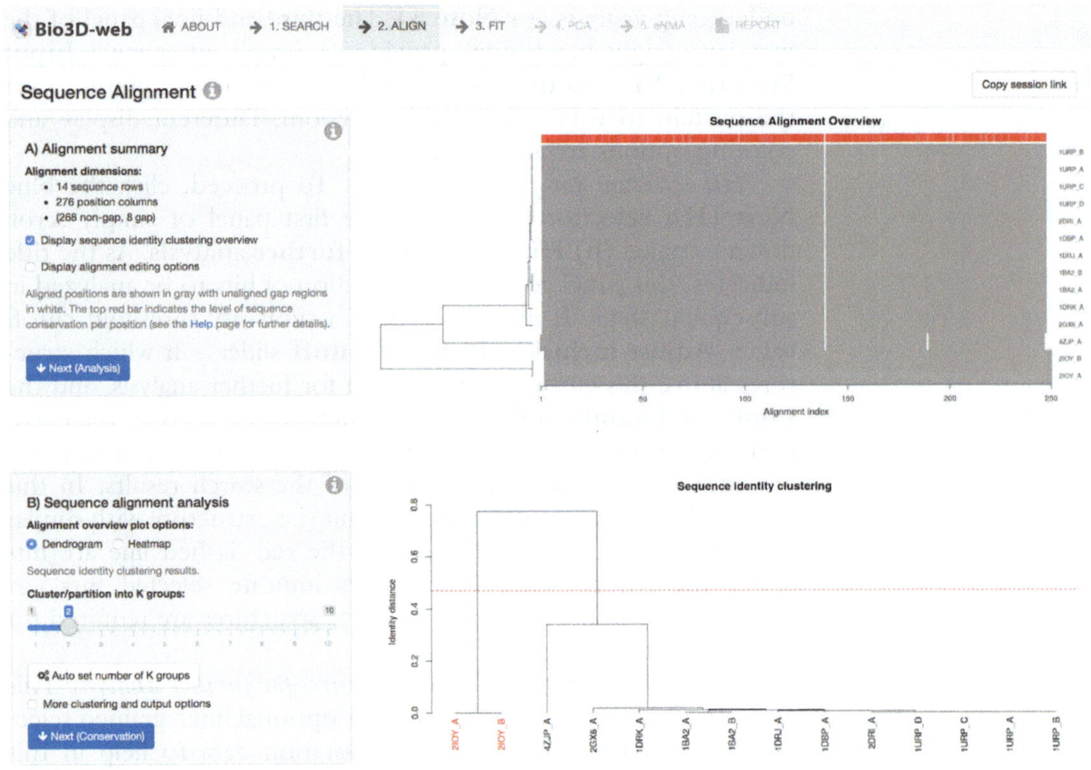

Fig. 2 Sequence alignment summary and analysis. (**a**) The first row of the ALIGN tab provides an alignment summary, as well as the option to include/exclude structures for further analysis. Note the information on the missing in-structure residues. (**b**) Clustering dendrogram of the pairwise sequence identities. Further sequence conservation analysis not shown

conserved positions and white indicating less conserved. Note that the sequences in this figure are ordered based on their similarity provided by the clustering dendrogram on the left-hand side (more on this below) (*see* **Note 7**). Provided functionality also allows the users to upload their own sequence alignment file in FASTA format. This is useful for the correction of potential alignment errors or for the investigation of alignments from other sources.

Sequence alignment analysis: This panel presents the results of structure clustering based on pairwise sequence identity as well as analysis of sequence conservation (*see* Fig. 2b). By default, a **dendrogram** (or tree diagram) representing the arrangement of clusters is shown. The *y*-axis of the dendrogram represents the distance (in terms of sequence identity) between the clusters. The cluster analysis shows that the sequences of the structures in the current analysis are very similar but can be divided into two major groups (indicated by black and red labels). Moving the **Cluster into K groups** slider will allow the user to set the number of cluster groups, and by clicking in the **More clustering options** button, the user can change the clustering method and obtain PDF and text outputs (*see* **Note 8**).

Residue conservation: In panel C (not shown in Fig. 2), the sequence conservation per residue position is displayed with respect to the aligned structure set (or optionally the PFAM database seed alignment set). A number of conservation scoring methods are available including entropy, similarity, and identity methods (*see* **Note 9**). Secondary structure elements are depicted in the marginal regions of the plot as black (helices) and gray (sheets) boxes. The residue numbers provided are obtained from the first structure in the ensemble.

Optional alignment display: The final sequence alignment is optionally shown with amino acid residues colored according to their physicochemical properties. Also note that conserved columns are depicted with an asterisk (*), while columns containing similar amino acid are marked with a hat (^) below the alignment.

When you are done inspecting the sequence analysis, proceed to the next step of the application by clicking on the **FIT** tab on the top of the page.

3.2.3 FIT: Structure Superposition and Analysis

When entering the **FIT** tab, the server will automatically start the process of superimposing all structures onto each other. By default, the program will identify the *invariant core*—a region with low structural variability within the ensemble—and superimpose this region. Additional superposition options are available including all C-alpha atoms.

Superposed PDB viewing options: The superimposed structures are shown in the first row of the **FIT** tab (*see* Fig. 3). Click and drag the mouse over the structures to rotate, and scroll to zoom. By default, the structures are colored according to the Residue index. Another useful coloring option is by their cluster membership, for example, **RMSD Cluster Groups** (these are calculated based on pairwise structural deviations discussed further below), i.e., structures with the same color share a similar conformation (*see* **Note 10**).

To visualize the invariant core (in which all structures are superimposed to), toggle the **Invariant core** radio button under **Structure color** list. This will color the region defined as the invariant core red, and all other residues are colored black. As you can see, the regions colored red show very little structural variability. Next, color by **Gap regions**. This will color all residues placed in a gap containing column (in the ALIGN tab) red. In this case, there are no gap regions to color (*see* **Note 11**).

Initial structure analysis: Scroll further down to the **Initial structure analysis** panel (Fig. 3). This panel provides basic analyses and plotting options of the structure data. This includes clustering of the structures based on all pairwise RMSD (root mean square deviation) values. By default, a **dendrogram** (or tree diagram) representing the arrangement of clusters is shown. The *y*-axis of

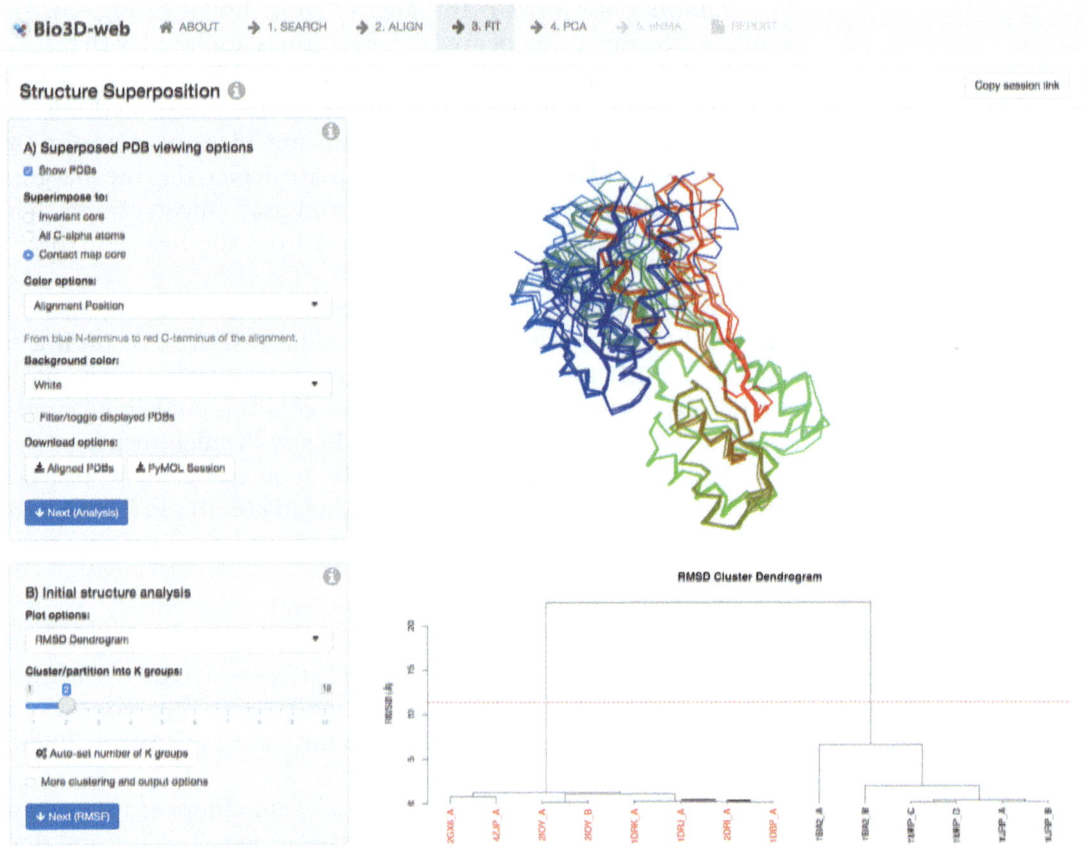

Fig. 3 Structure analyses of selected PDBs. (**a**) Visualization of all superimposed structures. (**b**) Clustering dendrogram based on the pairwise RMSD values. The label colors depict the two cluster memberships obtained by RMSD clustering. Further structural displacement analysis not shown

the dendrogram shows the distance (in Å) between the clusters. The cluster analysis shows that the structures can be divided into two major groups (indicated by black and red labels).

The full matrix of pairwise RMSD values can also be visualized as a **RMSD heatmap** representing the structural deviations using a color scale from white (dissimilar) to red (similar). Toggle the **Row side color by sequence identity clusters** checkbox to compare the clusters from the sequence and RMSD analysis. Notice that in this case, the sequence differences are not reflected in major structure differences. The next panel (C) of this tab shows the structural variation per residue positions in the structure ensemble.

Structural analysis summary: At the bottom of the **FIT** tab, a section with three panels provides additional data of the structure analyses. The first panel gives an overview of the residues comprising the invariant core (with residue identifiers belonging to the **Reference PDB**). The second panel (**RMSD summary**) displays the RMSD values between the reference PDB and every other PDB in the ensemble. Finally, the third panel shows the list of cluster representatives—one structure from each cluster with the minimal distance to all the other cluster members.

3.2.4 PCA: Principal Component Analysis

The **PCA** tab provides principal component analysis (PCA) of the structure data. PCA is a statistical approach used to transform a dataset down to a few important components that describe the directions where there is most variance. In terms of protein structures, PCA is used to capture major structural variations within an ensemble of structures. More explicitly, Bio3D-web utilizes PCA to provide a new condensed view of user-defined structural datasets. This condensed view is a reframing that retains the essence of the entire coordinate data. The new view is given in terms of what are known as principal components. These principal components are new directions in the data along which there is maximal variance—or more simply put, the directions where the structure set differs most (i.e., are most spread out). The whole idea of PCA is to find these new directions of maximal variation in the coordinate data and use them to better understand major conformational features of the dataset.

Principal Component Visualization: The first panel of the PCA tab (**Principal Component Visualization**) provides an interactive visualization of the principal components (PCs). By default, the PC describing the most of the structural variations (PC-1) is shown in the visualization window (Fig. 4a). Change the **Color options** to **Variability Per Position**. In this view, atoms are colored on a scale from blue to red, where red represents atoms showing large motion amplitudes and blue for more rigid atoms (*see* **Note 12**).

Conformer plot: The second panel of the PCA tab shows a *conformer plot*—a low-dimensional representation of the conformational variability within the ensemble of PDB structures (Fig. 4b). The plot is obtained by projecting the individual structures onto two selected PCs (e.g., PC-1 and PC-2). These projections display the inter-conformer relationships in terms of the conformational differences described by the selected PCs.

The plot shows that the RBP structures can be divided into two major groups along the two first PCs (Fig. 4b). To inspect which PDB IDs correspond to the different dots, scroll down to the **PCA conformer plot annotation** panel. Click on the row of a given PDB ID, and this will highlight the structures in the conformer plot above.

Toggle the **Interactive** plotting mode option in the **Conformer plot** panel. In this plot type, you can hover over any dot to get information on which PDB ID the dot represents. Note that red structures are ligand bound "closed" structures with black structures "open" unbound structures of RBP.

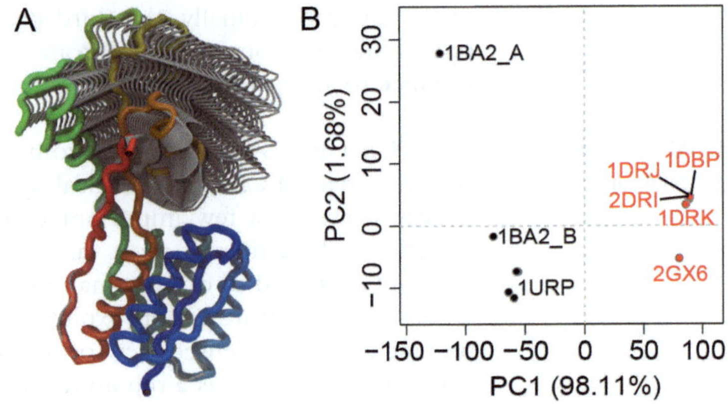

Fig. 4 The first principal component (PC-1) of the RBP structures reveals a closing motion and two distinct conformational clusters. (**a**) Visualization of PC-1 characterizing the major collective conformational variation. (**b**) The conformer plot of all available RBP structures. The conformer plot shows a two-dimensional representation of the conformational variability, where each point represents a structure and point color indicates the conformational cluster membership. Note red structures are ligand bound "closed" structures with black structures "open" unbound structures

Residue contributions: The final panel of the PCA tab shows the contribution of each residue to the individual PCs. The height of each bar represents the relative displacement of each residue described by a given PC. Toggle the **Show RMSF** checkbox to plot the RMSF profile in the same plot, and use the **Choose Principal Component** input field to plot the contributions of additional PCs.

3.2.5 eNMA: Ensemble Normal Mode Analysis

The penultimate tab of the app enables normal mode analysis (NMA) on selected structures of the ensemble [3]. This facilitates characterizing and comparing flexibility profiles of all selected structures. Traditional NMA application most often involves the analysis of only a single protein structure. As the normal modes are sensitive to the specific protein conformation for which they are calculated, the exclusion of alternative protein conformations provides only a limited picture of the overall flexibility of the protein under different conditions. A more complete picture of protein flexibility can be obtained with Bio3D-web by performing NMA across all structures in an ensemble in a way that facilitates the interpretation of structural similarity and dissimilarity trends. This allows a user to explore dynamic trends of all crystalized states in relation to each other without the conventional caveat of potentially overinterpreting the differences between extreme cases and a single artifactual structure. Furthermore, by carefully contrasting the fluctuation profiles, one can provide new information on state-specific global and local dynamics of potential functional relevance.

Filter structures: Prior to calculating the normal modes, we have added the option to reduce the size of the structure ensemble by filtering out structures of similar conformation (panel **Filter structures**). This is useful to reduce the computational load of the ensemble NMA approach and is used on our public server to allow expedient return of results. Set the **RMSD Cutoff** to 0. Observe that all structures have now been selected and are labeled in the cluster dendrogram on the right-hand side. Click the green **Run Ensemble NMA** button to start the calculation of the normal modes.

Normal Modes Visualization: Once the calculation of the normal modes is complete, multiple panels offering various types of analyses of the normal modes appear. The first panel (**Normal Modes Visualization**) offers an interactive visualization of the motions described by the normal modes. Increase the **Magnification factor** to amplify the motions (*see* **Note 12**).

Residue fluctuations: The next panel offers plotting of the NMA-derived fluctuation profiles (Fig. 5a). Here the lines in the plot are colored according to their cluster membership (*see* checkboxes **Cluster by**). Note that the number of cluster groups was specified on the previous pages. The fluctuation profiles of the two

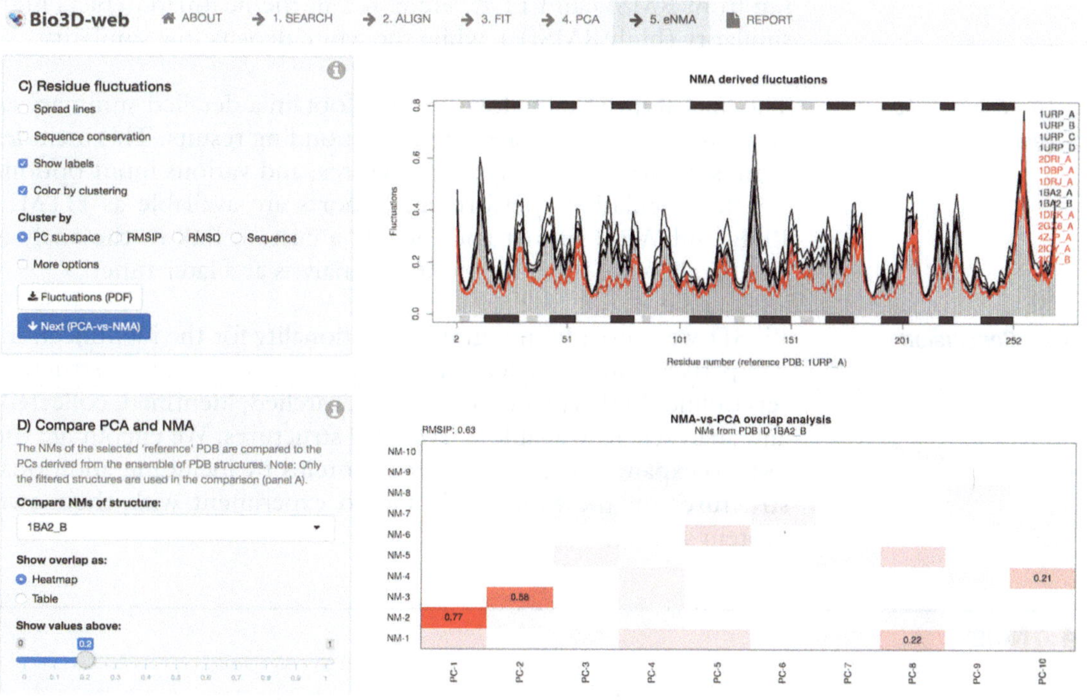

Fig. 5 Ensemble normal mode analysis (eNMA) of RBP structures. (**a**) This panel shows NMA-derived fluctuations of seven representative structures. The fluctuation profiles are colored according to the PCA-based clustering analysis. Note that the red ligand bound "closed" structures are predicted to be less flexible in certain regions. (**b**) Comparison of NMs and PCs showing the overlap between individual modes

groups of PDB structures reveal that the black cluster shows higher flexibility in specific areas than the red cluster; the latter is more restrained in its fluctuations.

Compare PCA and NMA: In this panel, the normal modes of the selected PDB structure are compared to the PCs derived from the ensemble of PDB structures (Fig. 5b). The values provided in the table correspond to the overlap (i.e., the dot product) between the mode vectors. The RMSIP value provides an overall score for the similarity between the PC and NM vectors. The heatmap reveals that PC-1 and NM-2 show a very high similarity with an overlap of 0.77. This indicates that the major conformational transition of this protein system is greatly facilitated by the structure.

Overlap analysis: This panel facilitates comparison between normal mode vectors and a vector describing the conformational difference between two structures. This enables the visualization of which modes contribute to a known functional conformational transition. An overlap of 1 corresponds to identical vectors, while an overlap of 0 corresponds to orthogonal vectors.

Clustering analysis based on NMA: In this panel, the structures grouped based on the similarity of the normal mode vectors using RMSIP as similarity measure. The clustering heatmaps in the final row allow a comparison between NM-based clustering and clustering from RMSD and PCA. Here, red in the heatmap depicts high similarity (high RMSIP), while the white depicts low similarity.

3.2.6 Summary Report Generation

The final **Report** tab allows users to obtain a detailed summary of all analyses performed and the corresponding results. This includes figures, tables of all analyzed structures, and various input options for the calculations performed. Reports are available as HTML, PDF, and Word format and include a custom link to the analysis session allowing users to revisit the analysis at a later time.

3.3 Conclusion

Bio3D-web provides integrated functionality for the identification, comparison, and detailed analysis of large user-defined structure sets online. In this protocol, we have searched, identified, collected, and analyzed all available *E. coli* RBP structures. We encourage the user to expand analysis to related proteins by including additional structures on the Search tab and to experiment with their own protein systems.

4 Notes

1. The Bio3D structural bioinformatics package contains extensive functionality for the analysis of biomolecular sequence and structure from both experiment and theory. For full details, please *see* http://thegrantlab.org/bio3d/.

2. We note that most user needs will be fulfilled by our main public server. However, for protecting IP and other reasons, users in industry and elsewhere may wish to set up their own Bio3D-web server.
3. Additional information and help on each tab and panel can be found by clicking the **About this tab** button on a given page and the small question marks in each panel.
4. The best place to find PDB structure codes is the RCSB PDB database where text searches can be used to locate structures of interest.
5. Note the **Limit to chain ID** dropdown selector in this section, which enables selecting the chain ID (in multichain PDB structures). In this particular case, the input PDB contains only one chain (chain "A"). Note also that there are options available for biounit and multichain analysis.
6. Note that at the time of writing, the first 11 entries (rows) in this table are colored blue. These are the PDBs selected for further analysis. You can (de)select any PDB entry in this table to include/exclude it for further analysis. For example, deselect the first entry by clicking, and notice that the blue circle in the panel b plot above also disappears.
7. The **Display alignment editing options** checkbox enables custom filtering of the structures. To remove a particular PDB from the sequence alignment and further analysis, click on it with the mouse pointer and hit the delete button on your keyboard. Notice that the alignment regenerates once your edit is performed. To include it again, click the **Reset alignment** button. This panel also contains the option to upload a manually corrected sequence alignment file.
8. Choose the **Heatmap** option to display a clustering dendrogram with an associated heatmap representing the pairwise sequence identity (red color corresponds to high identity and white to low identity). The colored boxes between the heatmap and the dendrogram correspond to the cluster membership of the structures.
9. To assess the level of sequence conservation at each position in an alignment, the "similarity," "identity," and "entropy" per position can be calculated. The "similarity" is defined as the average of the similarity scores of all pairwise residue comparisons for that position in the alignment, where the similarity score between any two residues is the score value between those residues in the BLOSUM62 substitution matrix. The "identity," i.e., the preference for a specific amino acid to be found at a certain position, is assessed by averaging the identity scores resulting from all possible pairwise comparisons at that position in the alignment, where all identical residue comparisons are

given a score of 1 and all other comparisons are given a value of 0. "Entropy" is based on Shannon's information entropy. Note that the returned scores are normalized so that conserved columns score 1 and diverse columns score 0.

10. To investigate only a subset of the structures, toggle the **Filter/toggle displayed PDBs** check box in the **PDBs Viewing Options** panel. A table of all the hits will now appear below the visualization window. Select PDB IDs you are interested in, and the visualizer will now only show these selected PDB IDs.
11. The ensemble of aligned PDBs can also be visualized in your favorite molecular viewer program (e.g., PyMOL or VMD) by downloading the aligned PDBs with the **Download Aligned PDBs** or **Download PyMOL session file** button. The latter option will generate a PyMOL session file with the structures aligned and colored according to the structure color options chosen in the panel. Click the button to download the file. Note that the file is zipped and you will have to unzip it before opening it in PyMOL.
12. A trajectory view of the motion described by the PCA or NMA can be obtained by clicking on the **Download PDB trajectory** or **PyMOL session** buttons. This gives you a multi-model PDB file to be opened in your favorite molecular viewer, e.g., PyMOL or VMD. The PyMOL session file provides the motions as a vector field.

Acknowledgments

We thank Shashank Jariwala, Hongyang Li, and Dr. Guido Scarabelli for extensive testing throughout development as well as the Bio3D user community for their feedback and comments that have improved this application.

References

1. Skjaerven L, Jariwala S, Yao XQ, Grant BJ (2016) Online interactive analysis of protein structure ensembles with Bio3D-web. Bioinformatics 32(22):3510–3512. https://doi.org/10.1093/bioinformatics/btw482. Epub 2016/07/18. PubMed PMID: 27423893; PMCID: PMC5181562
2. Grant BJ, Rodrigues AP, ElSawy KM, McCammon JA, Caves LS (2006) Bio3d: an R package for the comparative analysis of protein structures. Bioinformatics 22(21):2695–2696. Epub 2006/08/31. https://doi.org/10.1093/bioinformatics/btl461
3. Skjaerven L, Yao XQ, Scarabelli G, Grant BJ (2014) Integrating protein structural dynamics and evolutionary analysis with Bio3D. BMC Bioinformatics 15:399. https://doi.org/10.1186/s12859-014-0399-6. Epub 2014/12/11. PubMed PMID: 25491031; PMCID: PMC4279791
4. Dwyer MA, Hellinga HW (2004) Periplasmic binding proteins: a versatile superfamily for protein engineering. Curr Opin Struct Biol 14(4):495–504. Epub 2004/08/18. https://doi.org/10.1016/j.sbi.2004.07.004

Chapter 3

Using Dali for Protein Structure Comparison

Liisa Holm

Abstract

The exponential growth in the number of newly solved protein structures makes correlating and classifying the data an important task. Distance matrix alignment (Dali) is used routinely by crystallographers worldwide to screen the database of known structures for similarity to newly determined structures. Dali is easily accessible through the web server (http://ekhidna.biocenter.helsinki.fi/dali). Alternatively, the program may be downloaded and pairwise comparisons performed locally on Linux computers.

Key words Classification of protein folds, Database searching, Distance geometry, Pattern recognition, Protein structure alignment

1 Introduction

At the end of 2018, the Protein Data Bank (PDB) contained the structure of 300,000 protein chains. Nearly all proteins have structural similarities to other proteins. General similarities arise from principles of physics and chemistry that limit the number of ways in which a polypeptide chain can fold into a compact globule. Evolutionary relationships result in surprising similarities, which are even stronger than similarity due to convergence caused by physical principles. Comparing 3D structures may reveal biologically interesting similarities that are not detectable by comparing only sequences and may help to infer functional properties of hypothetical proteins. For example, the recent discovery of structural homology and a conserved Cys-Asp-His catalytic triad unified two previously uncharacterized effectors from *Legionella pneumophila* with the cycle inhibiting factor (cif) gene family, leading to mechanistic insights of host manipulation by this pathogenic bacterium [1].

Large proteins can be decomposed into semiautonomous, globular folding units called domains. Domains are often evolutionarily mobile modules and may carry specific biological functions. Because a common domain may be surrounded by

Zoltán Gáspári (ed.), *Structural Bioinformatics: Methods and Protocols*, Methods in Molecular Biology, vol. 2112, https://doi.org/10.1007/978-1-0716-0270-6_3, © Springer Science+Business Media, LLC, part of Springer Nature 2020

completely unrelated domains, most structure comparison methods search for local similarities. A structural alignment defines a set of one-to-one correspondences between Cα atoms in two proteins. This is analogous to sequence alignment, except that the notion of similarity is much more complex between three-dimensional objects than between linear sequences. A large variety of scoring functions for structural similarity have been proposed [2]. The most important categories are (1) scoring functions based on the root mean square deviation (RMSD) of rigid-body superimposition and (2) scoring functions allowing flexible superimposition or plastic deformations. Early works based on visual analysis of folds stressed the importance of plastic deformations in the evolution of protein structure. Dali's scoring function belongs to the latter category, and it has been shown to yield structural dendrograms that agree well with expert classifications [3–5].

The Dali method is based on a sensitive measure of geometrical similarities defined as a weighted sum of similarities of intramolecular distances [3]. Let's consider two proteins labeled A and B. The match of two substructures is evaluated using an additive similarity score S of the form:

$$S(\mathrm{A},\mathrm{B}) = \sum_{i\in\mathrm{core}}\sum_{j\in\mathrm{core}}\varphi(i,j) \tag{1}$$

where i and j label residues, core is the common substructure, and φ is a similarity measure based on some pairwise relationship, here on the similarity of intramolecular Cα–Cα distances. Unmatched residues do not contribute to the overall score. For a given functional form of $\phi(i,j)$, the largest value of S corresponds to the optimal set of residue equivalences. The similarity measure needs to balance two contradictory requirements: maximizing the number of equivalenced residues and minimizing structural deviations. The use of relative rather than absolute deviations of equivalent distances is tolerant to the cumulative effect of gradual geometrical distortions. In Dali, the residue-pair score φ has the form:

$$\varphi(i,j) = (\theta - \mathrm{diff}(i,j)) * \mathrm{env}\left(\mathrm{d}_{ij}^{*}\right) \tag{2}$$

where the first term of the multiplication is the relative distance difference compared to a similarity threshold θ and the second term is an envelope function which downweights pairs in the long-distance range. In Dali, the similarity threshold is set to $\theta = 0.2$. The envelope is a Gaussian function $\mathrm{env}(x) = \mathrm{e}^{-\left(\frac{x}{R_0}\right)^2}$ where $R_0 = 20$ Å, calibrated on the size of a typical domain. The relative distance difference $\mathrm{diff}(i,j) = \frac{|d_{ij}^{\mathrm{A}} - d_{ij}^{\mathrm{B}}|}{d_{ij}^{*}}$, where d_{ij}^{A} and d_{ij}^{B} are intramolecular Cα–Cα distances in structure A and B, respectively, and their average is $d_{ij}^{*} = \frac{d_{ij}^{\mathrm{A}} + d_{ij}^{\mathrm{B}}}{2}$. Inserting the values of the constants,

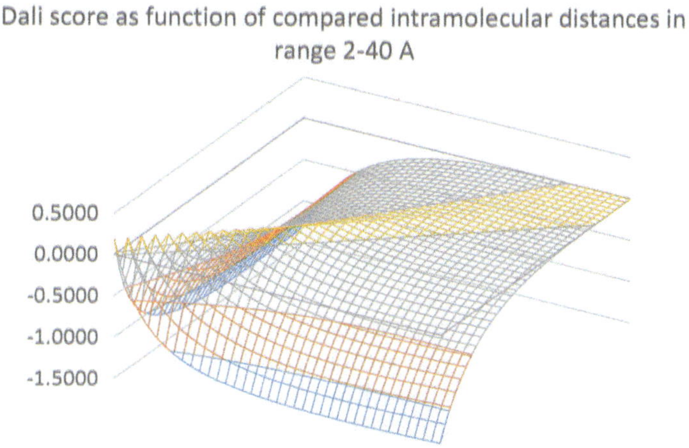

Fig. 1 Pairwise distances contribute a positive score (yellow color) when the relative deviation is less than 20%. An envelope function damps the contribution of longer distances

the resulting raw Dali score describing the structural similarity is given by

$$S(\mathrm{A},\mathrm{B}) = \sum\nolimits_{i\in\mathrm{core}}\sum\nolimits_{j\in\mathrm{core}}\left(0.2 - \frac{2|d_{ij}^{\mathrm{A}} - d_{ij}^{\mathrm{B}}|}{d_{ij}^{\mathrm{A}} + d_{ij}^{\mathrm{B}}}\right)\mathrm{e}^{-\left(\frac{d_{ij}^{\mathrm{A}}+d_{ij}^{\mathrm{B}}}{40\mathrm{A}}\right)^2} \quad (3)$$

The residue-pair scores (Fig. 1) can get both positive and negative values; therefore, the maximum of Eq. 3 corresponds to a local alignment. Hydrogen bonded backbone segments in helices and sheets have a distance of around 5 A. Here, absolute distance deviations up to 1 A generate positive scores, while larger deviations incur a steeply increasing penalty. At 10 A distance, as found between helices or sheets in tertiary contact, a deviation of up to 2 A still contributes positively, and larger deviations incur a mild penalty. The diameter of a typical domain is around 20 A. Beyond this distance, the score function is relatively insensitive to distance deviations. For example, two conformations of a two-domain structure, which are not superimposable as rigid bodies because of hinge rotation, can be structurally aligned by Dali since the similarity of local structure compensates for the downweighted deviations in interdomain distances.

For random pairwise comparison, the expected Dali score (Eq. 3) increases with the number of residues in the compared proteins. In order to describe the statistical significance of a pairwise comparison score S(A,B), we use the Z-score defined as

$$Z(\mathrm{A},\mathrm{B}) = \frac{S(\mathrm{A},\mathrm{B}) - m(L)}{\sigma(L)} \quad (4)$$

The relation between the mean score m, standard deviation σ, and the average length $L = \sqrt{L_A L_B}$ of two proteins was derived empirically from a large set of random pairs of structures. Fitting a polynomial gave the approximation:

$$\begin{aligned} m(L) &\approx 7.95 + 0.71L - 2.59 \cdot 10^{-4}L^2 - 1.92 \cdot 10^{-6}L^3, && \text{if } L \leq 400, \\ m(L) &= m(100) + L - 400, && \text{if } L > 400 \end{aligned} \quad (5)$$

For standard deviation, the empirical estimate was $\sigma(L) = 0.5 * m(L)$. The *Z*-score is computed for every possible pair of domains, and the highest value is reported as the *Z*-score of the protein pair [6]. Possible domains are determined by the Puu algorithm (*p*arser for protein *u*nfolding *u*nits), which recursively cuts a structure into smaller compact substructures at the weakest interface [7].

2 Materials

The Dali method is available as a web service at http://ekhidna.biocenter.helsinki.fi/dali. The standalone version can be downloaded from http://ekhidna.biocenter.helsinki.fi/dali/#download, which gives instructions for installation. The DaliLite.v4 package contains two Perl wrapper scripts along with installation instructions in the README file as well as source code and sample input/output files. The program is designed to run under Linux. Compiling the source code requires Fortran-90 (e.g., gfortran) and C compilers. Openmpi is optional. Standard Perl is required to execute the wrapper scripts. To install the programs, unpack the zip archive in a suitable directory, edit the path to the Dali home directory in the Makefile, and follow the instructions.

The installed package contains two Perl scripts:

1. A script **import.pl** which must be used to convert PDB files to Dali's internal data format. This script handles the input PDB files which might contain multiple chains, passes them to the DSSP program for extracting the coordinates and defining secondary structure elements, reads the output of DSSP, prepares a hierarchical tree of folding units, and outputs a data file for each chain in the input PDB file (*see* **Note 1**).
2. The script **dali.pl** performs pairwise comparisons of a list of query structures to a list of target structures. The lists of query and target structures must be provided by the user (*see* **Note 2**).

3 Methods

3.1 Input File

The input structure must be a PDB format text file. The PDB format consists of records (lines) where the first six characters are a keyword and data follows in fixed-width columns. Dali uses data from the COMPND and ATOM records. Only the first model of an NMR ensemble is read in. The input structure must have complete backbone atoms (C, CA, N, O); this requirement comes from the DSSP program used to parse PDB files. Though only CA coordinates are used in structural alignment, the DSSP step is necessary because also secondary structure assignments by DSSP are used as input to structure comparison. Chains shorter than 29 amino acids are ignored (*see* **Note 3**). The maximum throughput of the web server is 100–200 structure database searches per day. To apply the method on a larger number of structures, we advise the use of the standalone version.

3.2 Structure Data Parsing

The DSSP method [8] is used to parse Cα coordinates and to define secondary structure elements from the PDB file. The dsspCMBI implementation of DSSP is included in the standalone package. dsspCMBI is maintained at ftp://ftp.cmbi.ru.nl/pub/software/dssp/. The DSSP algorithm defines hydrogen bonds based on the dipole interaction of backbone amide and carbonyl groups. The interaction energy is modeled by a Coulomb potential between partial charges, which leads to a function of the angle and distance of the dipoles. Regular patterns of hydrogen bonds between runs of residues generate turns, helices, bridges, ladders, and sheets. Dali uses secondary structure elements (helices, beta strands) to simplify structural alignment. Alignment is further simplified by using a tree of compact substructures to guide alignment identifying first local matches and then solving a combinatorial problem in building up larger clusters of matching substructures. The tree is generated by the Puu program [7]. The underlying physical concept is maximal interactions within each unit and minimal interaction between units (domains). In a simple harmonic approximation, interdomain dynamics is determined by the strength of the interface and the distribution of masses. The most likely domain decomposition involves units with the most correlated motion or largest interdomain fluctuation time. The decomposition of a convoluted 3D structure is complicated by the possibility that the chain can cross over several times between units. Grouping the residues by solving an eigenvalue problem for the contact matrix reduces the problem to a one-dimensional search for all reasonable trial bisections. Recursive bisection yields a tree of putative folding units. Simple physical criteria are used to identify units that could exist by themselves.

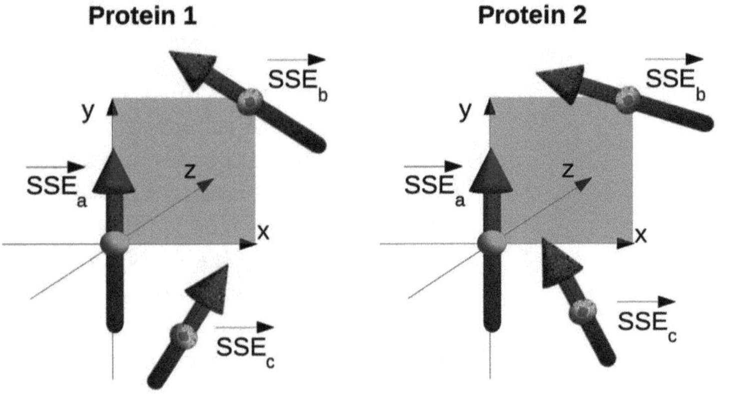

Fig. 2 Scheme of the coordinate system of the Wolf method. The first unit vector representing a secondary structure element (SSE_a) is aligned along the *y* axis with its midpoint at the origin. The midpoint of the second vector (SSE_b) is placed in the *xy* plane. In the example shown, the SSE_b vectors in the two proteins have similar orientations and midpoint positions. Figure redrawn based on [10]

3.3 Pairwise Comparison

Dali implements four structure alignment algorithms. In the standalone package, these are available through the serialcompare/mpicompare programs, though in practice they are invoked through the wrapper script dali.pl.

1. The Soap algorithm [9] is used to align structures with few (*see* **Note 4**) secondary structure elements. Soap minimizes a "soap film" metric between two Cα traces superimposed in 3D space. The minimal surface area between the virtual backbones of two proteins is determined numerically using an iterative triangulation strategy. The first protein is then rotated and translated in space until the smallest minimal surface is obtained. Such a technique yields the optimal structural superposition between two protein segments.
2. The Wolf algorithm is a very fast filter to identify obvious similarities [10]. It models secondary structure elements as vectors. Three points taken from an ordered pair of secondary structure elements (SSE) define an internal coordinate frame. Here, the midpoint of the first SSE is the origin, the vector representing the first SSE aligns with the *y* axis, and the midpoint of the second SSE is in the positive *z–y* half-plane (Fig. 2). Each database structure is presented in the "poses" defined by all possible frames of SSE pairs. Testing all frames of the query structure allows counting the number of matching SSEs at nearby positions in all possible "poses" by a fast lookup procedure (*see* **Note 5**). The result is a ranked list of database structures which can be used as a filter in a database search.

3. Parsi is a sensitive branch-and-bound alignment algorithm [11]. The algorithm is guaranteed to deliver an exact solution to the subproblem of ungapped alignment of secondary structure elements (SSEs), ignoring loops. Dali is based on a sum of pairs score. The score of an alignment involving n segments has n diagonal terms and $n(n - 1)$ off-diagonal terms in the summation. The off-diagonal dependencies pose a difficult combinatorial problem. The branch-and-bound algorithm overcomes the difficulty by initially pooling all possible segment-to-segment pairs as potential constituents of the optimal alignment. An upper bound of the total alignment score is given by the sum of the maxima of each of the n^2 terms independently of the others. More formally, the problem of optimizing the alignment score (Eq. 1) over all possible alignments $\mathrm{A} \rightarrow \mathrm{B}$ can be rewritten using an indicator function $\mathbf{1}_{\mathrm{A}\rightarrow\mathrm{B}}$ as

$$S^*(\mathrm{A},\mathrm{B}) = \max_{\mathrm{A}\rightarrow\mathrm{B}} \sum_{x=1}^{m} \sum_{y=1}^{m} \sum_{x'=1}^{n} \times \sum_{y'=1}^{n} \varphi(x \rightarrow x', y \rightarrow y) * \mathbf{1}_{\mathrm{A}\rightarrow\mathrm{B}}(x \rightarrow x') * \mathbf{1}_{\mathrm{A}\rightarrow\mathrm{B}}(y \rightarrow y') \tag{6}$$

$$\text{where } \mathbf{1}_X(a) = \begin{cases} 1, & \text{if } a \text{ is a member of set } X \\ 0, & \text{if } a \text{ is not a member of set } X \end{cases}$$

Here, the indicator function picks one-to-one correspondences defined by a given alignment $\mathrm{A} \rightarrow \mathrm{B}$, while all other terms are zero. The maximum is searched over all possible alignments of m residues in structure A and n residues in structure B, which is a hard combinatorial problem. A partition is a subset of the search space, which can contain many-to-many correspondences between residues in structure A and residues in structure B. An upper bound on the best one-to-one alignment score that is contained within a partition is given by

$$S^*(\text{partition}) \leq \sum_{x=1}^{m} \sum_{y=1}^{m} \max_{\substack{x'=1\ldots n \\ y'=1\ldots n}} \left(\varphi(x \rightarrow x', y \rightarrow y'') * \mathbf{1}_{\text{partition}}(x \rightarrow x') * \mathbf{1}_{\text{partition}}(y \rightarrow y')\right). \tag{7}$$

The search space is recursively partitioned to derive tighter upper bounds for the subspaces. A binary partition moves one particular query-target segment pairing to one subspace and excludes it from the other. The algorithm terminates when the partition that has the highest upper bound corresponds to unique pairings of all segments.

4. All alignments generated by methods 1–3 above use different objective functions that only approximate the Dali score or exclude loops from the alignment. All alignments generated by methods 1–3 are therefore refined using a Monte Carlo algorithm (Dalicon) that aims to maximize the Dali score over the whole structures [3].

Interestingly, Dali has been shown to generate close to optimal solutions on a benchmark of small proteins [12].

3.4 Web Server Methods

The web server and standalone version use the same algorithms for structure comparison. However, the web server has search and data visualization options which are not included in the standalone package. The web server supports four types of comparison:

(a) Search query structure against the Protein Data Bank using heuristics and a knowledge base of precomputed pairwise structure similarities.

(b) Compare query structure against a representative subset of the Protein Data Bank using systematic pairwise comparison.

(c) Perform pairwise comparison of a query structure against a set of target structures.

(d) Perform all against all comparison of up to 64 structures.

All methods are based on pairwise comparison. Methods **(b)**–**(d)** are available in the standalone version.

The search (method **(a)**) heuristically prunes the list of targets so that dissimilar target structures can be eliminated without explicit computation [13]. The elimination relies on a knowledge base of accumulated pairwise comparisons of structures in the PDB, which are represented as a graph. The nodes of the graph represent protein structures, and edges represent structural alignments. The idea is that once a strong similarity to the query structure has been found, other structural neighbors can be collected by walks through the graph, provided that structurally similar proteins form a connected component in the graph. A cascade of comparison methods is used to try and find a strong similarity from the query structure to known structures with little computational effort. The cascade starts with sequence comparison followed by Wolf or Soap. When a strong similarity is found, the search switches to "walking" based on transitive alignments. If no strong similarity was found, the query structure is compared against a representative subset of PDB using Parsi. Finally, a sequence search of the structurally most similar targets identifies homologs not caught by the previous steps. The *Z*-score threshold for extending the walk is adjusted dynamically during the search. Edges with lower *Z*-score than the threshold are effectively removed from the structural similarity graph. There are only empirical rules for setting the threshold. Initially, it is set to the square root of the Z-score for the comparison of the query structure to itself. Subsequently, it is increased if there are many higher scoring targets. Specifically, the aim is to report the 100 highest scoring PDB90 representatives (PDB structures with less than 90% sequence identity to each other). Because small domains obtain smaller Z-scores than large domains, we recommend cutting multidomain structures and searching each domain separately.

3.5 Interpretation of the Result

Like in sequence analysis, the goal of structural database searching is usually to identify homologous proteins which might provide clues to the function of the query protein. Homology means descent from a common ancestor. We can infer homology from sequence or structural similarities that are so strong they would not be expected to have arisen by chance. The boundary between homologous and unrelated proteins varies from one family to another, and there is no universally applicable Z-score cutoff to separate homologous from analogous (nonhomologous) structures. As a rule of thumb, a Z-score above 20 means the two structures are definitely homologous, between 8 and 20 means the two are probably homologous, between 2 and 8 is a gray area, and a Z-Score below 2 is not significant. The wide gray zone is because the size of the proteins influences Z-scores—small structures will tend to have small Z-scores, whereas a medium Z-score for very large structures need not imply a biologically interesting relationship. Fold type also has an effect—α/β proteins also usually have higher Z-scores than all β proteins. For example, all $(\beta\alpha)8$-barrel folds are unified at Z-scores above 10. In contrast, a small avian polypeptide (PDB code 1ppt) contains only one helix and a proline-rich loop and gets a Z-score under 8 even in comparison with itself. In view of the Z-score, it is much more improbable to observe 16 helices and strands in a similar packing arrangment than to find a similar arrangement of just a helix and a loop.

Other criteria than the mere Z-score are often required to make a convincing case for homology. Structural dendrograms are useful in locating the boundary between homologous and analogous folds, the idea being that homologous proteins should be monophyletic and functionally similar [4]. Dali generates structural dendrograms from the matrix of pairwise Z-scores by average linkage clustering. Branch lengths in the dendrogram represent distances, which are modeled ad hoc as the difference of Z-scores.

Dali web server results (Fig. 3) are linked to interactive sequence search and function assignment servers [14, 15]. The structural alignments can be visualized as stacked sequence logos, where the logos are generated from sequence neighbors of the target protein and the alignment of the logos is based on the structure comparison. In particular, enzyme superfamilies have sharp sequence signatures, but binding domains can have very little

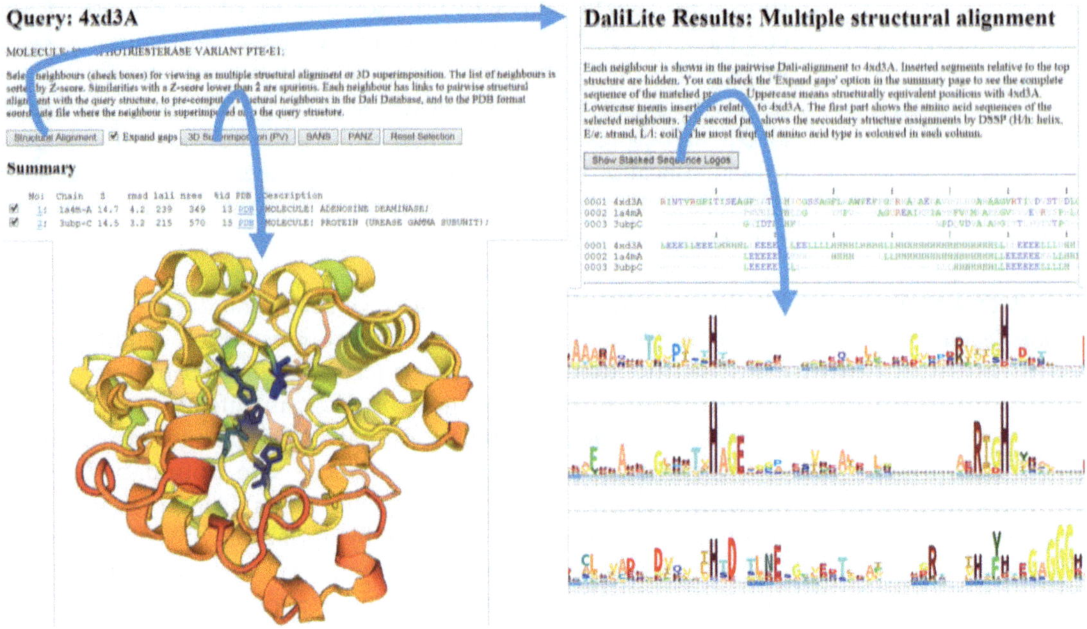

Fig. 3 Outputs of the Dali web server, clockwise from top left: summary of similar structures ordered by *Z*-score, structural alignment showing amino acid sequences and secondary structure assignments, stacked sequence logos highlighting conserved structurally equivalent positions, 3D structure view (here colored by sequence conservation)

sequence similarity. Without a sequence signature, it is harder to establish homology. Web server results additionally have a link labeled PDB for each target structure, which returns the coordinates superimposed onto the query structure for viewing in a molecular graphics program of your choice.

Besides the *Z*-score, Dali reports the RMSD and the number of equivalent residues (LALI), because they are traditional measures and often quoted as qualifiers of structural similarity. RMSD is a measure of the average deviation in distance between aligned Cα atoms in 3D superimposition. For sequences sharing 50% identity, this should be around 1.0. Dali maximizes a geometrical similarity score, which is defined in terms of similarities of intramolecular distances and is thus not primarily aiming to generate alignments with low RMSD. Numerous programs for structure comparison have been published over the last 30 years, based on a variety of similarity measures [2]. Consequently, method evaluations often assess the quality of structural alignments using a non-native yardstick, such as the popular RMSD measure. An alignment is "better" if it has both smaller RMSD and larger LALI. If both RMSD and LALI are smaller or both are larger, it is not possible to establish an order between the alignments.

3.6 Dali Comparison with the Locally Installed Standalone Version

The standalone version can be used for a PDB format file with one or multiple protein chains (identified by a letter in column 19). Compressed files with extension .gz and normal text files are accepted (*see* **Note 6**). There is generally no reason to change parameters from their default values which are hardcoded in the standalone version. In a preparatory step, protein structure data must be imported by invoking the wrapper script import.pl. Normally, the script can be invoked as

perl -I $DALI_HOME/bin $DALI_HOME/bin/import.pl pdb1xg8ent.gz 1xg8 $DALI_DAT/

where environment variables $DALI_HOME and $DALI_DAT point to the installation directory and data directory, respectively. The first argument is the PDB file name, and the second argument is the four-letter identifier to be used for the structure. One data file is generated for each chain the PDB file. Thus, the number of output files corresponds to the number of chains longer than 28 residues (*see* **Note 3**). The example above creates a file named 1xg8A.dat, where the chain identifier has been automatically appended to the four-letter structure identifier. The first line of a properly formed data file looks like Fig. 4 (*see* **Note 7**). If reading the coordinates failed, for any reason, you only find lots of zeros on the first line of the data file. Troubleshooting tips include installing the DSSP program from the distribution package and checking the format of the PDB entry (Fig. 5).

Pairwise structure comparison is performed by invoking the wrapper script dali.pl as

perl -I $DALI_HOME/bin $DALI_HOME/bin/dali.pl
--query query.list --db target.list
--dat1 $DALI_DAT --dat2 $DALI_DAT

```
>>>> 1xg8A  108    7    3    4  EHEHEEH
     |   |    |    |    |    |  L------  sequential order secondary structure elements
     |   |    |    |    |    L---------  number of beta-strands (E)
     |   |    |    |    L--------------  number of helices (H)
     |   |    |    L-------------------  number of secondary structure elements
     |   |    L------------------------  number of residues
     |   L-----------------------------  chain identifier (one character, appended automatically)
     L---------------------------------  PDB identifier (four characters)
```

Fig. 4 First line of a data file

```
ATOM      1  N   GLY A   1       2.296  -9.636  18.253  1.00  0.00           N
ATOM      2  CA  GLY A   1       1.470  -9.017  17.255  1.00  0.00           C
ATOM      3  C   GLY A   1       0.448  -9.983  16.703  1.00  0.00           C
ATOM      4  O   GLY A   1       0.208 -11.066  17.345  1.00  0.00           O
```

Fig. 5 Example of coordinate data in PDB format. The format has its roots in the era of punch cards when data fields had a fixed width and position

```
# Job: test
# Query: 1pptA
# No:  Chain   Z    rmsd lali nres  %id PDB  Description
   1:  1bba-A  3.6  1.8   33   36   39   MOLECULE: BOVINE PANCREATIC POLYPEPTIDE;
         |      |    |     |    |    L- percent identity of aligned amino acids
         |      |    |     |    L------ number of residues in target structure
         |      |    |     L----------- number of aligned C-alpha atoms
         |      |    L----------------- root-mean-square deviation of aligned C-alpha atoms
         |      L---------------------- Z-score
         L----------------------------- target structure

# Structural equivalences
   1: 1ppt-A 1bba-A      1 -  33 <=>    1 -  33   (GLY    1 - ARG   33 <=> ALA    1 - ARG   33 )
      query  target    \___________________/    \___________________________________________/
                        query   <=>   target                query    <=>    target
                        sequential numbering                PDB residue numbering

# Translation-rotation matrices
-matrix  "1ppt-A 1bba-A  U(1,.)   0.631906 -0.761372 -0.144939          -0.890845"
-matrix  "1ppt-A 1bba-A  U(2,.)   0.512616  0.550832 -0.658642         -10.882093"
-matrix  "1ppt-A 1bba-A  U(3,.)   0.581308  0.341902  0.738366           4.946664"
                                 \____________________________/        \_________/
                                       rotation matrix U            translation vector T
```

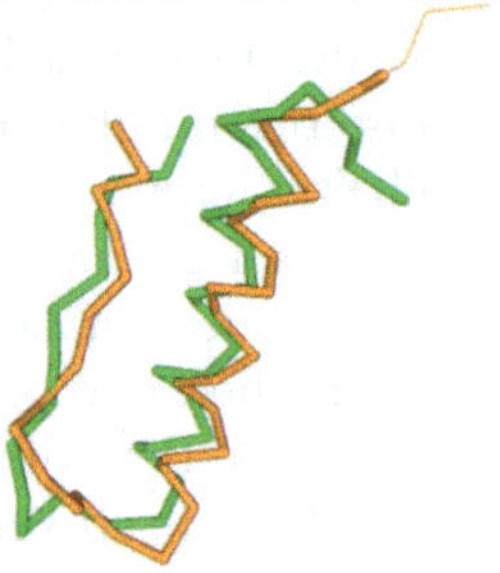

Fig. 6 (Top) Example output from pairwise comparison. Comments have been added in *italics*. (Bottom) Cα traces of 1pptA (green) and 1bbaA (orange) in structural superimposition. Unaligned residues are shown by a thin line

where query.list and target.list are text files containing one 5-letter structure identifier per line. All query and target structures must have been imported beforehand using the script import.pl (*see* above). The output is a text file xxxxY.txt for each structure xxxxY in query.list (Y is the chain identifier, while xxxx is the PDB entry identifier). The output is sorted in decreasing order of the *Z*-score (*see* **Note 8**). The results are presented in three blocks (Fig. 6). The summary block lists statistical parameters for the matched target structures. The alignment block lists the aligned segments. The last block gives the translation-rotation matrices for rigid-body superimposition of target structures onto the query structure. $UY + T$ is the best approximation of X, where X and Y are the (x,y,z) coordinates of the query and target structure, respectively, U is the rotation matrix, and T is the translation vector [16].

All-against-all structure comparison of a set of structures is invoked with the matrix option as

```
perl -I $DALI_HOME/bin $DALI_HOME/bin/dali.pl
  --query query.list --matrix --dat1 $DALI_DAT
```

This option generates additional outputs named "ordered" and "newick_unrooted", which contain a matrix of pairwise *Z*-scores and a dendrogram in Newick format. Note that the matrix of *Z*-scores represents similarities between structures, whereas branch lengths of the Newick tree represent distances. Branch lengths are modeled ad hoc as the difference of *Z*-scores.

The wrapper scripts generate a number of intermediate results in the current work directory. The lock file dali.lock is created at the start of the job and is deleted automatically, when it completes successfully. If a file named dali.lock is present, you cannot start another Dali job in the same directory.

4 Notes

1. Dali handles each chain separately. Structure identifiers have a fixed length of five characters, where the last character is the chain identifier. Quaternary structure comparisons are not possible at present.
2. The dali.pl script has two parameters for data directories (DALIDATDIR_1 and DALIDATDIR_2). All query structures must be imported to DALIDATDIR_1. All target structures must be imported to DALIDATDIR_2. DALIDATDIR_1 and DALIDATDIR_2 can be identical, but usually DALIDATDIR_2 contains public structures imported from the Protein Data Bank (PDB), and DALIDATDIR_1 contains private structures.
3. The parameter $MINLEN in the Perl module mpidali.pm is set to 29 by default. The insulin peptide is accepted, shorter chains are rejected.
4. The parameter $MINSSE in the Perl module mpidali.pm is set to 3 by default. This means that structures with two or fewer SSEs are compared using the Soap method and structures with three or more SSEs are compared using Wolf or Parsi.
5. The Wolf algorithm uses three parameters. There is generally no reason to change the defaults. The parameters rcut and maxiter control the iterative refinement, which cycles between superimposition and alignment. The alignment favors the pairing of C-alpha atoms from the query and target structure gets a positive score if their positional deviation is smaller than rcut, which is 4 A by default. Maxiter limits iterations to 10. The parameter neiborcutoff says that internal coordinate frames are generated using every pair of SSE vectors whose midpoints are closer than 12 A.
6. Structures of the Protein Data Bank can be mirrored using the following command:

rsync -rlpt -v -z --delete --port=33444 rsync.rcsb.org::ftp_data/structures/divided/pdb/ $MIRRORDIR

where environment variable $MIRRORDIR is the top level of the local structure data directory.

7. The standalone version works on common Linux operating systems with the exception of Debian, which generates erroneous output from the DSSP program.

8. The amount of output from structure comparison is limited by the parameter $zcut in the Perl module mpidali.pm. $zcut is the minimum *Z*-score (default 2.0).

References

1. Valleau D, Quaile AT, Cui H, Xu X, Evdokima E, Chang C, Cuff ME, Urbanus ML, Houliston S, Arrowsmith CH, Ensminger AW, Savchenko A (2018) Discovery of ubiquitin deamidases in the pathogenic arsenal of *Legionella pneumophila*. Cell Rep 23:568–583
2. Hasegawa H, Holm L (2009) Advances and pitfalls of protein structural alignment. Curr Opin Struct Biol 19:381–389
3. Holm L, Sander C (1993) Protein structure comparison by alignment of distance matrices. J Mol Biol 233:123–138
4. Dietmann S, Holm L (2001) Identification of homology in protein structure classification. Nat Struct Biol 8:953–957
5. Fox NK, Brenner SE, Chandonia JM (2014) SCOPE: structural classification of proteins—extended, integrating SCOP and ASTRAL data and classification of new structures. Nucleic Acids Res 42:D304–D309
6. Holm L, Sander C (1998) Dictionary of recurrent domains in protein structures. Proteins 33:88–96
7. Holm L, Sander C (1994) Parser for protein folding units. Proteins 19:256–268
8. Kabsch W, Sander C (1983) Dictionary of protein secondary structure: pattern recognition of hydrogen bonded and geometrical features. Biopolymers 22:2577–2637
9. Falicov A, Cohen FE (1996) A surface of minimum area metric for the structural comparison of proteins. J Mol Biol 258:871–892
10. Holm L, Sander C (1995) Fast protein structure database searches at 90% reliability. ISMB 3:179–187
11. Holm L, Sander C (1996) Mapping the protein universe. Science 273:595–602
12. Wohlers I, Andonov R, Klau GW (2013) DALIX: optimal DALI protein structure alignment. IEEE/ACM Trans Comput Biol Bioinform 10:26–36
13. Holm L, Kääriäinen S, Rosenström P, Schenkel A (2008) Searching protein structure databases with DaliLite v.3. Bioinformatics 24:2780–2781
14. Somervuo P, Holm L (2015) SANSparallel: interactive homology search against Uniprot. Nucleic Acids Res 43:W24–W29
15. Törönen P, Medlar A, Holm L (2018) PANNZER2: a rapid functional annotation webserver. Nucleic Acids Res 46:W84–W88
16. Kabsch W (1978) A discussion of the solution for the best rotation to relate two sets of vectors. Acta Crystallogr A 34:827–828

Chapter 4

Assessing Protein Function Through Structural Similarities with CATH

Natalie L. Dawson, Christine Orengo, and Zoltán Gáspári

Abstract

The functional diversity of proteins is closely related to their differences in sequence and structure. Despite variations in functional sites, global structural similarity is a valuable source of information when assessing potential functional similarities between proteins. The CATH database contains a well-established hierarchical classification of more than 430,000 protein domain structures and nearly 95 million protein domain sequences, with integrated functional annotations for each represented family. The present chapter provides an overview of the main features of CATH with emphasis on exploiting structural similarities to obtain functional information for proteins.

Key words Structure similarity, Structure classification, Functional assignment, Protein function, Protein structure

1 Introduction

A long-standing observation in structural biology is that protein structure is more conserved than sequence [1–4]. It is also apparent that evolutionarily related proteins with similar folds often exhibit similar functions [5, 6]. CATH [7] is a free, publicly available resource that identifies protein domains within proteins from the Protein Data Bank [8] and classifies them into evolutionary related groups according to sequence, structure, and function information. The CATH structural classification was established in the mid-1990s, and the latest release (version 4.2) contains 434,857 structural domains. The sister resource, Gene3D [9], adds in additional protein domain sequences with no known structure, which brings the current total number of domains in CATH-Gene3D up to 95 million. Extensive search and browsing features are provided through the web interface (www.cathdb.info), and a suite of open-source tools is available through GitHub (https://github.com/UCLOrengoGroup/cath-tools).

Zoltán Gáspári (ed.), *Structural Bioinformatics: Methods and Protocols*, Methods in Molecular Biology, vol. 2112, https://doi.org/10.1007/978-1-0716-0270-6_4, © Springer Science+Business Media, LLC, part of Springer Nature 2020

CATH uses a hierarchical classification scheme where the units compared and classified are structural domains. Domains, defined here as globular structural domains capable of semi-independent folding, are extracted from experimentally determined protein structures available in the Protein Data Bank (PDB) archive [8]. Various structure-based algorithms (SSAP [10], CATHEDRAL [11]) and sequence-based algorithms (Needleman-Wunsch-based sequence alignments, jackhmmer from the HMMER3 suite [12], Profile Comparer [13], HHsearch [14]) are used to assess the similarity of domains to each other and to recognize protein homologues.

In addition to the structural classification, a functional subclassification is also provided in the form of Functional Families (FunFams) [15]: homologous superfamily subgroups whose relatives have similar functions. FunFams illustrate the functional diversity of proteins with similar folds and help to assess the biological role of related domains. To generate FunFams, relatives from each superfamily are first grouped into a tree of clusters using a profile-based, hierarchical agglomerative clustering method [16]. The FUNFHMMer algorithm [17] then determines where to cut this tree by detecting changes in specificity-determining residues [18] between two nodes (i.e., multiple sequence alignments) and calculating which nodes are considered to be functionally related. The FunFams have been independently validated by the international experiment, CAFA (Critical Assessment of Function Annotation) [19]. The CAFA experiment has found the FunFams to be highly competitive in assigning functional annotations to sequences of unknown function. The FunFams have also been shown to be useful in repurposing drugs [20] and for selecting suitable structural templates when building 3D models [21].

CATH is updated on a daily basis, which contains the latest domain boundary definitions, superfamily assignments, names of each node in the hierarchy, and sequence family annotations for all nonredundant sequence representatives (clusters at 35% sequence identity). Full releases on the other hand, denoted CATH-Plus, include additional detailed functional annotations (e.g., GO terms, EC terms) along with other types of derived data (e.g., FunFams, multiple sequence/structure alignments, structural clusters, superfamily superpositions).

In 2017, CATH was awarded the title of Core Data Resource (CDR) by ELIXIR. It is one of only 18 CDRs across Europe and 1 of only 9 outside of the EBI [22].

In this chapter, we provide an overview of the main features of the CATH server and demonstrate its usage with examples.

2 Methods and Materials for Accessing CATH Data

2.1 The CATH Hierarchy

The fundamental unit of CATH classification is the protein domain. A domain is derived from a chain in a PDB entry. The domain identifier has the form "*ppppcnn*," where "*pppp*" is the four-character PDB identifier, "*c*" is the one-character chain id, and "*nn*" is the two-digit number of the given domain within the chain as identified by CATH. For example, the identifier 2bwbA00 represents the sole domain found in chain A of the PDB structure 2BWB, which corresponds to the C-terminal region of the RuvA DNA helicase.

The domains are classified in a nine-level hierarchy (i.e., C.A.T.H.S.O.L.I.D). The four main levels (C.A.T.H), from top to bottom, are as follows: Class (C), based on the secondary structure content; Architecture (A), which considers the spatial arrangement of these elements; Topology (T) (or fold group) that further divides structures based on the connections between secondary structure elements and their 3D arrangement; and the level Homologous superfamily (H) that groups evolutionarily related domains together. Within superfamilies, domains are further divided according to the level of their sequence similarity, and relatives are clustered at increasing levels of sequence similarity to produce the SOLID sequence clusters: 35% (S), 60% (O), 95% (L), and 100% (I). The individual domains (D) are found at the bottom of the hierarchy.

2.2 The CATH Web Interface

The CATH web interface is available at www.cathdb.info (Fig. 1). This page offers multiple entry points to the CATH analysis pipeline. Here we will primarily focus on structure-based queries. Submitting a structure evidently carries information on the sequence, so it is worth to stress that in such a case both sequence- and structure-based queries can be initiated.

2.2.1 Using a Text Search to Find a Protein in CATH

The most straightforward query is to use a PDB identifier. In the search field on the CATH home page, the user can submit a PDB id and get the information about the given structure. On the results page, matching superfamilies, domains, and PDB hits will be reported, with the best matches displayed. Further matching entries can be explored by clicking on the "View all entries" button. If multiple CATH domains are identified in the query structure, all of these will be listed in the "Matching CATH Domains" section, and each superfamily is listed in the "Matching CATH Superfamilies." Clicking on a domain, superfamily, or PDB id will take the user to the web page of that entry for more information. For example, a PDB entry page lists information from the PDB file, experimental data details from the PDBe Prints widget, the chain

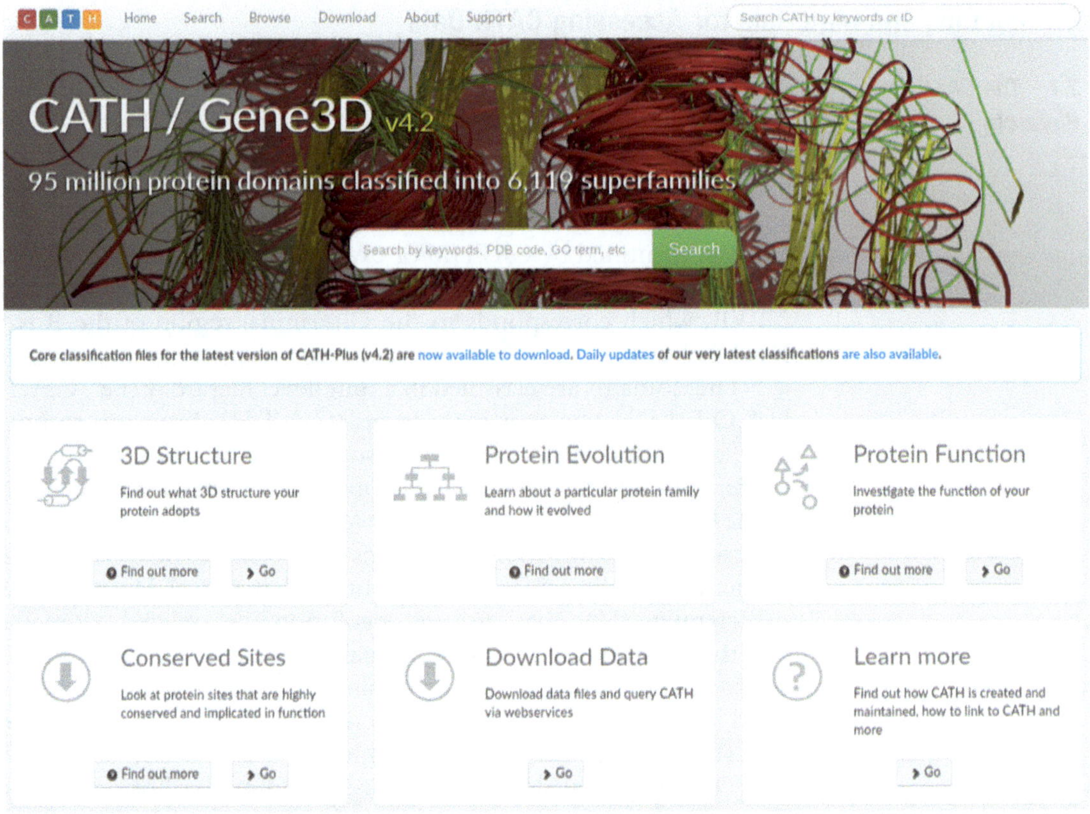

Fig. 1 The CATH home page, available at www.cathdb.info

ids, the CATH domains identified within these chains, and the status of the domain processing within the CATH hierarchy.

2.2.2 Using a Structure-Based Search to Identify Related Proteins

Users can submit a structure of interest to CATH to identify related proteins. From the CATH home page (www.cathdb.info), the user can either select the "3D Structure" entry point by clicking on the button "Go" in the corresponding region of the page or select the "Search" button at the top of the home page. Three search options will be offered, one based on text/identifiers, the second on sequence, and the third on structure search. Selecting the "Search by Structure" tab, a PDB file can be uploaded using the "Choose File" button. By clicking "Submit," the server identifies the PDB details and list of chains. It also automatically carries out a fast sequence similarity search to determine any protein sequence matches in CATH (*see* Subheading 2.2.3 for details on the sequence-based search tool). For each polypeptide chain identified in the submitted PDB structure, a structural scan can be submitted by clicking on the red "Submit structure" button. A structure similarity search is performed using the CATHEDRAL structure comparison pipeline [11]. The query protein chain is searched against a library of classified structural domains in CATH, and high-scoring matches are used to predict domain boundaries. CATHEDRAL first searches the query protein structure using GRATH [23] to rapidly identify fold matches. SSAP [10] is then run on the fold matches to more accurately identify structural domain matches. SSAP is an efficient structure alignment program capable of identifying structural similarities between proteins regardless of their sequence. Structure alignment is more computationally demanding than sequence alignment but allows the identification of homologous proteins with similar, conserved structure but divergent sequence.

2.2.3 Using a Sequence-Based Search to Identify Related Proteins

Users can submit a sequence of interest to identify related proteins. The tool can be found by selecting the "Search" button at the top of the home page and then selecting the "Search by Sequence" tab. Pasting the FASTA sequence and clicking "Search" submits the query. The progress of the scan will be reported and a tick will appear at each of the stages upon completion. When the scan is finished, the CATH structural domain and CATH functional family matches found can be viewed using the "Found *N* matches." Clicking on the "Found *N* matches" button on the right brings the user to the corresponding domain or FunFam summary page(s).

2.2.4 Homologous Superfamily Data

Each homologous superfamily has its own summary page that describes the main characteristics of the family (Fig. 2). Three diagrams provide information on the diversity of members of the superfamily represented by: Gene Ontology (GO) terms [24, 25] and Enzyme Commission (EC) numbers [26] and organisms. Naturally, EC diversity applies only to members having enzymatic function and represented in the EC database.

A diagram on the sequence-structure diversity of all CATH superfamilies is also shown, where the position of the given superfamily is highlighted in red in relation to all the others.

Selecting "Functional Families" on the left-side menu provides a list of FunFams associated with members of the superfamily. Depending on the size and functional diversity of the superfamily, the number of FunFams can vary greatly. Each FunFam has an id and a structural representative assigned where available. Clicking on the name of the FunFam, a summary page is displayed with GO, EC, and species diversity. Further details can be viewed by selecting the "Alignment," "GO," "EC," and "Taxonomy" tabs. For example, the FunFam multiple sequence alignment can be viewed and downloaded under the "Alignment" tab (Fig. 3). Sequence conservation patterns are color-coded within the alignment from blue (i.e., no residue conservation) to red (i.e., the residue is completely conserved). The color-coded alignment positions are mapped onto the representative structure, where available.

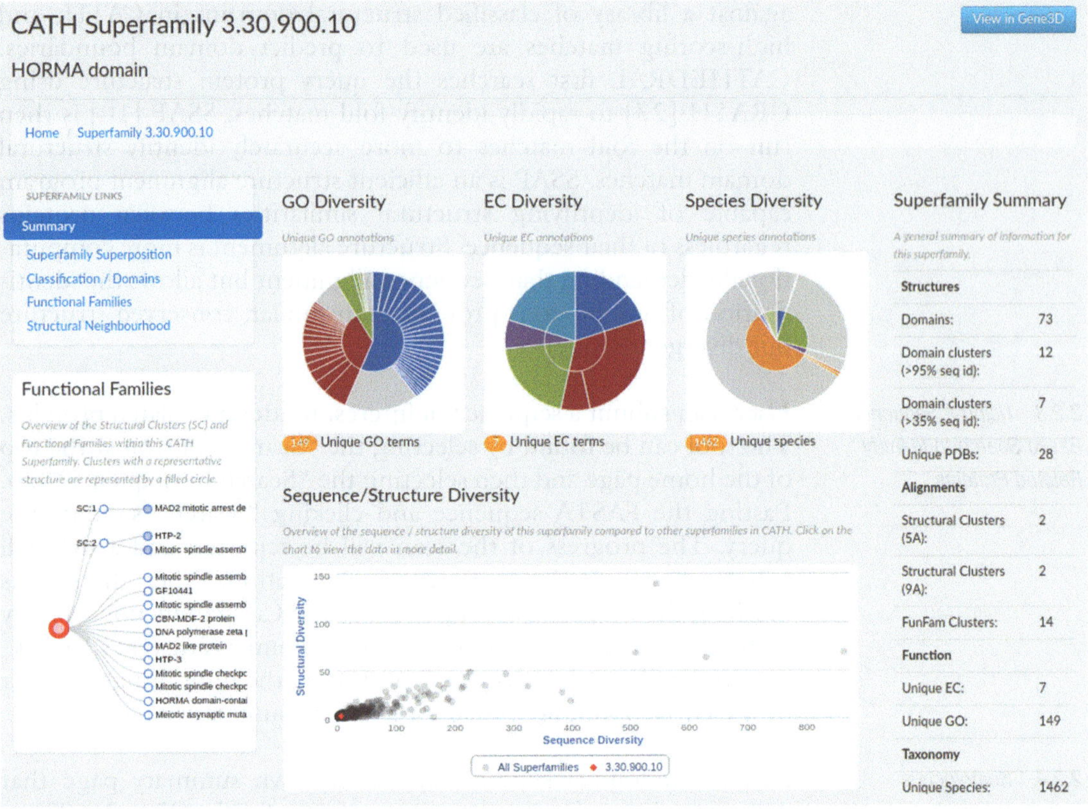

Fig. 2 Example of the summary superfamily page provided for each superfamily. Through the menu on the left-hand side, users can access superfamily superposition data, domain SOLID classification data, functional family data, and structural neighborhood data. Below this menu is an overview of the structural clusters and functional families in the selected superfamily. In the center of the page, users are provided with information on the GO terms, EC numbers, and species diversity. There is a sequence/structure diversity plot that places the selected superfamily (red dot) in the context of all other CATH superfamilies (gray dots). The right-hand side of the page is a summary of the superfamily statistics

2.3 Downloading CATH Data

CATH offers a suite of files for download via the FTP site, along with a README file containing detailed data descriptions. Data is available for each CATH-Plus release, as well as the daily updates. A link is also provided to the latest release of CATH-Plus for ease of use. The daily release provides putative domain boundary annotations, name descriptions for each CATH hierarchy node, and all domain ids within each superfamily S35 cluster (*see* Subheading 2.1 for S35 description). The CATH-Plus release provides the same classification of data-based files as the daily release, as well as sequence-based data (e.g., FunFam HMMs, superfamily domain FASTA files) and nonredundant ASTRAL-like sequence datasets at 20% and 40% sequence identity levels.

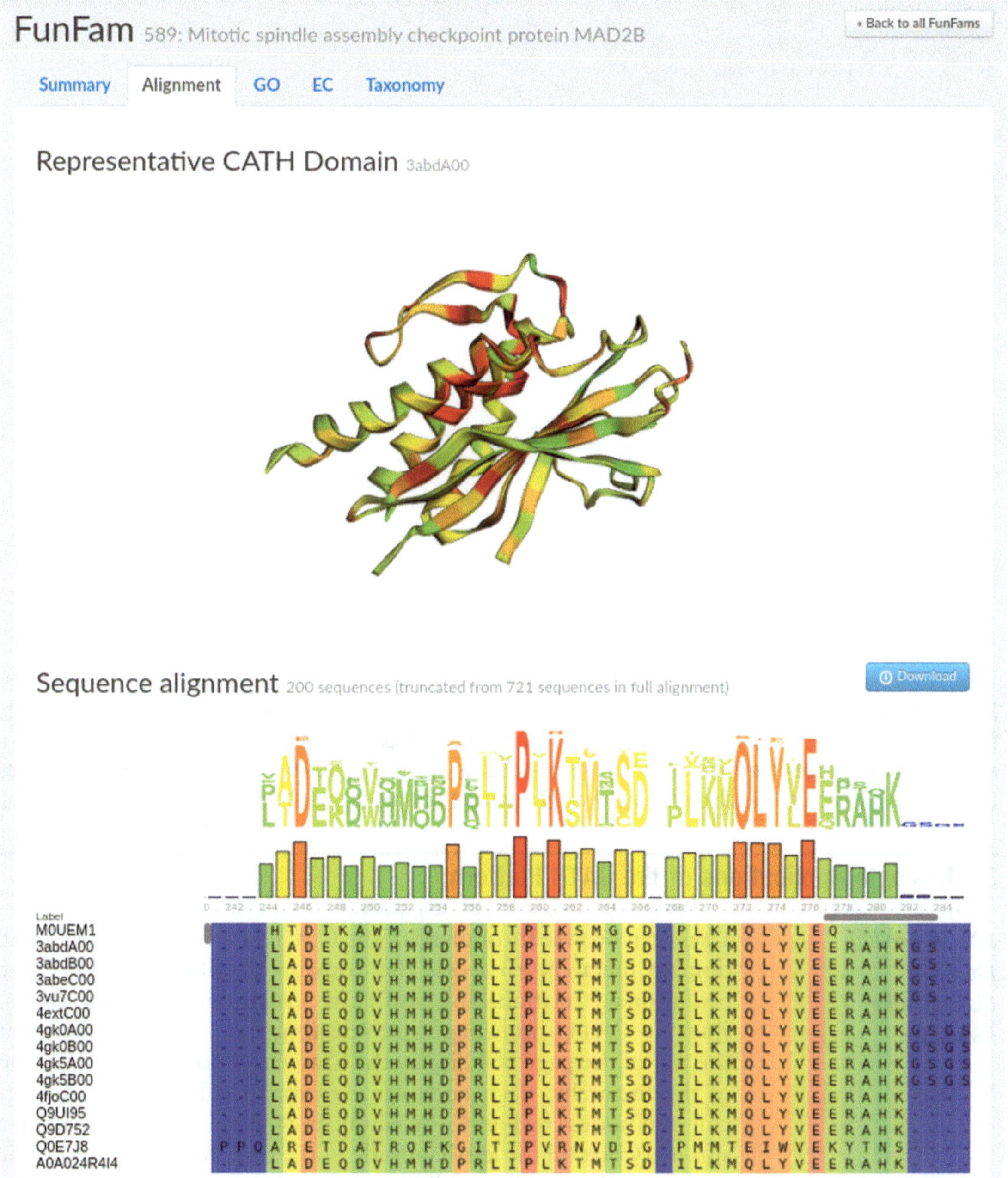

Fig. 3 Example of the information found under the "Alignment" tab of a FunFam web page. Where a 3D structural representative domain is available for a FunFam, the interactive structure is displayed together with the FunFam alignment (truncated to up to 200 sequences for visualization purposes). Alignment positions are colored according to conservation scores, from blue (no conservation) to red (complete conservation). The conservation score-based coloring is mapped onto the representative structure, illustrating the mapping from sequence to structure

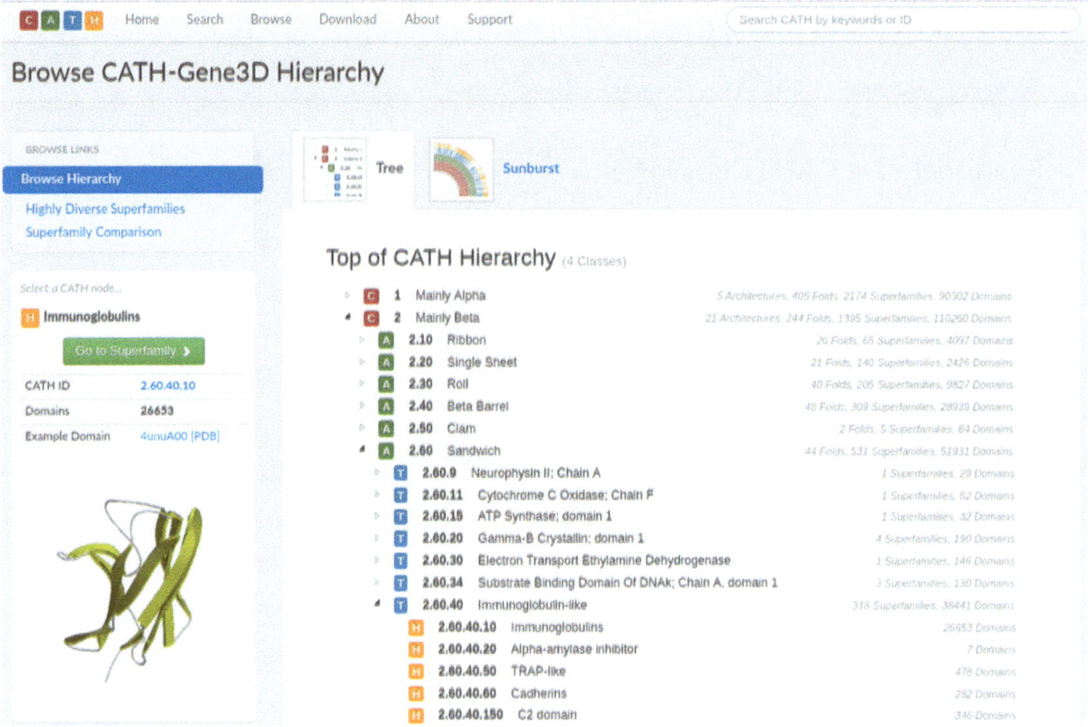

Fig. 4 Browsing the CATH-Gene3D hierarchy. The user can navigate across all of the nodes in the hierarchy to find information on structural properties, evolutionary relationships, and statistics on node counts. When a superfamily node is selected, such as the Immunoglobulins (CATH id: 2.60.40.10), the user can easily access the superfamily page and is also provided with an example domain id and structure

3 Example Applications of the CATH Platform

3.1 Browsing the CATH Classification Hierarchy

Users can browse through all of the C.A.T.H levels of the structural domain hierarchy to access the domains classified in CATH, the evolutionary relationships, and their structural properties. Accessing the CATH homepage and selecting the “Browse” button at the top of the page will display the Class-level nodes at the very top of the hierarchy. Each node can be expanded to display all of its children nodes (*see* Fig. 4). Selecting any CATH node will load up a summary box on the left-hand side of the page with statistics on the current node and all its children nodes. An example domain is also provided, where the user can click through to the domain summary page and/or download its PDB file.

3.2 Identifying Structural Relatives of the Protein Atg101

As the first example application, we will use structures from the HORMA domain family. This family contains three known subgroups that are dissimilar in sequence. The earliest identified members of the family are HOP1, REV7, and MAD2, proteins involved in the regulation of cell division and DNA repair [27]. The protein

Atg13, a key factor in autophagy initiation, was shown to possess a HORMA domain that was not identified based on its sequence alone [28]. Another autophagy-related protein, Atg101, itself a binding partner for Atg13, was also predicted to adopt the HORMA fold [29], which was later confirmed experimentally [30]. In this example, we show how these proteins are represented in CATH, with emphasis on the added value of structure-based comparisons, a key feature of the resource allowing practical investigation and demonstration of the "structure is more conserved than sequence" principle.

We will use the PDB structure 5XUY [31] as our query, a heterodimer of Atg13–Atg101, containing four chains. Chains A and C correspond to Atg13 and chains B and D to Atg101. Performing an id-based search reveals that as of writing this chapter, 5XUY is not yet integrated in CATH. To replicate this search, enter the PDB id in the search box and press the green "Search" button. The results display that no CATH superfamilies, domains, or PDB structures associated with this ID are present in the latest full release.

Next, we will use the sequence search tool to identify domain and functional family matches to the query sequence. Here, the FASTA sequence for chain A of 5XUY was entered into the "Search by Sequence" tab (Fig. 5), which resulted in one domain match and no functional family matches. Clicking on "Found 1 matches" will display that the match is to domain 5c50B00. Selecting the domain ID brings the user to the domain summary page (Fig. 6).

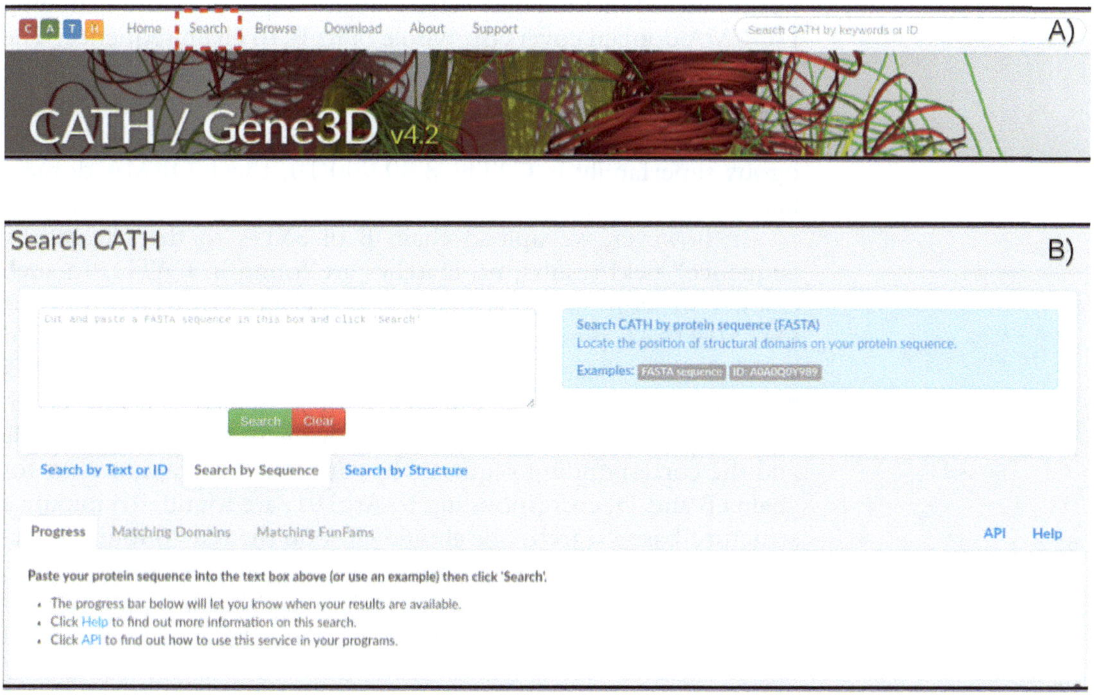

Fig. 5 How to navigate from the home page (**a**) to the sequence search tool (**b**)

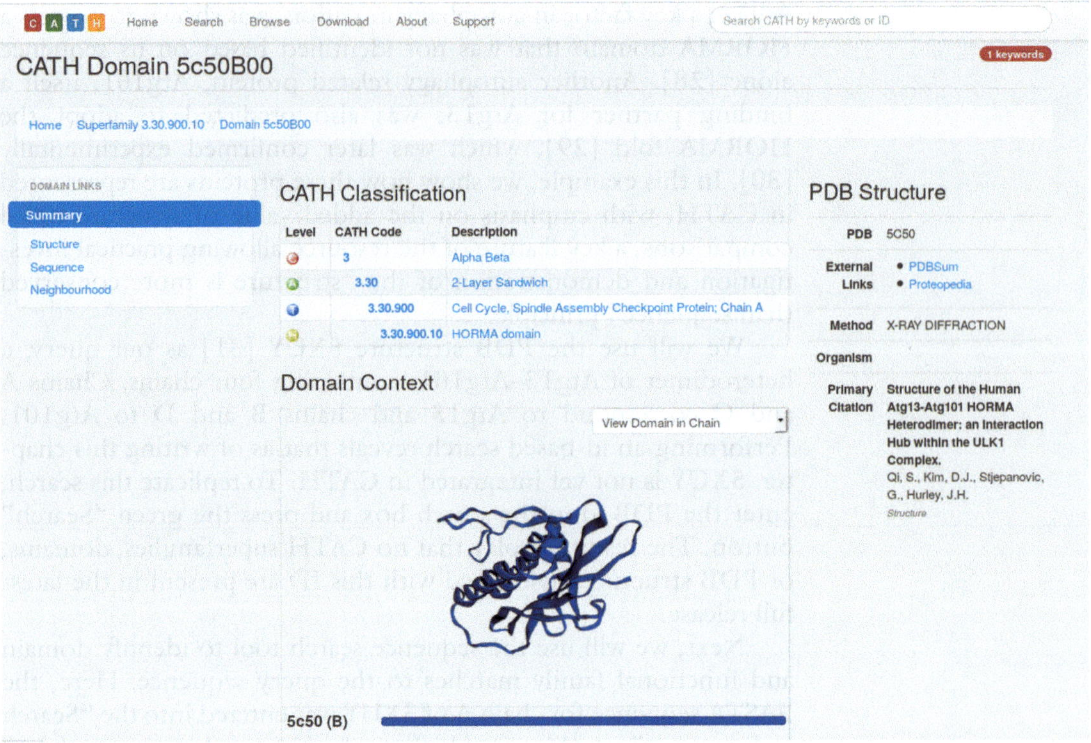

Fig. 6 CATH domain summary page for 5c50B00

In the interactive structure display panel, several views can be selected. By default the "View domain in chain" mode is active, a feature added in the latest release, 4.2. In our case, the matching HORMA domain covers the whole of its PDB chain sequence. The PDB structure that this match belongs to (PDB id 5C50) is another Atg13–Atg101 complex. Chain B of 5C50 corresponds to Atg13, and this chain comprises a single domain (5c00B00) whose homologous superfamily is CATH 3.30.900.10, the HORMA domain family.

If, however, we upload chain B of 5XUY to the "Search by Sequence" field results, no matches are found in CATH. In such cases, a structure-based search can be performed, which uses the CATHEDRAL pipeline (*see* Subheading 2.2.2). The user should select the tab "Search by Structure," choose a PDB file, and click "Submit." A summary with details of the PDB file will be displayed, below which a brief summary of the chains found in the PDB file and the corresponding sequence hits are shown. Again, no hits for chains B and D, corresponding to Atg101, are found. To initiate a structure-based search, one should click on the red "Submit Structure" button, in the example on that in the second row, corresponding to chain B. The server displays the "Scan submitted" message and offers to monitor the progress of the scan. Note that such a scan is generally slower than sequence-based searches. The monitoring page refreshes itself regularly, and when the search is done, the user can click on the "View" button to display the results.

The results page shows a diagram plus a list of the identified structures with their SSAP scores and RMSDs. In the example, all the hits are from the Architecture level 3.30, firmly positioning the domain among two-layer sandwich alpha-beta proteins. Closer inspection of the hits reveals that only two of those with an SSAP score above 70, the first and third one, cover the full sequence; the others are partial matches. Indeed, these two hits correspond to the HORMA family. This result shows that we have successfully identified structural domains belonging to a relative of Atg101.

Clicking on the CATH code of the HORMA domain superfamily, the user arrives at the summary page for the family, in which in the left-side menu the user can select several views like structure superposition, and the structural classification of the domains in the family.

There are a number of GO terms associated with the HORMA family, the most specific ones associated with functions of its earliest known members, HOP1, REV7, and MAD2 (Fig. 7).

3.3 Functional Analysis of Guanylate Kinase-Like Proteins

Guanylate Kinase-like (GK) proteins are a large family consisting of two main branches with different functions, namely, an enzymatic one (GK^{enz}) catalyzing phosphate group transfer and a peptide binding (GK^{dom}). The more ancient enzymatic function is present in all major clades of organisms, whereas protein-binding GK domains are characteristic for animals only. MAGUK proteins (Membrane-Associated GUanylate Kinases) are one class of proteins with a protein-binding GK domain. All GK proteins exhibit similar folds, and their evolutionary relationship is detectable based on their sequence. In a study addressing the emergence of novel protein functions, it was found that a single Ser → Pro mutation is capable of transforming a potent GK enzyme into a functional peptide-binding domain [33].

In this example, we will use such a single-mutant protein as our query, derived from a yeast guanylate kinase enzyme but possessing peptide binding activity, represented in the PDB with id 4F4J. Initiating a general text search from the CATH home page results in hits to two superfamilies, two domains, and one PDB structure.

The two superfamily hits reflect that the GK domain is represented by two types of structural domains in CATH: the guanylate kinase phosphate binding domain (3.30.63.10) and the P-loop containing nucleotide triphosphate hydrolase family (3.40.50.300). As its name suggests, the former one is specific for GK domains, whereas the latter is a 100 times larger, widespread superfamily that, according to the sequence-structure diversity

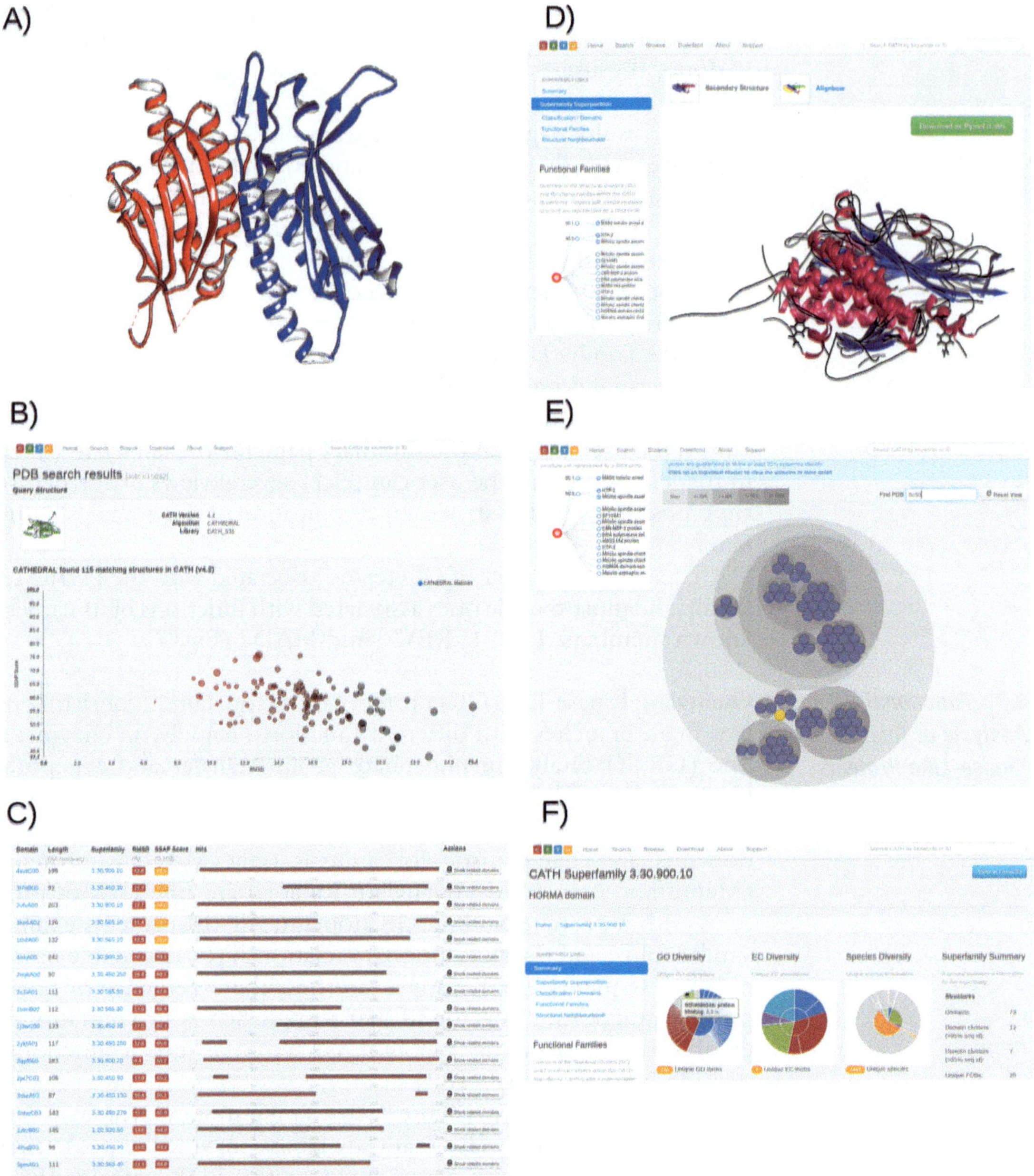

Fig. 7 Structure of the Atg13–Atg101 heterodimer and information on the HORMA domain superfamily (3.30.900.10) in CATH. (**a**) The Atg13 (chain A, red)–Atg101 (chain B, blue) complex in the PDB structure 5XUY. Figure prepared with UCSF Chimera [32]. (**b**, **c**) Diagram and list of the top hits obtained with the Structure search feature of CATH for chain B of 5XUY as query. (**d**) Superfamily superposition of members of the HORMA superfamily in CATH. (**e**) Classification of HORMA domains in the CATH database with domain 5c50B00 highlighted. (**f**) Diversity of the HORMA superfamily, with the GO term "protein binding" highlighted

diagram, is the structurally most diverse superfamily in CATH. The functional diversity of the domains in family 3.30.63.10 including the peptide binding activity is illustrated in Fig. 8.

Fig. 8 Information about the protein 4F4J in CATH. (**a**) The two domain superfamily hits obtained for the search with PDB ID "4F4J": Guanylate kinase phosphate binding domain (3.60.63.10) and P-loop containing nucleotide triphosphate hydrolases (NTPases) (3.40.50.300). (**b**) Visualization of PDB id 4F4J, chain A, using the 3D structure viewer. The phosphate-binding domain is colored blue and the position of Pro35 is highlighted. Note that this domain is inserted into the sequentially noncontinuous P-loop NTPase domain, shown in red. (**c**) Functional diversity of the domains in the superfamily 3.60.63.10. Note the peptide binding function (highlighted). (**d**) Sequence-structure diversity of the domains in family 3.40.50.300. Note that this superfamily has the largest structural diversity in CATH

In this example of a complex functional unit, the GK domains are represented in CATH as belonging to two different superfamilies, reflecting the composite nature and high functional diversification of one of its building blocks, the P-loop NTPase domain [34]. In this specific case, the clue on the function can be found in the annotation of the structurally and functionally less diverse unit, the guanylate kinase phosphate-binding domain, which, among other functions, is associated with peptide binding characteristic of GK^{dom} proteins.

4 Concluding Remarks

The CATH resource, established in the mid-1990s, provides publicly available high-quality information on the classification of protein domains. This chapter describes methods (e.g., using structure-based and sequence-based tools) that allow users to research for a protein(s) of interest, together with examples of how to apply these methods. The CATH resource is particularly valuable in providing an overview of current known structural folds in the protein structure universe, predicting the possible functions of a protein sequence, predicting structural domains in a protein sequence, providing evolutionary relationship insights, and providing highly curated gold standard datasets: e.g., fold libraries for protein structure prediction methods.

Acknowledgment

N.L.D. acknowledges funding from the Wellcome Trust (Award number: 104960/Z/14/Z).

References

1. Chothia C, Lesk AM (1986) The relation between the divergence of sequence and structure in proteins. EMBO J 5:823–826
2. Holm L, Sander C (1996) Mapping the protein universe. Science 273:595–602
3. Rost B (1999) Twilight zone of protein sequence alignments. Protein Eng Des Sel 12:85–94
4. Illergård K, Ardell DH, Elofsson A (2009) Structure is three to ten times more conserved than sequence—a study of structural response in protein cores. Proteins 77:499–508
5. Todd AE, Orengo CA, Thornton JM (2001) Evolution of function in protein superfamilies, from a structural perspective. J Mol Biol 307:1113–1143
6. Lee D, Redfern O, Orengo C (2007) Predicting protein function from sequence and structure. Nat Rev Mol Cell Biol 8:995–1005
7. Sillitoe I et al (2019) CATH: expanding the horizons of structure-based functional annotations for genome sequences. Nucleic Acids Res 47:D280–D284
8. wwPDB consortium (2019) Protein Data Bank: the single global archive for 3D macromolecular structure data. Nucleic Acids Res 47: D520–D528
9. Lewis TE et al (2018) Gene3D: extensive prediction of globular domains in proteins. Nucleic Acids Res 46:D1282
10. Orengo CA, Taylor WR (1996) SSAP: sequential structure alignment program for protein structure comparison. Methods Enzymol 266:617–635
11. Redfern OC, Harrison A, Dallman T, Pearl FMG, Orengo CA (2007) CATHEDRAL: a fast and effective algorithm to predict folds and domain boundaries from multidomain protein structures. PLoS Comput Biol 3:e232
12. HMMER. Available from: http://hmmer.org/ . Accessed July 15 2019
13. Madera M (2008) Profile Comparer: a program for scoring and aligning profile hidden Markov models. Bioinformatics 24:2630–2631
14. Steinegger M et al (2019) HH-suite3 for fast remote homology detection and deep protein annotation. https://doi.org/10.1101/560029
15. Das S et al (2015) Functional classification of CATH superfamilies: a domain-based approach for protein function annotation. Bioinformatics 31:3460–3467
16. Lee DA, Rentzsch R, Orengo C (2010) GeMMA: functional subfamily classification

within superfamilies of predicted protein structural domains. Nucleic Acids Res 38:720–737

17. Das S et al (2015) CATH FunFHMMer web server: protein functional annotations using functional family assignments. Nucleic Acids Res 43:W148–W153
18. Capra JA, Singh M (2008) Characterization and prediction of residues determining protein functional specificity. Bioinformatics 24:1473–1480
19. Zhou N et al (2019) The CAFA challenge reports improved protein function prediction and new functional annotations for hundreds of genes through experimental screens. bioRxiv 653105. https://doi.org/10.1101/653105
20. Moya-García A et al (2017) Structural and functional view of polypharmacology. Sci Rep 7:10102
21. Lam SD, Das S, Sillitoe I, Orengo C (2017) An overview of comparative modelling and resources dedicated to large-scale modelling of genome sequences. Acta Crystallogr D Struct Biol 73:628–640
22. Blomberg N, Oliveira A, Mons B, Persson B, Jonassen I (2015) The ELIXIR channel in F1000Research. F1000Res 4:ELIXIR-1471
23. Harrison A et al (2003) Recognizing the fold of a protein structure. Bioinformatics 19:1748–1759
24. Ashburner M et al (2000) Gene ontology: tool for the unification of biology. The Gene Ontology Consortium. Nat Genet 25:25–29
25. The Gene Ontology Consortium (2019) The gene ontology resource: 20 years and still going strong. Nucleic Acids Res 47: D330–D338
26. International Union of Biochemistry and Molecular Biology. Nomenclature Committee, Webb EC (1992) Enzyme nomenclature 1992: recommendations of the Nomenclature Committee of the International Union of Biochemistry and Molecular Biology on the nomenclature and classification of enzymes. Academic Press, London
27. Almutairi ZM (2018) Comparative genomics of HORMA domain-containing proteins in prokaryotes and eukaryotes. Cell Cycle 17:2531–2546
28. Jao CC, Ragusa MJ, Stanley RE, Hurley JH (2013) A HORMA domain in Atg13 mediates PI 3-kinase recruitment in autophagy. Proc Natl Acad Sci U S A 110:5486–5491
29. Hegedűs K, Nagy P, Gáspári Z, Juhász G (2014) The putative HORMA domain protein Atg101 dimerizes and is required for starvation-induced and selective autophagy in Drosophila. Biomed Res Int 2014:1–13
30. Michel M et al (2015) The mammalian autophagy initiator complex contains 2 HORMA domain proteins. Autophagy 11:2300–2308
31. Kim B-W et al (2018) The C-terminal region of ATG101 bridges ULK1 and PtdIns3K complex in autophagy initiation. Autophagy 14:2104–2116
32. Pettersen EF et al (2004) UCSF Chimera—a visualization system for exploratory research and analysis. J Comput Chem 25:1605–1612
33. Johnston CA, Whitney DS, Volkman BF, Doe CQ, Prehoda KE (2011) Conversion of the enzyme guanylate kinase into a mitotic-spindle orienting protein by a single mutation that inhibits GMP-induced closing. Proc Natl Acad Sci U S A 108:E973–E978
34. Leipe DD, Koonin EV, Aravind L (2003) Evolution and classification of P-loop kinases and related proteins. J Mol Biol 333:781–815

Chapter 5

Protein Thermal Stability Engineering Using HoTMuSiC

Fabrizio Pucci, Jean Marc Kwasigroch, and Marianne Rooman

Abstract

The rational design of enzymes is a challenging research field, which plays an important role in the optimization of a wide series of biotechnological processes. Computational approaches allow screening all possible amino acid substitutions in a target protein and to identify a subset likely to have the desired properties. They can thus be used to guide and restrict the huge, time-consuming search in sequence space to reach protein optimality. Here we present HoTMuSiC, a tool that predicts the impact of point mutations on the protein melting temperature, which uses the experimental or modeled protein structure as sole input and is available at the dezyme.com website. Its main advantages include accuracy and speed, which makes it a perfect instrument for thermal stability engineering projects aiming at designing new proteins that feature increased heat resistance or remain active and stable in nonphysiological conditions. We set up a HoTMuSiC-based pipeline, which uses additional information to avoid mutations of functionally important residues, identified as being too well conserved among homologous proteins or too close to annotated functional sites. The efficiency of this pipeline is successfully demonstrated on *Rhizomucor miehei* lipase.

Key words Protein melting temperature, Protein design, Thermal stability, Statistical potentials, Artificial neural network, Lipase

1 Introduction

In the last decades, lots of efforts have been devoted to analyze the molecular mechanisms associated to protein thermal stability, since their understanding is fundamental not only for advancing the theoretical comprehension of the protein folding process but also for potential applications to a wide series of biological processes that range from drug design to the synthesis of new protein nanomaterials. The design of new enzymes that remain active and stable at temperatures well above or below their physiological temperature can also lead to the improvement of the efficiency of catalytic processes while reducing their economic costs and their environmental impact [1–3].

Different experimental and computational approaches have been developed and largely used to enhance protein thermal resistance [4–6] such as:

Zoltán Gáspári (ed.), *Structural Bioinformatics: Methods and Protocols*, Methods in Molecular Biology, vol. 2112,
https://doi.org/10.1007/978-1-0716-0270-6_5, © Springer Science+Business Media, LLC, part of Springer Nature 2020

- **Directed evolution methods** in which randomly distributed mutations are introduced in the target protein and are followed by screening and selection steps.
- **Rational protein design** in which the understanding of the protein structure/function relationships is the key ingredient.
- **Semi-rational approaches** that combine the benefits of the directed evolution and the rational design methods.

Despite impressive current achievements, the development of a systematic and cheap way to optimize a target protein remains a challenging goal. This is primarily due to the huge size of the sequence space to be explored and to the incomplete knowledge of the molecular mechanisms involved.

Here we present some advances in the computational protein design field by describing the HoTMuSiC software [7] that we recently developed. HoTMuSiC is an efficient tool that, given the experimental or modeled three-dimensional (3D) structure of the target protein as input, screens the protein sequence and predicts the impact of all possible single amino acid substitutions on the protein melting temperature in just a few minutes. Its performance makes it a perfect instrument to design proteins with improved thermal stability and to guide mutagenesis experiments that are usually expensive and time-consuming. Moreover, due to its speed, it can also be employed in large-scale investigations aimed at gaining insights into the relationship between protein thermal stability and natural evolution and into the adaptation mechanisms to extreme environmental conditions [8].

In the next sections, we briefly introduce the key ingredients and model structure of HoTMuSiC, show how it can be fruitfully applied for protein design applications, and discuss its performances in detail.

2 HoTMuSiC Key Instrument: The Statistical Potentials

The key instrument used in the construction of HoTMuSiC is the statistical potential formalism, known to be quite efficient when applied to a wide range of problems including protein structure prediction, protein design, and protein-ligand scoring.

These knowledge-based, effective potentials are derived from frequencies of sequence and structure elements in a nonredundant dataset of well-resolved protein 3D structures [9]. More precisely, let c be a structure element, consisting of the distance between two residues, the solvent accessibility of a residue, its backbone torsion angle, or combinations thereof, and let s be a sequence element consisting of the amino acid type of one or two residues. The free energy contribution of the association (c,s) is ruled by the Boltzmann law:

$$\Delta W(c,s) = -k_B\, T\, \mathrm{Log}\left[\frac{P(c,s)}{P(c)P(s)}\right] \quad (1)$$

where $P(c)$, $P(s)$, and $P(c,s)$ are the relative frequencies of c, s, and (c, s) in the dataset, k_B is the Boltzmann constant, and T is the absolute temperature.

In order to construct energy functions that describe the impact of the temperature on the amino acid interactions, we introduced temperature-dependent statistical potentials [10, 11]. They are extracted from datasets of proteins with known structures and specified thermal stability properties. The potentials derived from mesostable proteins, denoted as $\Delta W(c,s)^{\mathrm{meso}}$, describe the interactions at low T, while those derived from thermostable or hyperthermostable proteins, $\Delta W(c,s)^{\mathrm{thermo}}$, represent the interactions at high temperatures.

Using these new potentials, the T-dependence of several types of amino acid interactions were unraveled [10]. Moreover, these potentials were successfully applied to the prediction of the protein stability curve as a function of the temperature [11, 12].

The potentials that are used in HoTMuSiC's model are the standard potentials $\Delta W(c,s)$ and the T-dependent potentials $\Delta W(c, s)^{\mathrm{meso}}$ and $\Delta W(c,s)^{\mathrm{thermo}}$, for various structure and sequence elements (c,s).

3 HoTMuSiC Harmony: The Artificial Neural Networks

The statistical potentials were combined to predict how the melting temperature T_m changes upon amino acid substitution:

$$\Delta T_m = T_m(\mathrm{mutant}) - T_m(\mathrm{wild\ type}) \quad (2)$$

We set up two different models to compute ΔT_m. The HoTMuSiC model requires as sole input the 3D structure of the wild-type protein and is based on a linear combination of the standard statistical potentials $\Delta W(c,s)$ for different sequence and structure elements (c,s). The T_m-HoTMuSiC model requires in addition the melting temperature of the wild-type protein and employs a combination of standard and T-dependent potentials $\Delta W(c,s)$, $\Delta W(c, s)^{\mathrm{thermo}}$, and $\Delta W(c,s)^{\mathrm{meso}}$.

Both models include three additional terms—a constant term **1** and two terms $\Delta V_{\pm}$ that represent the difference in volume between the wild-type and mutant residues—and describe the creation of stress or holes in the protein structure.

All these terms are weighted by sigmoid functions of the solvent accessibility A, which smoothly connect the surface to the protein interior while allowing different behaviors in these regions.

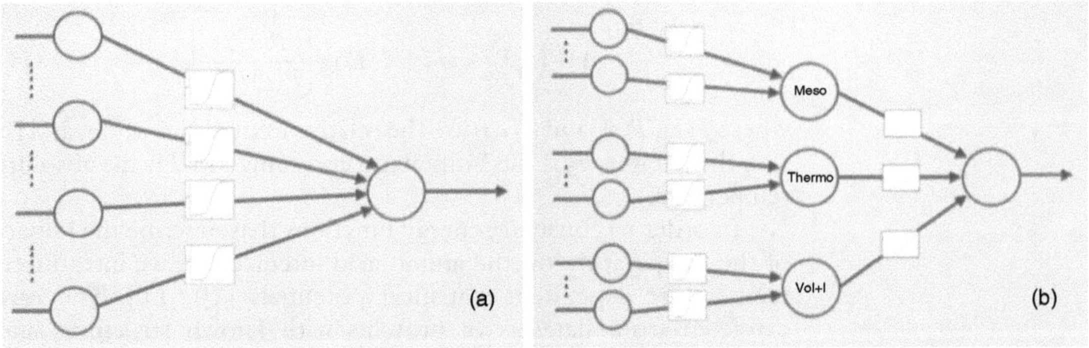

Fig. 1 Schematic representation of the ANNs used for the parameter identifications. (**a**) HoTMuSiC: two-layer ANN, consisting of a perceptron with sigmoid activation functions and 12 input neurons encoding 9 potentials, 2 volume terms and a constant term; (b) T_m-HoTMuSiC: three-layer ANN, consisting of 3 perceptrons with sigmoid weights; the first two perceptrons have 5 input neurons each, encoding 5 mesostable and 5 thermostable potentials, respectively; the third perceptron has 3 neurons for the volume and constant terms. The outputs of these three perceptrons (Meso, Thermo, and Vol + I) are the inputs of another perceptron with polynomial weight functions of the wild-type T_m value

Indeed, according to the type of potential, the mutational impact can be stronger at the surface or in the protein core.

Each sigmoid is a function of four parameters, which were identified using a cost function of the difference between experimental and computed ΔT_m values on a learning dataset. We used for that purpose the T1626 set [13] that contains 1626 mutations inserted in about 90 proteins of known X-ray structure of resolution below 2.5 Å, collected by literature screening.

For minimizing the cost function, artificial neural networks (ANN) were considered. For HoTMuSiC, a standard single-layer ANN was used (Fig. 1a), in which each neuron is associated with one input term ($\Delta W(c,s)$, ΔV, $\mathbf{1}$) and the activation functions are sigmoid functions of A.

The ANN of T_m-HoTMuSiC is composed of three layers, where the additional, hidden, layer gets activated by functions of the T_m of the wild-type protein and confers more weight to mesostable or thermostable perceptrons according to the thermal properties of the wild-type protein (*see* Fig. 1b and [7] for details).

A standard back-propagation algorithm was employed in the training of the neural network. To test the predictor, we performed fivefold cross-validation using an early stopping technique to avoid overfitting.

4 HoTMuSiC Sound: The Results

4.1 Performances

To validate the method, we first applied the two HoTMuSiC models to the T1626 learning dataset using a fivefold cross-validation procedure and compared their performances with another ΔT_m predictor developed in the literature, i.e., AutoMute [14]. The results are reported in [7]. Here, in addition, we compared the HoTMuSiC performances with popular $\Delta\Delta G$ prediction methods, namely, PoPMuSiC [15], FoldX [16], and Rosetta [17]. Indeed, there is the tendency in the literature to use thermodynamic stability prediction methods to compute thermal stability changes even though the two quantities $\Delta\Delta G$ and ΔT_m are only partially correlated [13]. These comparisons give us an idea of the additional error that one makes by using $\Delta\Delta G$ predictors to predict ΔT_m. The results are given in Table 1.

To extend the performance analysis and get more robust results, we also did a manual literature search and used dedicated annotation servers such as ProTherm [18] and Brenda [19] to collect mutations that are not in T1626, i.e., newly characterized substitutions inserted in experimentally solved protein structures, and mutations inserted in structures that are not yet solved but were modeled by homology modeling. In this way, we obtained a second dataset, called T526, containing 526 mutations in 42 experimental and 58 modeled protein structures.

As seen in Table 1, T_m-HoTMuSiC is slightly more accurate than HoTMuSiC, which is normal as it relies on additional information, i.e., the experimental T_m of the wild type. Both HoTMuSiC models outperform the other predictors, especially on the T526 test set, which is not part of any of the predictors' learning sets. This could indicate that some of these predictors suffer from overfitting problems. Finally, the performance of all the tested $\Delta\Delta G$ predictors is significantly lower. We may thus conclude that using $\Delta\Delta G$ predictors to predict ΔT_m leads to a drop of the performance, of at least 0.1 in correlation coefficient.

4.2 Webserver

HoTMuSiC and T_m-HoTMuSiC are freely available for academic use on the webserver dezyme.com. The site contains an introductory and an explanatory page (**Software** and **Help**) and three working pages:

- **My Data** privately stores the user's personal files, containing protein structures in PDB [20] format or lists of mutations.
- **Query**: On the "HoT" query page, the user must choose either a PDB code that is automatically retrieved from the PDB data bank [20] or a private structure file stored in his My Data page. He must also choose among three options:

Table 1
Performances of the predictors on the T1626 [13] and T526 datasets as evaluated by the Pearson correlation coefficients *R* between the predictor output and the experimental ΔT_m values

Predictors	T1626 *R*	T526 *R*
***T_m*-HoTMuSiC**	**0.61**	**0.59**
HoTMuSiC	0.59	0.56
AutoMute 2.0	0.54	0.22
PoPMuSiC v3.0	-0.50	-0.35
FoldX	-0.42	-0.27
Rosetta	-0.47	-0.26

The three predictors on a blue background are ΔT_m predictors and the three on a green background are $\Delta\Delta G$ predictors; the latter show negative correlations as, by convention, negative $\Delta\Delta G$ and positive ΔT_m values are stabilizing

1. *Systematic* mutation option in which all the possible amino acid substitutions for each residue in the protein sequence are computed.
2. Uploading a mutation *File* containing a list of single-site mutations.
3. Giving *Manually* a list of single-site mutations.

The last option gives the choice between HoTMuSiC and T_m-HoTMuSiC. When selecting the latter, the T_m of the wild type must be specified.

- The **Results** page contains all the predictions performed by the user, which can easily be downloaded. If the systematic mutation mode is chosen, the following files are available:

– [pdbname].hot contains the predicted ΔT_m value for all possible single-site substitutions inserted in the protein of structure [pdbname] as well as other biophysical characteristics of the mutated residue, i.e., its solvent accessibility and its secondary structure.

– [pdbname].hots contains information at the residue level: the mean ΔT_m of all substitutions at each position and the sum of all positive and of all negative ΔT_m values.

– [pdbname].html shows a histogram picture in which the ΔT_m sum of all stabilizing mutations at each position is indicated. Figure 2 contains an example of this output.

– [pdbname.zip] is an archive containing the three abovementioned files, ready to download.

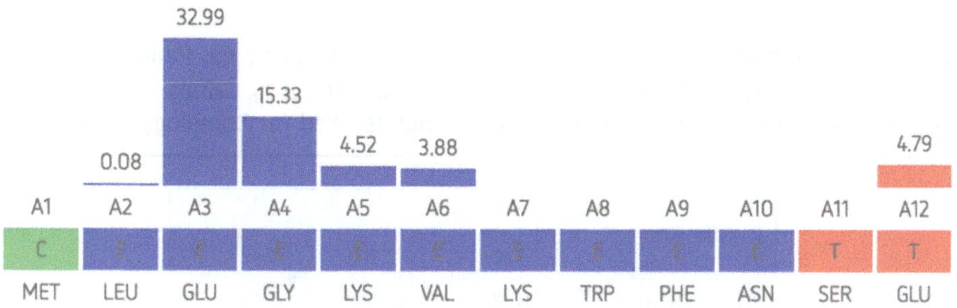

Fig. 2 Example of the webserver's histogram results for the systematic scanning of a target protein. The bars represent the sum of the positively predicted ΔT_m values at each sequence position; these sums are indicated above the bars (in °C). The largest bars represent residues whose mutations are likely to thermostabilize the protein and on which protein engineering experiments should focus first

4.3 Application to Modeled Structures

The quality of the input structures is of fundamental importance for stability predictions. Indeed, the more precise the structure in terms of resolution, the more accurate the predictions. This point is often overlooked when the performances of predictors are discussed. However, it will probably become even more important in the near future since, for example, a growing number of structures will be resolved with cryo-electron microscopy techniques, which usually have lower resolutions (at least for now) than those obtained with X-ray crystallography.

In the newly developed dataset T526, we have also included structures obtained via homology modeling using the Swiss-Model server [21], in view of analyzing the robustness of HoTMuSiC with respect to structural inaccuracies. We compared the performance on proteins for which we have a good-resolution X-ray structure or a modeled structure obtained with a template of which the sequence identity (SI) with respect to the target is either greater than 98% or between 22% and 98%.

As seen from Table 2, the proteins whose X-ray structure is available or can reliably be modeled from templates with SI $\geq$ 98% show good performances with a linear correlation coefficient of about 0.60, whereas the proteins that are modeled from templates SI $<$ 98% (with a mean $<$SI$>$ of 69%) have, as expected, lower but still reasonably good performances measured by a correlation coefficient $R \approx 0.45$.

Another important aspect that impacts the prediction accuracy is the experimental technique used to determine protein structures and the associated resolution. Clearly, the best results are obtained with X-ray structures of resolution of 2.0–2.5 Å at most, as expected from the above analysis on modeled structures.

These results show that HoTMuSiC can be reliably applied not only to good-resolution structures but also to low-resolution and modeled structures. This further broadens the field of applicability of the method.

Table 2
Performances of the predictors on the T526 datasets as evaluated by the Pearson correlation coefficients *R* between the predictor output and the experimental ΔT_m values. "≥98%" and "<98%" indicate the sequence identity with respect to the template used for homology modeling

Predictors	*R*		
	X-ray	≥98%	<98%
T_m-HoTMuSiC	0.59	0.62	0.45
HoTMuSiC	0.56	0.58	0.46

4.4 Protein Design with HoTMuSiC Pipeline

In order to optimize the thermal stability of a target protein to allow it, for example, to remain stable and active at temperatures higher than the physiological temperature, one has to select mutations that increase the protein melting temperature without affecting the protein function. We constructed a systematic pipeline for that purpose, which includes the HoTMuSiC tool but also the ConSurf software [22] that computes the evolutionary conservation score of each residue, as well as annotations from UniProt [23]. This pipeline aims at selecting combinations of point mutations likely to achieve a substantial increase of the protein thermoresistance without affecting the function of the target protein. In what follows, we describe step by step how this pipeline works, as illustrated in Fig. 3:

1. In a first stage, the impact of all possible point substitutions on the melting temperature of a target protein is evaluated with HoTMuSiC and/or T_m-HoTMuSiC in the systematic mode (*see* Subheading 4.2), using as input the (experimental or modeled) protein structure in PDB format. If multiple PDB structures of the target protein are available, HoTMuSiC is applied to all of them (with usually a cutoff of about 2.5 Å on the X-ray resolution), and the mean ΔT_m of each substitution is computed. The mutations are then classified according to their mean ΔT_m values.
2. The following step consists of running ConSurf [22] to estimate the evolutionary conservation score (from 0 to 9) of each amino acid of the target sequence aligned to its homologous sequences. At this level, we performed the first selection by imposing that the mutated positions have a conservation score of at most 7. This filters out mutant residues that are too conserved and thus probably too important for functional or structural reasons.
3. The next step is to retrieve from UniProt [23] the available annotations about catalytic residues and binding sites, and to

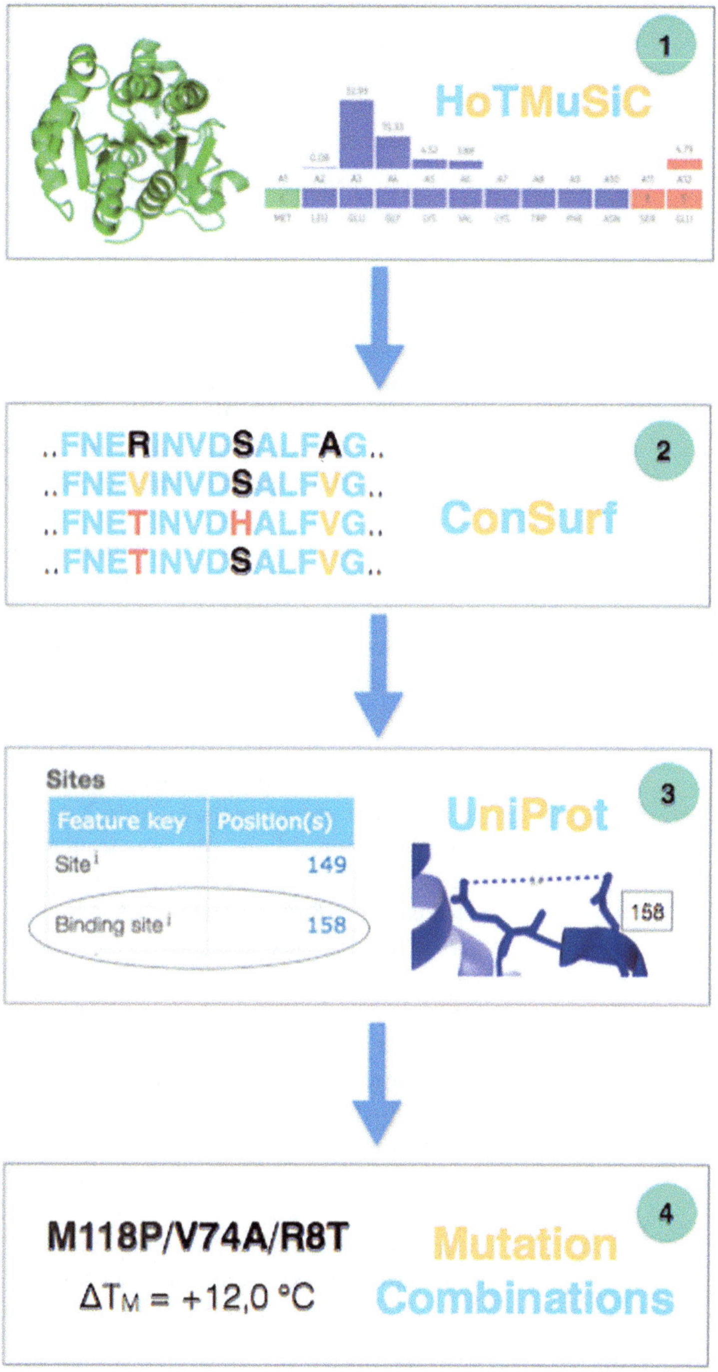

Fig. 3 Schematic picture of the HoTMuSiC-based pipeline for protein design

compute the distance between these sites and all the protein residues. All positions that are closer than 7 Å from one of the functional sites were overlooked. This ensures that the selected mutations do not touch or influence the catalytic activity or the binding to other biomolecules.

4. The remaining list thus contains point mutations that have a high chance to improve the thermoresistance of the target protein without affecting its function. Often, however, one cannot reach the required increase in melting temperature by just a single mutation. The strategy is then to combine several point mutations. For this purpose, we made the approximation that if two mutated sites are separated by a spatial distance of more than 10 Å, the stabilization effect is additive. This final criterion led us to select groups of two, three, or more mutations that optimize the target protein.

4.5 HoTMuSiC-Based Pipeline Applied to Rhizomucor Miehei *Lipase*

To illustrate how the HoTMuSiC pipeline can be used to optimize a protein, we applied it to thermally stabilize the lipase from *Rhizomucor miehei* (RML). This enzyme catalyzes a wide range of reactions such as the hydrolysis of oil, the esterification of fatty acids, and the transesterification and alcoholysis of glycerides. Therefore, it received a lot of attention, and different studies tried to improve its thermal stability to increase the efficiency of these biocatalytic reactions [24].

There are several available X-ray structures of the enzyme, and we considered here the two structures with PDB code 3TGL and 4TGL, which have a resolution of 1.9 and 2.6 Å, respectively. RML folds into a ß-sheet surrounded by helices (Rossmann fold) and has a melting temperature of 58.7 °C [25]. It contains a Ser-His-Asp trypsin-like catalytic triad, in which the active serine is buried under a short helical lid that undergoes conformational changes (*see* Fig. 4a). By exposing or protecting the catalytic pocket, the lid movement controls the enzymatic activity [25].

The first step to predict thermostabilizing mutations consists in running HoTMuSiC in the systematic mode on the two PDB structures of RML to compute the ΔT_m of all possible point mutations. These values, stored in the 3TGL.hot and 4TGL.hot files on the Results page, were first averaged for each individual mutation and then ranked decreasingly according to their average ΔT_m value.

The top 15 mutations in the list are given in Table 3. Note that glycine and proline substitutions were excluded from this list since their mutations are likely to induce changes in conformation which are not taken into account in the prediction model.

In the next steps, the residue conservation across homologous sequences was evaluated using an in-house implementation of the ConSurf algorithm, and the spatial distance between the mutated residues and the catalytic triad Ser-His-Asp was computed. The mutations inserted at highly conserved positions (ConSurf

Table 3
The 15 point mutations in RML that are predicted as the most thermostabilizing by HoTMuSiC

Mutations	ΔT_m^{pred} (°C)	ΔT_m^{exp} (°C)	ConSurf	Distance (Å)
E221V	3.2	-	9	6.3
E221I	2.8	-	9	6.3
Q174F	2.5	-	6	5.3
Q174Y	2.3	-	6	5.3
E230F	2.0	1.6	3	8.5
E230I	2.0	5.7	3	8.5
E230Y	1.8	2.1	3	8.5
E230V	1.8	5.4	3	8.5
Q176F	1.8	-	8	5.2
E221L	1.8	-	9	6.3
Q174W	1.8	-	4	5.3
Q176Y	1.8	-	8	5.2
E230L	1.7	5.4	3	8.5
A64I	1.7	-	8	8.0
E230W	1.7	2.7	3	8.5

The predicted and experimental [25] ΔT_m values are reported in columns 2 and 3. The ConSurf conservation scores and the spatial distance from the catalytic residues, computed between average side chain centroids, are shown in columns 4 and 5. The substitutions that are dropped on the basis of their conservation scores and of their distance to the catalytic triad are on a yellow and green background, respectively

score ≥ 8) or that are too close to the catalytic site (distance$\leq$7 Å) were then removed, as they could impact the protein function or other biophysical characteristics that we do not want to modify.

The substitutions filtered out due to their high conservation scores or their proximity to the catalytic triad are indicated in Table 3. Among the 15 top mutations predicted in **step 1**, we kept only the 6 mutations of Glu at position 230.

To get a larger number of candidate mutations, we have to relax the first, ΔT_m-based, criterion. In this way we obtained the 15 most stabilizing mutations that satisfy all our selection filters, shown in Table 4.

As a final step of our pipeline, we determined subsets of mutations of residues whose relative distances are equal to 10 Å or more and are thus assumed to be independent (Fig. 4). For example, this procedure yields the sets of multiple mutations:

- E230I/S103W/K53I
- E230F/R68V/K53F
- E230M/K106I/N63V

Table 4
Final list of proposed mutations in RML, which satisfy all our selection criteria

Mutations	ΔT_m^{pred} (°C)	ΔT_m^{exp} (°C)	ConSurf	Dist (Å)
E230F	2.0	1.6	3	8.5
E230I	2.0	5.7	3	8.5
E230Y	1.8	2.1	3	8.5
E230V	1.8	5.4	3	8.5
E230L	1.7	5.4	3	8.5
E230W	1.7	2.7	3	8.5
S103W	1.7	–	3	8.5
R68V	1.6	–	6	17.0
E230M	1.6	1.3	3	8.5
K53I	1.5	–	2	13.3
K106I	1.5	–	4	13.2
K53F	1.4	–	2	13.3
K106F	1.3	–	4	13.2
N63V	1.3	–	6	7.4
N63I	1.3	–	6	7.4

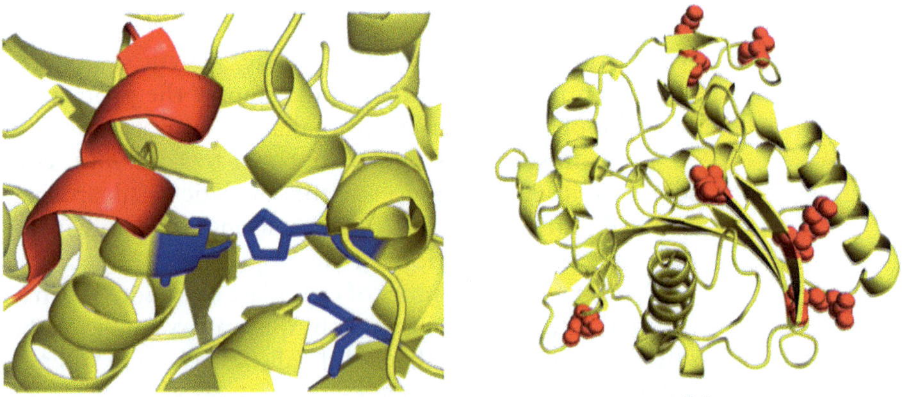

(a) (b)

Fig. 4 Three-dimensional structure of *Rhizomucor miehei* lipase (PDB code 3TGL). (**a**) Zoom on the Ser-His-Asp catalytic triad (in blue sticks) with the helical lid (in red) that undergoes conformational changes and modulate the protein activity. (**b**) Residues to be mutated (in red spheres) for the thermostabilization of the lipase

Finally, we compared our computational predictions on lipase with the available experimental data reported in [25]. The prediction score of HoTMuSiC on this set of 36 point mutations was found to outperform the competitor methods reported in Table 1 and to reach quite a good accuracy evaluated by a root mean square deviation between predicted and experimental ΔT_m values of 1.7 °C and by a linear correlation coefficient of 0.6. This score is the same as the one obtained from the other tests shown in Tables 1 and 2.

Moreover, four sets of multiple mutations have been experimentally shown to lead to a substantial increase of the RML thermoresistance [25]. If we assume that the effect of mutations with a sufficient spatial separation is additive, these multiple mutations are also very well predicted by HoTMuSiC. The final score of our pipeline, when the 4 multiple mutations are added to the 36 point mutations, reaches a linear correlation coefficient higher than 0.8.

Finally note that the predicted ΔT_m values tend to be lower than the experimental values. Indeed, as we already noted [26], the training datasets are dominated by destabilizing mutations and this induces biases toward these types of mutations and leads to an underestimation of the predicted stabilization effects.

5 Conclusion

In this chapter, we presented recently developed protein thermal stability predictors, and their application to efficiently optimize targeted proteins. We focused more specifically on the HoTMuSiC predictor, which predicts the ΔT_m values of all the possible point mutations in a medium-size protein in a few minutes, on the basis of its experimental or modeled 3D structure. Our tool outperforms the other available ΔT_m predictors, as well as $\Delta\Delta G$ predictors when applied to thermal stability even though they are designed for thermodynamic stability.

The fastness of HoTMuSiC allows scanning and predicting all possible substitutions inserted in a target protein. It can thus guide mutagenesis experiments aimed to improve the thermoresistance of proteins and be employed in the optimization of a wide series of biotechnological processes.

To optimize the selection of mutations that need to be tested experimentally, we complemented the HoTMuSiC results with additional information on the residue conservation among homologous proteins and the distance from annotated functional sites. This novel pipeline is designed to filter out mutations that are likely to affect functionally or structurally important residues. The point mutations that satisfy the criteria are then combined into subsets of non-interacting mutations that are assumed to be independent.

These subsets are identified to strongly enhance the effect on thermal stability.

The HoTMuSiC pipeline was applied to the thermal stabilization of lipase from *Rhizomucor miehei*. Several series of multiple mutations were predicted for this enzyme. For those mutations whose ΔT_m values were measured experimentally, the prediction score was shown to be high.

Finally note that the potentiality of HoTMuSiC is not restricted to the protein design field. Due to its speed, it can also be applied on a proteomic scale to gain important theoretical insights into the thermal and evolutionary adaptation of proteins to extreme environments.

Acknowledgments

We thank Raphael Bourgeas for the help in implementing HoTMuSiC on the webserver and César Ngabo for his contribution to the HoTMuSiC pipeline. We acknowledge support from the Fund for Scientific Research (FNRS) through a PDR research project. F.P. and M.R. are FNRS postdoctoral researcher and research director, respectively.

References

1. Pucci F, Rooman M (2017) Physical and molecular bases of protein thermal stability and cold adaptation. Curr Opin Struct Biol 42:117–128
2. Kumar S, Tsai CJ, Nussinov R (2000) Factors enhancing protein thermostability. Protein Eng 13:179–191
3. Razvi A, Scholtz JM (2006) Lessons in stability from thermophilic proteins. Protein Sci 15:1569–1578
4. Fowler DM, Fields S (2014) Deep mutational scanning: a new style of protein science. Nat Methods 11:801–807
5. Jäckel C, Kast P, Hilvert D (2008) Protein design by directed evolution. Annu Rev Biophys 37:153–173
6. Chica R, Doucet N, Pelletier JN (2005) Semi-rational approaches to engineering enzyme activity: combining the benefits of directed evolution and rational design. Curr Opin Biotechnol 16:378–384
7. Pucci F, Bourgeas R, Rooman M (2016) Predicting protein thermal stability changes upon point mutations using statistical potentials: introducing HoTMuSiC. Sci Rep 6:23257
8. Pucci F, Rooman M (2016) Improved insights into protein thermal stability: from the molecular to the structurome scale. Philos Trans A Math Phys Eng Sci 374:2080
9. Dehouck Y, Gilis D, Rooman M (2006) A new generation of statistical potentials for proteins. Biophys J 90:4010–4017
10. Folch B, Dehouck Y, Rooman M (2010) Thermo- and mesostabilizing protein interactions identified by temperature-dependent statistical potentials. Biophys J 98:667–677
11. Pucci F, Dhanani M, Dehouck Y, Rooman M (2014) Protein thermostability prediction within homologous families using temperature-dependent statistical potentials. PLoS One 9:e91659
12. Pucci F, Kwasigroch JM, Rooman M (2017) SCooP: an accurate and fast predictor of protein stability curves as a function of temperature. Bioinformatics 33:3415–3422
13. Pucci F, Bourgeas R, Rooman M (2016) High-quality thermodynamic data on the stability changes of proteins upon single-site mutations. J Phys Chem Ref Data 45:023104
14. Masso M, Vaismann II (2010) AUTO-MUTE: web-based tools for predicting stability changes in proteins due to single amino acid replacements. Protein Eng Des Sel 23:683–687

15. Dehouck Y, Grosflis A, Folch B, Gilis D, Bogaerts P, Rooman M (2009) Fast and accurate predictions of protein stability changes upon mutations using statistical potentials and neural networks: PoPMuSiC-2.0. Bioinformatics 25:2537–2543
16. Guerois R, Nielsen JE, Serrano L (2002) Predicting changes in the stability of proteins and protein complexes: a study of more than 1000 mutations. J Mol Biol 320:369–387
17. Kellogg EH, Leaver-Fay A, Baker D (2010) Role of conformational sampling in computing mutation-induced changes in protein structure and stability. Proteins 79:830–838
18. Bava KA, Gromiha MM, Uedaira H, Kitajima K, Sarai A (2004) ProTherm, version 4.0: thermodynamic database for proteins and mutants. Nucleic Acids Res 32:D120–D121
19. Placzek S (2017) BRENDA in 2017: new perspectives and new tools in BRENDA. Nucleic Acids Res 45:D380–D388
20. Berman HM et al (2000) The Protein Data Bank. Nucleic Acids Res 28:235–242
21. Waterhouse A et al (2018) SWISS-MODEL: homology modelling of protein structures and complexes. Nucleic Acids Res 46:W296–W303
22. Ashkenazy H, Abadi S, Martz E, Chav O, Mavrose I, Pupko T, Ben-Tai N (2016) ConSurf 2016: an improved methodology to estimate and visualize evolutionary conservation in macromolecules. Nucleic Acids Res 44: W344–W350
23. The UniProt Consortium (2017) UniProt: the universal protein knowledgebase. Nucleic Acids Res 45:D158–D169
24. Rodrigues RC, Fernandez-Lafuente R (2010) Lipase from Rhizomucor miehei as a biocatalyst in fats and oils modification. J Mol Catal B Enzym 66:15–32
25. Li G, Fang X, Su F, Chen Y, Xu L, Yan Y (2018) Enhancing the thermostability of Rhizomucor miehei lipase with a limited screening library by rational-design point mutations and disulfide bonds. Appl Environ Microbiol 84: e02129–e02117
26. Pucci F, Bernaerts KV, Kwasigroch JM, Rooman M (2018) Quantification of biases in predictions of protein stability changes upon mutations. Bioinformatics 34:3659–3665

Chapter 6

Contact Area-Based Structural Analysis of Proteins and Their Complexes Using CAD-Score

Kliment Olechnovič and Česlovas Venclovas

Abstract

Quantifying discrepancies between computationally derived and native (reference) structures is an essential step in the development and comparison of protein modeling and protein-protein docking methods. Measuring conformational differences of proteins or protein complexes is also important in other areas of structural biology such as molecular dynamics and crystallography. There are multiple scores to do that. However, nearly all of them, whether superposition-based (e.g., RMSD) or superposition-free, use distances to measure similarity. CAD-score is conceptually different as it uses physical contacts represented as contact areas. Such representation makes it possible to quantify differences of both structures and surfaces (e.g., protein-protein interfaces and binding sites) using the same framework. A number of studies have found CAD-score to be among the most robust scores. The method is implemented both as a web server and as standalone software available at http://bioinformatics.lt/software/cad-score. Here, we describe how to use the standalone CAD-score software for comparison and analysis of protein structures, interfaces, and binding sites.

Key words Protein structure, Protein-protein interactions, Voronoi tessellation, Interatomic contacts, Contact area, Global similarity score, Local similarity score

1 Introduction

Comparison of different structures (conformations) for the same protein or protein complex is a common task in both computational and experimental structural biology. For example, measuring discrepancies between computational models and corresponding native (reference) structures is at the heart of development and comparison of protein structure prediction and/or refinement methods. Other common uses include comparison of experimental structures solved in different crystal forms, at different temperature or pH, with and without bound ligand, etc. Analysis of a molecular dynamics simulation also involves comparison of structures obtained along the simulation course.

Zoltán Gáspári (ed.), *Structural Bioinformatics: Methods and Protocols*, Methods in Molecular Biology, vol. 2112, https://doi.org/10.1007/978-1-0716-0270-6_6, © Springer Science+Business Media, LLC, part of Springer Nature 2020

Over the years, multiple scores have been developed for performing such comparisons. Some of the scores, such as RMSD [1], GDT-TS [2, 3] or TM-score [4], are based on global structure superposition. Others, like Local Distance Difference Test (LDDT) [5], are superposition-free and focus on local deviation. Despite some differences, the majority of such methods use distances to derive a similarity score. Contact Area Difference (CAD) score [6] is conceptually different as it uses areas of physical contacts to quantify differences between the reference structure (target) and the one being evaluated (model). CAD-score is superposition-free measure and can be used for the evaluation of both local and global structural similarity. Moreover, since CAD-score is based on contact areas, it can be directly applied not only for structures but also for surfaces such as protein-protein interfaces or protein binding sites. A recent comprehensive analysis revealed a number of advantages of CAD-score over various other scores [7]. For example, CAD-score shows robust performance on structures displaying large local deviations and multidomain proteins with flexible linkers, the cases presenting a serious problem for superposition-based global scores. Another important advantage of CAD-score is that it strongly favors models with realistic stereo-chemical features, the property that might be particularly important for the analysis of homology modeling and refinement results.

2 CAD-Score Definition

2.1 Contacts

Contacts in CAD-score are derived from protein structure represented as a set of atomic balls, each ball having a van der Waals radius depending on the atom type. A ball can be assigned a region of space that contains all the points that are closer (or equally close) to that ball than to any other. Such a region is called a Voronoi cell, and the partitioning of space into Voronoi cells is called Voronoi tessellation [8]. Two adjacent Voronoi cells share a set of points that form a surface called a Voronoi face (Fig. 1a, b). A Voronoi face can be viewed as a geometric representation of a contact between two atoms; the area of the Voronoi face corresponds to the contact area. Voronoi cells of atomic balls are constrained inside the boundaries defined by the solvent-accessible surface as described in the recent paper [9].

The resulting constrained Voronoi faces can be combined into residue-residue contacts involving either all atoms (Fig. 1c) or only a subset, for example, side chain atoms (Fig. 1d). For practical purposes, three standard subsets of residue atoms are defined (A, "all atoms"; S, "side chain atoms"; and M, "main chain atoms") resulting in six nonredundant categories of residue-residue contacts (A-A, A-S, S-S, A-M, M-M, M-S). The most useful categories are those that include side chain-side chain interactions, namely, A-A, A-S, and S-S.

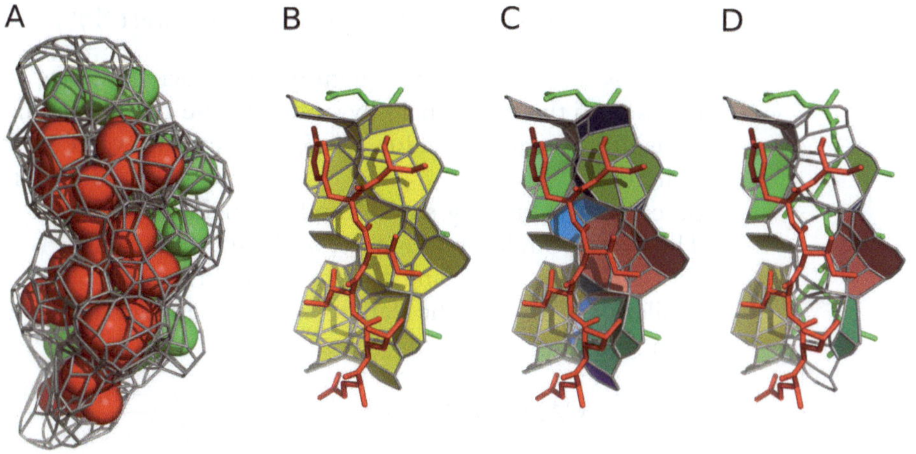

Fig. 1 Example of interatomic and inter-residue contacts, with two groups of atoms distinguished by red and green coloring. (**a**) Voronoi cells of atomic balls. (**b**) Interatomic contacts between two groups of atoms. (**c**) Grouping of interatomic contacts into inter-residue contacts. (**d**) Contacts between residue side chains

2.2 Structure Scores

Given reference structure T (target) and structure to be compared, M (model), let G denote the set of all the pairs of residues (i,j) that have a nonzero contact area $T_{(i,j)}$ in the target structure. Then for every residue pair $(i,j) \in G$, the corresponding contact area $M_{(i,j)}$ in the model is calculated. $M_{(i,j)}$ is assigned zero if there is no contact between residues i and j in the model or if either residue (i or j) is missing from the model. The CAD-score for the model structure is then defined as:

$$\text{CAD-score}(G) = 1 - \frac{\sum \min\left(|T_{(i,j)} - M_{(i,j)}|, T_{(i,j)}\right)}{\sum T_{(i,j)}} \quad (1)$$

Values of Eq. 1 are always within the [0,1] range. If model and target structures are identical, CAD-score(G) = 1. At the other extreme, if not a single contact is reproduced with sufficient accuracy, CAD-score(G) = 0.

Scores for individual residues are calculated by applying Eq. 1 to a residue-specific subset of G. Thus, the score for residue i equals CAD-score(G_i), where G_i is a set of all pairs $(i,j) \in G$. Per-residue scores can be smoothed along the sequence using a sliding window technique.

2.3 Scores for Interfaces

A straightforward way to compare the inter-chain interfaces of two protein complexes is to apply Eq. 1 to a set of inter-chain contacts. Let I and J denote the sets of interface residues of the first and the second subunits (chains), respectively, in the target protein complex. Then the set of target interface contacts $G_{I,J}^{\text{iface}}$ and the interface similarity score CAD-score^{iface} are defined as:

$$G_{I,J}^{\text{iface}} = G \cap (I \times J) \quad (2)$$

$$\text{CAD-score}^{\text{iface}}(I, J) = \text{CAD-score}(G_{I,J}^{\text{iface}}) \tag{3}$$

It is also possible to quantify how each interface residue is exposed to the other chain by summing the corresponding contact areas. For a specific residue $i \in I$, the exposure value in the target structure is $T_i = \sum_{(i,j)\in G_{I,J}^{\text{iface}}} T_{(i,j)}$. The set of T_i values for all $i \in I$ describes the binding site of the first chain in the target structure. The corresponding binding site in the model structure is defined in the same way, but using the model interface contacts. Then the similarity score of the target and the model binding sites is computed:

$$\text{CAD-score}^{\text{site}}(I) = 1 - \frac{\sum \min(|T_i - M_i|, T_i)}{\sum T_i} \tag{4}$$

Values of Eq. 4 can range from 0 (completely different binding site) to 1 (binding site with the same exposure of residues, but not necessarily the exact same inter-chain contacts).

Less detailed and, therefore, less stringent similarity measures can be defined using total interface contact areas (Eq. 5) and total binding site areas (Eq. 6):

$$\text{CAD-score}^{\text{iface-area}}(I, J) = \min\left(1, \frac{\sum M_{(i,j)}}{\sum T_{(i,j)}}\right) \tag{5}$$

$$\text{CAD-score}^{\text{site-area}}(I) = \min\left(1, \frac{\sum M_i}{\sum T_i}\right) \tag{6}$$

These latter two similarity measures essentially look whether the interface (binding site) corresponds to the same surface patch without paying attention to the exact contribution by individual residues.

2.4 CAD-Score Web Server

The CAD-score web server is accessible without any restrictions at the following URL: http://bioinformatics.ibt.lt/cad-score. It provides a simple and intuitive graphical user interface for running the original ("classic") implementation of CAD-score [6]. The server outputs tables of scores and provides interactive plots for exploring local contact differences. The CAD-score web server has an online tutorial, and in addition there is a separate paper devoted entirely to the description of the server [10]. Therefore, the focus of this chapter is solely on the standalone CAD-score software, which offers maximal flexibility in structural analyses.

2.5 Standalone CAD-Score Software

At present, there are two distinct software implementations of the CAD-score method. In the first, "classic" implementation (https://bitbucket.org/kliment/cadscore), contacts are constructed for every atom by subdividing the expanded atom sphere according to the Voronoi neighbors. In the more recent implementation, which is a part of a larger package called Voronota (https://bitbucket.org/kliment/voronota), contacts are derived directly from the Voronoi faces, as described in Subheading 2.1. Although the actual values for contact areas and similarity scores in these two implementations differ, the two versions correlate very strongly.

The "classic" implementation has been tested in CASP [11] and CAMEO [12] projects. Recently, it has also been extensively compared to other reference-based similarity measures [7]. On the other hand, the new implementation uses more intuitive and symmetric definition of contacts, making it especially suitable for analysis and comparison of interfaces. The new implementation has been extensively tested and employed in the comparison and clustering of protein-protein interfaces [13]. The scores defined by Eqs. 4–6 are attainable only through the new implementation.

This chapter describes the use of new CAD-score implementation. However, the software can always be run in the "classic" mode, which is enabled by simply using the **`--old-regime`** flag in the command line.

3 CAD-Score Usage

3.1 Installation

The latest version of CAD-score is implemented as the **`voronota-cadscore`** script, which is a part of the Voronota package. The package can be downloaded from https://bitbucket.org/kliment/voronota/downloads and installed (or run without installing) on any modern Linux or macOS system (also, *see* **Notes 1** and **2**). Ubuntu 18.04 and newer Voronota can be downloaded and installed with a single command: **`sudo apt install voronota`**.

3.2 Global Scoring of 3D Structures

For a basic yet realistic example, let us use a dataset from the CASP12 experiment. CASP12 target and model structures are available correspondingly from "targets" and "predictions" folders at http://predictioncenter.org/download_area/CASP12/. Let us consider the heterodimeric target structure "T0921-T0922.pdb" and its models. For clarity, let us rename "T0921-T0922.pdb" to "target.pdb" and rename the model files "TS188_1" and "TS208_1" to "model1.pdb" and "model2.pdb," respectively. These target and model structures already have the same residue numbering and the same chain naming, key requirements for proper use of CAD-score (*see* **Notes 3** and **4** for more details on how the **`voronota-cadscore`** script reads and interprets input PDB files). Below is an example of the global CAD-score calculation for "model1.pdb":

```
voronota-cadscore -t "target.pdb" -m "model1.pdb"

target.pdb model1.pdb AA 212 0.358224 13334.2 8287.38
```

The same result can be presented as a table with a header that explains the values; output can be aligned by passing it to the standard **column** command (also, *see* **Note 5**):

```
voronota-cadscore -t "target.pdb" -m "model1.pdb" --output-header | column -t

target_file model_file query_code residues score    target_area model_area
target.pdb  model1.pdb AA         212      0.358224 13334.2     8287.38
```

"query_code" indicates the category of residue-residue contacts as described in Subheading 2.1. "residues" is the number of target residues that were included in the evaluation. "score" is the global CAD-score value calculated by Eq. 1. "target_area" and "model_area" are total sums of considered contact areas for the target and the model.

3.3 Using Query Codes

Different query codes can be requested using the **--contacts-query-by-code** option, and all possible query codes may be used at once with the **--use-all-query-codes** flag (also, *see* **Note 6**):

```
voronota-cadscore -t "target.pdb" -m "model1.pdb" --use-all-query-codes \
--output-header | column -t

target_file model_file query_code residues score    target_area model_area
target.pdb  model1.pdb AA         212      0.358224 13334.2     8287.38
target.pdb  model1.pdb AS         212      0.268364  9736.82    5192.76
target.pdb  model1.pdb SS         194      0.19971   4908.5     2055.24
target.pdb  model1.pdb AM         212      0.404073  8425.71    5708.45
target.pdb  model1.pdb MM         212      0.479336  3597.39    2736.15
target.pdb  model1.pdb MS         212      0.266683  4828.31    2707.13

voronota-cadscore -t "target.pdb" -m "model1.pdb" --contacts-query-by-code "SS" \
--output-header | column -t

target_file model_file query_code residues score   target_area model_area
target.pdb  model1.pdb SS         194      0.19971 4908.5      2055.24
```

3.4 Caching and Reusing Contacts

Calculated contacts may be cached in a specified directory to be reused when possible. Reading contacts from a cache directory is much faster than recomputing them. In the Bash script below, the contacts for "target.pdb" are calculated only once when scoring the first model, stored in the "tmp" directory, and reused when scoring the second model:

```
for model in "model1.pdb" "model2.pdb"
do
  voronota-cadscore --cache-dir "tmp" -t "target.pdb" -m "$model"
done

target.pdb model1.pdb AA 212 0.358224 13334.2 8287.38
target.pdb model2.pdb AA 212 0.448115 13334.2 9738.18
```

3.5 Tolerating Non-matching Sequences

In order to compare contacts, the CAD-score method assigns a unique identifier to every contact. A contact identifier is a pair of residue identifiers. By default, a residue identifier is comprised of the chain name, the residue sequence number, the insertion code (if present), and the residue name. If for some residue in the target structure there is no residue in the model with the exact same identifier, the CAD-score algorithm considers the residue to be completely absent from the model structure. However, this rule can be softened by using the **--ignore-residue-names** flag. It forces the software to ignore residue names when matching residue identifiers. For example, it allows comparison of wild-type structures with their mutants. Using this flag, in principle, structures with entirely different sequences can be compared (also, *see* **Note 7**). This possibility was not tested for global structure scoring, but it was shown to be very useful for comparison of inter-chain interfaces of homologous protein complexes [13]. For closely related protein complexes, the standard CAD-score definition may be used, but as relationships become more distant, similarities between protein-protein interfaces can be effectively assessed only using less stringent CAD-score variants defined by Eqs. 4–6.

3.6 Focused Scoring

The CAD-score software allows the user to specify which contacts to include in the evaluation. In other words, it is possible to restrict the G parameter for Eq. 1. This is done using the **--contacts-query** option as shown in examples below (also, *see* **Note 8**):

```
#assessing contacts between chains A and B
voronota-cadscore -t "target.pdb" -m "model1.pdb" --cache-dir "tmp" --output-header \
--contacts-query "--match-first c<A> --match-second c<B>"

target_file model_file query_code residues score    target_area model_area
target.pdb  model1.pdb AA         54       0.178782 898.033     356.19

#assessing contacts between two residue sets in chain B
voronota-cadscore -t "target.pdb" -m "model1.pdb" --cache-dir "tmp" --output-header \
--contacts-query "--match-first c<B>&r<39:51> --match-second c<B>&r<39:66,75:87>"

target_file model_file query_code residues score    target_area model_area
target.pdb  model1.pdb AA         29       0.390721 834.729     586.403
```

Let us dissect the second example. The argument to the **--contacts-query** option is a string describing two constraints. The first constraint, specified as **--match-first c<B>&r<39:51>**, means that one side of any included contacts must be a residue that is from chain B and has a sequence number in the 39–51 range. The second constraint, specified as **--match-second c<B>&r<39:66,75:87>**, means that the other side of any included contacts must be a residue that comes from chain B and has a sequence number in either 39–66 or 75–87 range. The second constraint can be rewritten using both "**&**" (logical and) and "**|**" (logical or) operators: **--match-second c<B>&r<39:66>|c<B>&r<75:87>**.

There are more possibilities for specifying contact queries. They can be explored using the graphical contact query generator `support/generate-arguments-for-query-contacts.html` that is included in the Voronota package.

3.7 Scoring of Interfaces and Binding Sites

When scoring inter-chain interfaces, the calculation of binding site similarity score, as defined by Eq. 4, can be enabled using the **--enable-site-based-scoring** flag. This adds additional values to the output as shown in the following example:

```
voronota-cadscore -t "target.pdb" -m "model1.pdb" --cache-dir "tmp" \
--contacts-query "--match-first c<A> --match-second c<B>" \
--enable-site-based-scoring --output-header

target_file model_file query_code residues score    target_area model_area
target.pdb  model1.pdb AA         54       0.178782 898.033     356.19

            site_residues site_score site_target_area site_model_area
            31            0.337558   898.033          675.073
```

In the above example, the binding site is defined by the interface residues of chain A because chain A was indicated with **`--match-first`**. Swapping "`c<A>`" and "`c<B>`" in the contact query forces the evaluation of a different binding site, the one in chain B:

```
voronota-cadscore -t "target.pdb" -m "model1.pdb" --cache-dir "tmp" \
--contacts-query "--match-first c<B> --match-second c<A>" \
--enable-site-based-scoring --output-header

target_file model_file query_code residues score    target_area model_area
target.pdb  model1.pdb AA         54       0.178782 898.033     356.19

     site_residues site_score site_target site_model_area
     23            0.557076   898.033     795.095
```

In both of the above examples, the "`score`" values (the values of Eq. 3) are the same, while the "`site_score`" values (the values of Eq. 4) are different. They may differ radically if, for example, in a model of a protein heterodimer only the binding site in chain A, but not the one in chain B, appears at the dimer interface.

The value of Eq. 5 is not present in the output of **voronota-cadscore**, but it can be easily calculated from the "`model_area`" and the "`target_area`" values: CAD-score$^{\text{iface-area}}$ = min(1, model_area/target_area). Similarly, the value of Eq. 6 can be calculated from the "`site_model_area`" and the "`site_target_area`" values: CAD-score$^{\text{site-area}}$ = min(1, site_model_area/site_target_area).

Instead of specifying the exact interacting regions of an interface, it is possible to ask for all the inter-chain interactions that can be found in the target structure. This is done using the **`--contacts-query-inter-chain`** flag. It also allows calculating the binding site similarity score, where the binding site is a union of all the found interface residues (in other words, the union of all the chain-specific binding sites):

```
voronota-cadscore -t "target.pdb" -m "model1.pdb" --cache-dir "tmp" \
--contacts-query-inter-chain --enable-site-based-scoring --output-header

target_file model_file query_code residues score    target_area model_area
target.pdb  model1.pdb AA         54       0.178782 898.033     356.19

   site_residues site_score site_target_area site_model_area
   54            0.447317   1796.07          1470.17
```

To attain comprehensive understanding of structural differences when assessing multimeric models, it is advisable to look at both structure and interface-related scores. For example, let us look at Table 1 with different CAD-score values for dimeric structures

Table 1
Structure, interface, and binding site evaluation for two models of a dimeric structure using CAD-score

Score description		Formula	model1.pdb	model2.pdb
Structure	Score	CAD-score(G)	0.358	0.448
	Score for chain A	CAD-score(G_A)	0.376	0.424
	Score for chain B	CAD-score(G_B)	0.375	0.593
Inter-chain interface	Score	CAD-score^{iface}(A,B)	0.179	0.070
	Area score	CAD-score$^{\text{iface-area}}$(A,B)	0.397	0.189
Binding site	Score for chain A	CAD-score^{site}(A)	0.338	0.266
	Area score for chain A	CAD-score$^{\text{site-area}}$(A)	0.751	0.750
	Score for chain B	CAD-score^{site}(B)	0.557	0.528
	Area score for chain B	CAD-score$^{\text{site-area}}$(B)	0.885	0.760

"model1.pdb" and "model2.pdb." The second model is better according to the global quality of both the overall structure and its individual chains. However, the inter-chain interface and binding sites are better predicted in the first model. Different levels of detail in the representation of binding sites help to further understand the differences. The binding site in chain A is more accurate in the first model according to detailed representation of contacts, but overall binding site areas are of approximate accuracy in both models. In contrast, the accuracy of binding site in chain B in both models is comparable according to the detailed representation, but the overall area of the binding site is better reproduced in the first model.

3.8 Evaluation of Homo-Oligomeric Models

Comparing homo-oligomeric structures often presents an additional challenge, because the correspondence of the chain names in the model to the chain names in the target may not be optimal. Different arrangements of model chain names may lead to different similarity scores, and the optimal arrangement is the one that results in the highest similarity score. The CAD-score software can rearrange model chain names for higher global scores, this feature is turned on with the **--remap-chains** flag, and the resulting rearrangement can be recorded using the **--remap-chains-output** option.

For example, let us consider the homotrimeric target structure "T0860o.pdb" and its model "T0860TS203_1o." Below are the inter-chain interface scoring results without and with rearranging the model chain names:

```
#without rearranging model chain names
voronota-cadscore -t "T0860o.pdb" -m "T0860TS203_1o" --cache-dir "tmp" \
--contacts-query-inter-chain --output-header | column -t

target_file model_file    query_code residues score     target_area model_area
T0860o.pdb  T0860TS203_1o AA         159      0.0336622 3201.16     147.817

#with rearranging model chain names
voronota-cadscore -t "T0860o.pdb" -m "T0860TS203_1o" --cache-dir "tmp" \
--remap-chains --remap-chains-output "remapping.txt" \
--contacts-query-inter-chain --output-header | column -t

target_file model_file    query_code residues score   target_area model_area
T0860o.pdb  T0860TS203_1o AA         159      0.40027 3201.16     2025.39

cat "remapping.txt"

A C
B B
C A
```

This example shows how the scores can go from poor to reasonable ones after simply rearranging the model chain names.

3.9 Residue-Level Local Scoring

Per-residue scoring can be performed at the same time as the global or the focused scoring. One way to output residue scores is by writing them in place of the B-factor values for the target and/or the model coordinates in the PDB format:

```
voronota-cadscore -t "target.pdb" -m "model1.pdb" --cache-dir "tmp" \
--output-residue-scores-pdb-t "model1_local_scores_on_target.pdb" \
--output-residue-scores-pdb-m "model1_local_scores_on_model.pdb"

target.pdb model1.pdb AA 212 0.358224 13334.2 8287.38
```

The above command writes score values only for the evaluated residues. The produced PDB files can be displayed and colored, for example, in PyMol [14]. Note that PyMol interprets missing B-factor values as zeros. A possible visualization of local scores for two models is shown in Fig. 2. Below is a PyMol script that displays a structure using a color gradient (red-white-blue colors for worst-medium-best scores):

```
load model1_local_scores_on_target.pdb
spectrum b, red_white_blue, all, 0, 1
```

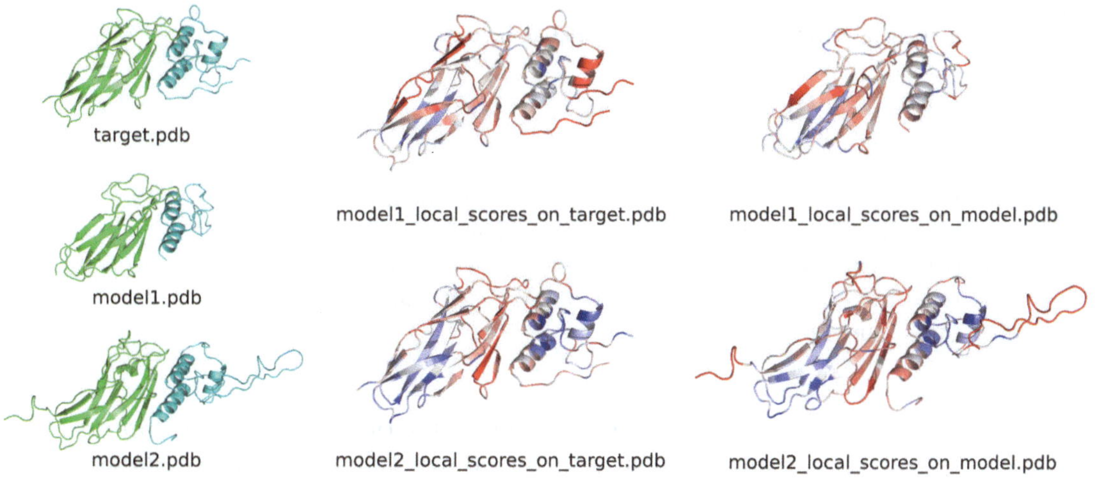

Fig. 2 Example of structure coloring by local CAD-score values, done using PyMol. Blue-red coloring corresponds to high-low scores

When performing focused scoring (e.g., interface scoring), it may be helpful to write a default B-factor value (e.g., 99) in the output PDB files for residues that were not evaluated. One way to do this is to use the **--input-filter-query** option as shown below:

```
voronota-cadscore -t "target.pdb" -m "model1.pdb" --cache-dir "tmp" \
--input-filter-query "--set-adjuncts score=99" \
--contacts-query-inter-chain \
--output-residue-scores-pdb-t "model1_interface_local_scores_on_target.pdb" \
--output-residue-scores-pdb-m "model1_interface_local_scores_on_model.pdb"

target.pdb model1.pdb AA 212 0.358224 13334.2 8287.38
```

This sets the B-factor values of the not scored residues to 99. Then the coloring of the scored and not scored residues can be controlled separately; an example PyMol script is shown below:

```
load model1_interface_local_scores_on_target.pdb
select scored_residues, b<99
spectrum b, red_white_blue, scored_residues, 0, 1
color gray, (not scored_residues)
```

3.10 Detailed Analysis of Contacts

The **voronota-cadscore** script produces similarity scores, but does not output the raw contact data that the scores are derived from. The contacts for a single structure can be produced using the **voronota-contacts** script. Below is an example with a minimal set of options:

```
#calculate contacts
voronota-contacts -i "model.pdb" > contacts.txt

#print first five lines of the output
cat contacts.txt | head -5 | column -t

c<A>r<16>a<1>R<THR>A<N> c<A>r<138>a<914>R<ILE>A<CA> 0.963464 5.6312  . .
c<A>r<16>a<1>R<THR>A<N> c<A>r<138>a<917>R<ILE>A<CB> 3.00998  5.38076 . .
c<A>r<16>a<1>R<THR>A<N> c<A>r<138>a<920>R<ILE>A<CD1> 3.67065  4.46106 . .
c<A>r<16>a<1>R<THR>A<N> c<A>r<139>a<921>R<THR>A<N>  0.11352  5.17191 . .
c<A>r<16>a<1>R<THR>A<N> c<solvent>                   37.4663  5.9     . .
```

The first two columns of the output contain descriptors of the contacting atoms, the third column contains contact areas (in squared angstroms), and the fourth one contains distances between the centers of the atoms. The remaining two columns contain additional contact-related tags (labels) and values. A single dot "." is printed when there are no tags or values to display.

For a more convenient field-based parsing (e.g., with the **awk** tool), the output can be further passed to the **voronota expand-descriptors** command that transforms each descriptor of an atom into a space-separated list of seven values (chain name, residue sequence number, insertion code, atom serial number, alternative location indicator, residue name, atom name); dots are printed in place of unavailable values:

```
voronota-contacts -i "model.pdb" \
| voronota expand-descriptors | head -5 | column -t

A 16 . 1 . THR N A       138 . 914 . ILE CA  0.963464 5.6312  . .
A 16 . 1 . THR N A       138 . 917 . ILE CB  3.00998  5.38076 . .
A 16 . 1 . THR N A       138 . 920 . ILE CD1 3.67065  4.46106 . .
A 16 . 1 . THR N A       139 . 921 . THR N   0.11352  5.17191 . .
A 16 . 1 . THR N solvent . . . . .           37.4663  5.9     . .
```

Atom-level contact can be summarized as residue-level ones:

```
voronota-contacts -i "model.pdb" --contacts-query "--inter-residue" \
| head -5 | column -t

c<A>r<16>R<THR> c<A>r<17>R<ALA>  26.3112 1.33342 central.
c<A>r<16>R<THR> c<A>r<18>R<LYS>  1.39527 4.28654 . .
c<A>r<16>R<THR> c<A>r<138>R<ILE> 12.6313 4.05087 central.
c<A>r<16>R<THR> c<A>r<139>R<THR> 30.6752 3.25293 central.
c<A>r<16>R<THR> c<solvent>       156.25  5.69    . .
```

The possibilities for the **--contacts-query** option are the same as for the analogous option of the **voronota-cadscore** script (some examples are presented in Subheading 3.6). Possible querying parameters can be viewed by running the **voronota query-contacts --help** command. For example, using **--contacts-query "--inter-residue --no-same-chain --no-solvent"** limits output to the contacts between residues of different chains without including contacts with the solvent.

The **voronota-contacts** command allows producing a script for drawing contacts in PyMol. In the example below, a script to display contacts between chains A and B in yellow color is generated:

```
#calculate contacts and generate a drawing script
voronota-contacts --cache-dir "tmp" -i "model.pdb" \
--contacts-query "--match-first c<A> --match-second c<B>" \
--output-drawing "draw_interface_AB.py" \
--drawing-parameters "--drawing-name interface_AB --default-color 0xFFFF00" \
> contacts.txt

#launch PyMol with the structure and the drawing
pymol model.pdb draw_interface_AB.py
```

The use of the **--cache-dir** option allows to generate several drawings for different queries without recomputing contacts every time. In order to load several drawings into PyMol, the names of the drawings should be distinct: providing the **--drawing-name** parameter is advised; otherwise, the name is set to "contacts." An example of multiple drawings in one scene is shown in Fig. 3, where interface contacts for different pairs of chains are displayed in distinct colors.

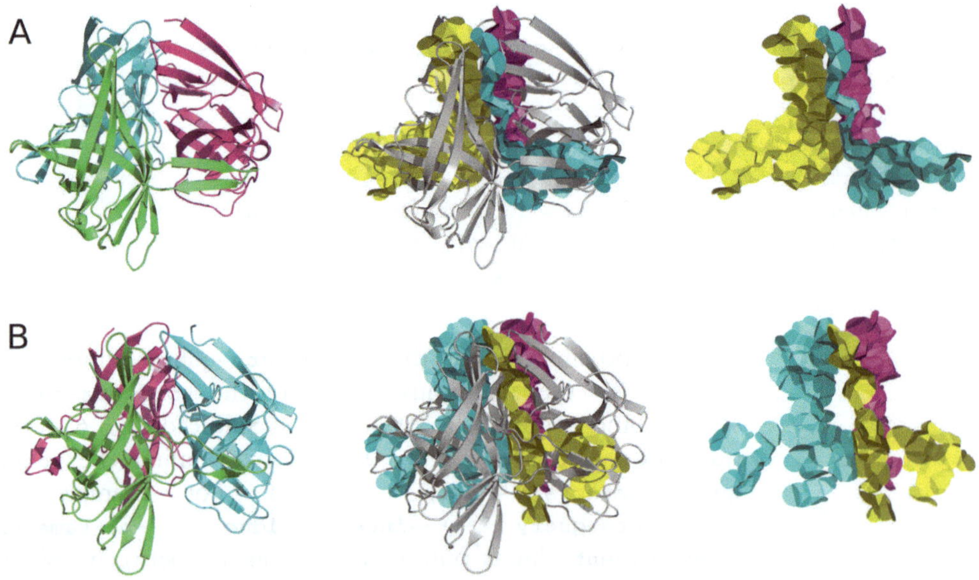

Fig. 3 Inter-chain interface contacts drawn in PyMol for two homotrimeric structures: (**a**) CASP12 target structure "T0860.pdb"; (**b**) model structure "T0860TS203_1o.pdb"

4 Notes

1. On macOS, it is advised to use Voronota 1.19 or a newer version, because earlier versions were not tested on macOS.
2. On Windows 10, the most convenient way to run CAD-score is through the Windows Subsystem for Linux.
3. The **`voronota-cadscore`** script reads both ATOM and HETATM records from PDB files. The script ignores TER records and determines chains just by the one-letter chain names from the ATOM or HETATM records.
4. If an input PDB file contains multiple MODEL blocks, then, by default, the **`voronota-cadscore`** script reads only the first MODEL block. This behavior can be changed with the **`--multiple-models`** option. It forces the script to treat input files as PDB biological assemblies (complexes assembled from the chains in every encountered MODEL block). This alters the internal representation of chain names: MODEL 1 chain names are left unchanged, and the names of the chains from the subsequent MODEL blocks are augmented with block numbers (e.g., chain “A” from MODEL 2 is renamed to “A2,” chain “A” from MODEL 3 is renamed to “A3,” and so on).
5. Command execution examples are presented as for the Bash shell that is the default shell for most Linux and macOS distributions.
6. Symbol “\” in a command example indicates that the command continues in the next line; “\” is not needed if a command is written in one line.
7. The **`voronota-cadscore`** script does not automatically align sequences or renumber residues in target and model structures. The correspondence between residues is determined simply based on their numbering and chain assignments in PDB files.
8. In most cases, it is necessary to enclose the argument to the **`--contacts-query`** option in quotes. Quotes are required for any argument that contains spaces or other special symbols (like “<,” “>,” “\&,” and “|”).

Acknowledgment

This work was supported by the Research Council of Lithuania [S-MIP-17-60].

References

1. Kabsch W (1976) A solution for the best rotation to relate two sets of vectors. Acta Crystallogr A 32(5):922–923. https://doi.org/10.1107/S0567739476001873
2. Zemla A, Venclovas Č, Moult J, Fidelis K (1999) Processing and analysis of CASP3 protein structure predictions. Proteins (Suppl 3):22–29
3. Zemla A, Venclovas Č, Moult J, Fidelis K (2001) Processing and evaluation of predictions in CASP4. Proteins (Suppl 5):13–21. https://doi.org/10.1002/prot.10052
4. Zhang Y, Skolnick J (2004) Scoring function for automated assessment of protein structure template quality. Proteins 57(4):702–710. https://doi.org/10.1002/prot.20264
5. Mariani V, Biasini M, Barbato A, Schwede T (2013) lDDT: a local superposition-free score for comparing protein structures and models using distance difference tests. Bioinformatics 29(21):2722–2728. https://doi.org/10.1093/bioinformatics/btt473
6. Olechnovič K, Kulberkytė E, Venclovas Č (2013) CAD-score: a new contact area difference-based function for evaluation of protein structural models. Proteins 81 (1):149–162. https://doi.org/10.1002/prot.24172
7. Olechnovič K, Monastyrskyy B, Kryshtafovych A, Venclovas Č (2018) Comparative analysis of methods for evaluation of protein models against native structures. Bioinformatics 35:937. https://doi.org/10.1093/bioinformatics/bty760
8. Olechnovič K, Venclovas Č (2014) Voronota: a fast and reliable tool for computing the vertices of the Voronoi diagram of atomic balls. J Comput Chem 35(8):672–681. https://doi.org/10.1002/jcc.23538
9. Olechnovič K, Venclovas Č (2017) VoroMQA: assessment of protein structure quality using interatomic contact areas. Proteins 85 (6):1131–1145. https://doi.org/10.1002/prot.25278
10. Olechnovič K, Venclovas Č (2014) The CAD-score web server: contact area-based comparison of structures and interfaces of proteins, nucleic acids and their complexes. Nucleic Acids Res 42(Web Server issue): W259–W263. https://doi.org/10.1093/nar/gku294
11. Kryshtafovych A, Monastyrskyy B, Fidelis K (2014) CASP prediction center infrastructure and evaluation measures in CASP10 and CASP ROLL. Proteins 82(Suppl 2):7–13. https://doi.org/10.1002/prot.24399
12. Haas J, Roth S, Arnold K, Kiefer F, Schmidt T, Bordoli L, Schwede T (2013) The Protein Model Portal—a comprehensive resource for protein structure and model information. Database (Oxford) 2013:bat031. https://doi.org/10.1093/database/bat031
13. Dapkūnas J, Timinskas A, Olechnovič K, Margelevičius M, Dičiūnas R, Venclovas Č (2017) The PPI3D web server for searching, analyzing and modeling protein-protein interactions in the context of 3D structures. Bioinformatics 33(6):935–937. https://doi.org/10.1093/bioinformatics/btw756
14. Schrödinger, LLC (2015) The PyMOL molecular graphics system, version 1.8

Chapter 7

A Comprehensive Computational Platform to Guide Drug Development Using Graph-Based Signature Methods

Douglas E. V. Pires, Stephanie Portelli, Pâmela M. Rezende, Wandré N. P. Veloso, Joicymara S. Xavier, Malancha Karmakar, Yoochan Myung, João P. V. Linhares, Carlos H. M. Rodrigues, Michael Silk, and David B. Ascher

Abstract

High-throughput computational techniques have become invaluable tools to help increase the overall success, process efficiency, and associated costs of drug development. By designing ligands tailored to specific protein structures in a disease of interest, an understanding of molecular interactions and ways to optimize them can be achieved prior to chemical synthesis. This understanding can help direct crucial chemical and biological experiments by maximizing available resources on higher quality leads. Moreover, predicting molecular binding affinity within specific biological contexts, as well as ligand pharmacokinetics and toxicities, can aid in filtering out redundant leads early on within the process. We describe a set of computational tools which can aid in drug discovery at different stages, from hit identification (EasyVS) to lead optimization and candidate selection (CSM-lig, mCSM-lig, Arpeggio, pkCSM). Incorporating these tools along the drug development process can help ensure that candidate leads are chemically and biologically feasible to become successful and tractable drugs.

Key words Graph-based signatures, mCSM, Mutation, Protein-ligand, Interatomic interactions, Docking, Drug development

1 Introduction

Structure-guided drug development uses knowledge of the three-dimensional structure of the biological target to more efficiently guide the design of small molecule binders. While it has become an integral strategy for both lead generation and optimization, the application of computational tools to take advantage of the explosion in structural information has often required specialist knowledge and resources and in some cases has been limited to commercial software.

Zoltán Gáspári (ed.), *Structural Bioinformatics: Methods and Protocols*, Methods in Molecular Biology, vol. 2112,
https://doi.org/10.1007/978-1-0716-0270-6_7, © Springer Science+Business Media, LLC, part of Springer Nature 2020

Using the concept of graph-based signatures, we have developed a robust, user-friendly, and freely accessible platform to analyze protein structures and interactions [1–12] and guide disease characterization [13–28] and drug development [29–32]. These include methods to perform virtual screening (EasyVS), score protein-small molecule docking solutions (CSM-lig [3]), look at all the molecular interactions being made (Arpeggio [7]), identify mutations that are likely to affect compound binding (mCSM-lig [5]), and characterize the pharmacokinetic and toxicity properties of the proposed molecules (pkCSM [33, 34]). These have been successfully employed in a number of drug development projects [30–32, 35–37] and together comprise a powerful platform that allows users to enhance their structure-guided drug development efforts (Fig. 1). Here we discuss how this platform can be leveraged to guide drug development.

2 Materials

Here we present four structure-based tools to help guide drug development. For each method, users are required to provide:

1. **Wild-type protein structure in PDB format**: For all methods, a wild-type structure in the Protein Data Bank [38] format must be provided to perform the analysis. This can be an experimentally solved structure previously deposited into the Protein Data Bank (www.rcsb.org or http://www.ebi.ac.uk/pdbe/) or a model, for instance, obtained by comparative homology modeling. We have previously shown that homology models built using templates down to 25% sequence identity do not significantly affect the accuracy of the methods [9, 10]. For Arpeggio, CSM-lig, and mCSM-lig, the protein structure file needs to include the ligand of interest, either already present in the experimental structure or computationally docked into the binding site. PDB structures are required to have a valid chain identifier (*see* **Note 1**), a single conformation (multiple occupancies need to be filtered out; *see* **Note 2**), and a single model, in case of NMR structures (*see* **Note 3**).
2. **Three-letter code of the ligand of interest**: When a structure of a protein-ligand complex is provided to the predictive web servers (CSM-lig and mCSM-lig), users will be asked to provide a three-letter code that identifies the residue ID for that ligand within the PDB file, according to the PDB format standards. In addition to the three-letter code, CSM-lig also requires the canonical SMILES of the compound of interest for additional property calculations. Several tools are available to aid users to convert between small molecule formats. These include stand-alone packages such as OpenBabel [39] and Avogadro [40].

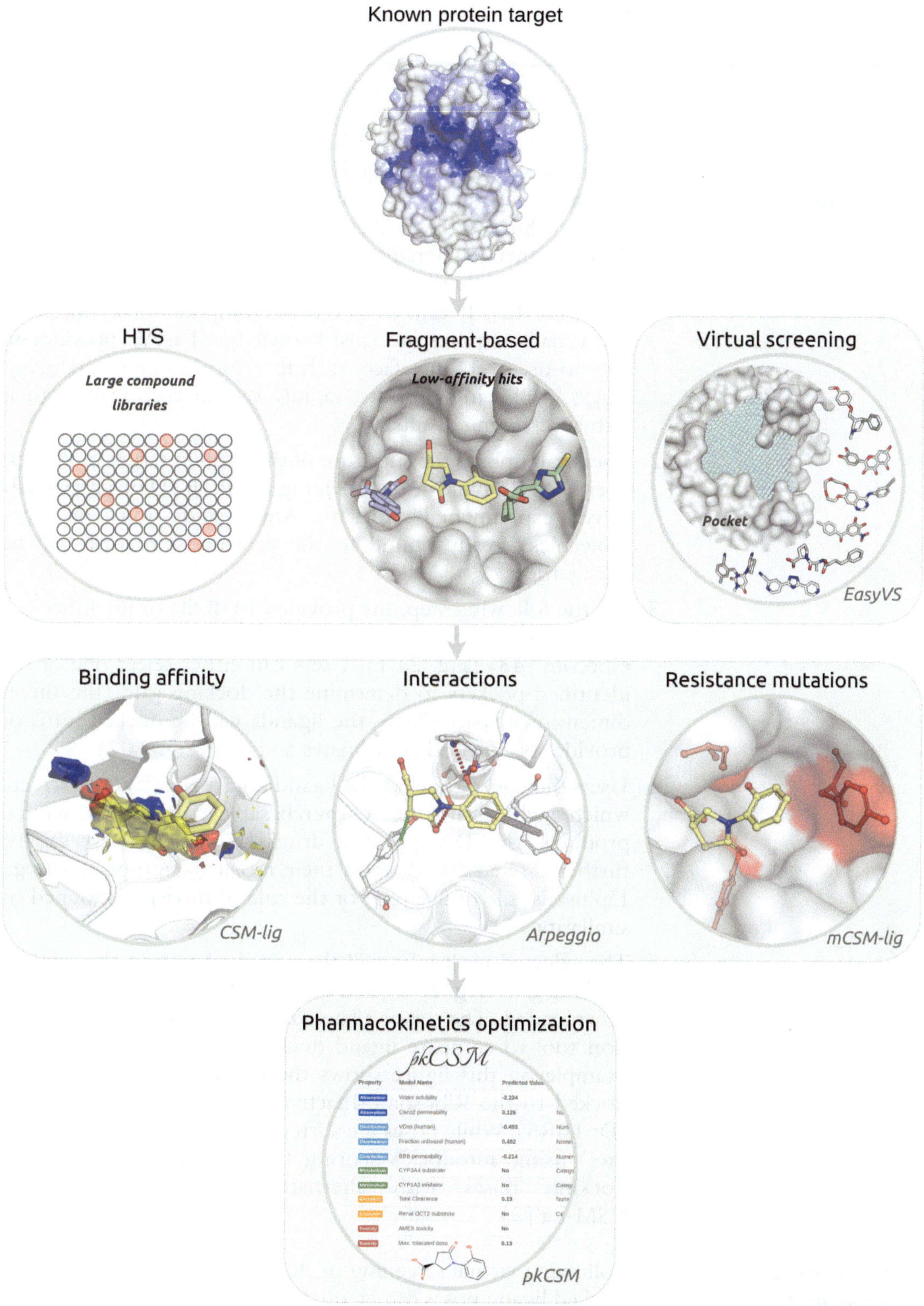

Fig. 1 A structure-based computational platform to guide drug development. To complement and support traditional experimental approaches, including high-throughput screening (HTS) and fragment-based drug discovery, this in silico platform supports hit identification via virtual screening, methods to better understand protein–small molecule interactions, affinity and effects of mutations, as well as the optimization of pharmacokinetic properties

3 Methods

3.1 Performing Automated Docking with EasyVS

1. Virtual screening is a powerful, high-throughput technique for computationally screening large libraries of small molecules (often in the order of millions) in order to identify those ligands which are most likely to bind to a drug target protein. When compared to traditional screening methods, this leads to significantly higher hit rates that can proceed to lead optimization [41, 42]. It can, however, be computationally intensive and usually requires specialist knowledge. EasyVS provides an easy-to-use web interface at http://biosig.unimelb.edu.au/easyvs/, allowing users to rapidly set up and analyze their virtual screening results.
2. Users can upload the structure of the protein target of interest as either a PDB file or by providing the PDB ID of a previously solved experimental structure. Any ligands, ions, or water molecules already bound to the provided structure will be disregarded.
3. On the following step, the provided PDB file or identifier will be processed, and pockets will be automatically detected using Ghecom [43] (Fig. 2a-1). Users can either select one of the identified pockets to determine the docking grid (the three-dimensional space where the ligands will be docked into) or provide specific grid coordinates and size (Fig. 2a-2).
4. Users then need to select the ligand library they want to screen, which includes libraries of purchasable compounds, natural products, or FDA-approved drugs (Fig. 2b). These can be further filtered based upon their molecular properties (e.g., Lipinski's rule of five [44] or the rule of three) or grouped by similarity.
5. The selected molecules will then be docked into the selected docking grid (Fig. 2c-1), and the top 20 poses per ligand can be downloaded. The server also provides an interactive visualization tool to compare ligand docking poses (Fig. 2c-2). The example on this figure shows the docking poses for ligands docked to the Ribosome-Inactivating Protein Ricin A (PDB ID: 1BR5). While poses are sorted by predicted affinity (kcal/mol) using autodock's scoring function, users can evaluate docking poses with alternative approaches, such as CSM-lig [3].

3.2 Predicting Protein-Small Molecule Affinity with CSM-lig

1. Following virtual screening or docking, the affinity of the top docked ligand poses can be quantified using CSM-lig. This is a machine learning-based tool which acts as a scoring function and enables the numerical affinity comparison between poses. It is implemented via an easy-to-use web interface at http://biosig.unimelb.edu.au/csm_lig, which is compatible with most operating systems and browsers.

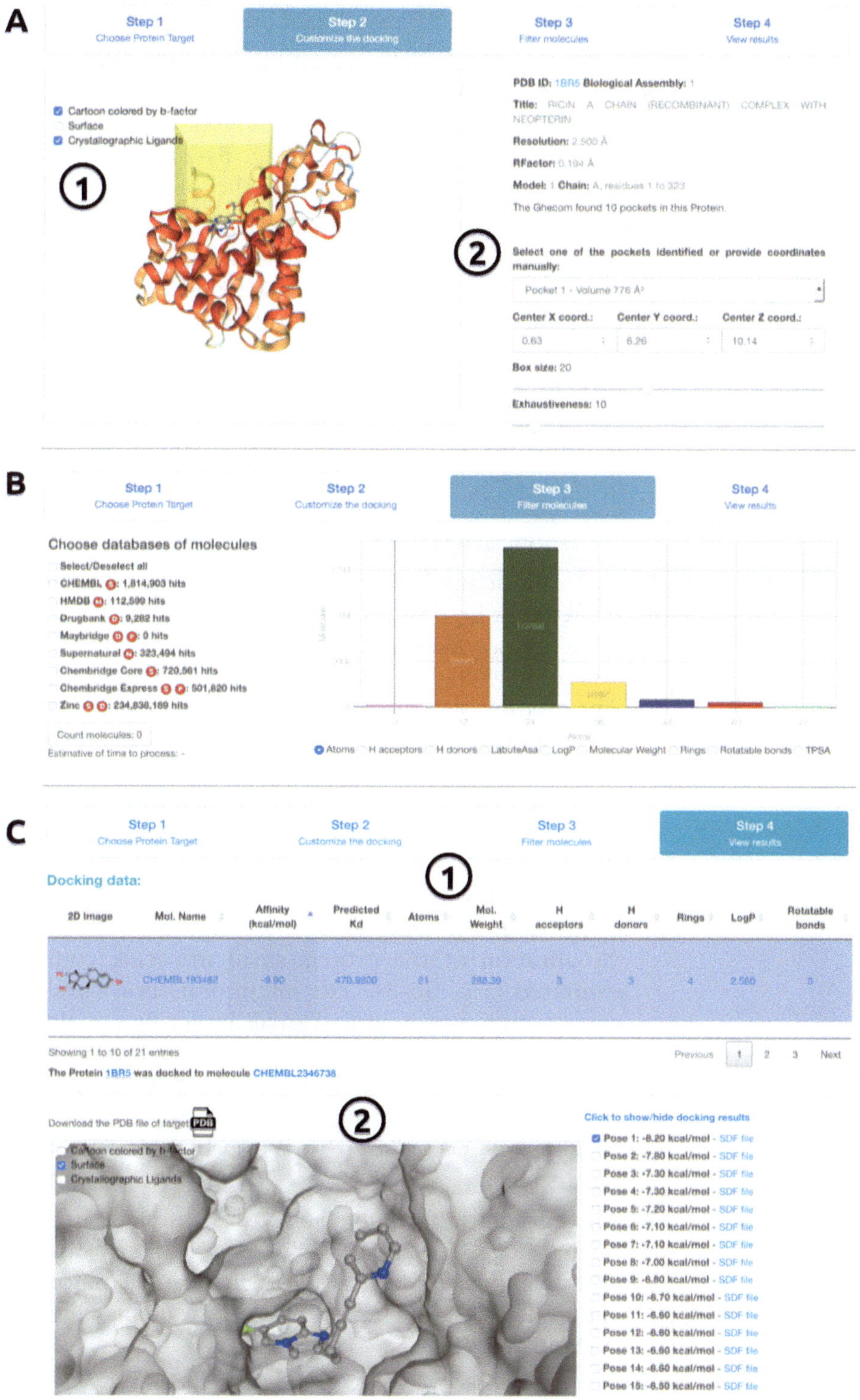

Fig. 2 Automated docking with EasyVS. After choosing a target of interest, EasyVS will automatically identify pockets (**a-1**) and allow user to further customize the docking protocol (**a-2**). A range of ligand libraries can be selected for docking (**b**), including FDA-approved drugs, purchasable compounds, and natural products, which can be further filtered based on physicochemical properties. Docking results are shown in tabular format (**c-1**), depicting ligands, their properties, and docking scores. An interactive viewer allows users to inspect the best poses for each ligand (**c-2**)

2. By selecting the "Predict" tab, users are presented with two job options, "Single Structure" and "Multiple Structures."
3. For "Single Structure" prediction, provide (Fig. 3a-1) the protein-small molecule complex you would like to evaluate the pose of in PDB format (Fig. 3a-2), the three-letter code for the small molecule (as in the provided PDB file) and (Fig. 3a-3) and the SMILES string of the small molecule.
4. Alternatively, for "Multiple Structures," provide two files. The first file (Fig. 3a-4) is a compressed zip file with all protein-small molecule PDB files you would like to evaluate. These could be, for instance, different poses or conformations for a given protein-ligand complex or multiple different complexes. The second (Fig. 3a-5) is a tab-separated file with the following information for each uploaded complex in the .zip file: (a) structure file name (file in PDB format), (b) three-letter code for the small molecule (as in the structure file), and (c) canonical SMILES for the small molecule.
5. The output prediction page for the "Single Structure" jobs depicted in Fig. 1b presents (Fig. 3b-1) the predicted affinity (as $-\log_{10}(\mathit{affinity})$ in molar, meaning a compound with an affinity predicted as 1 nM would have a predicted value of 9). The example presented in the figure and the web server shows the affinity prediction for the ligand Zanamivir bound to human sialidase-2 (PDB ID: 2F0Z). For this complex, CSM-lig generates a score of 12.6, denoting very high affinity (larger numbers denote higher affinity). A depiction figure of the small molecule is shown, together with calculated properties, including molecular weight (in Da) and partition coefficient (log *P*), among others (Fig. 3b-2). An interactive visualization of the protein-small molecule complex is also exhibited (Fig. 3b-3). The interatomic non-covalent interactions between protein and small molecule are also calculated and are available as a downloadable Pymol [45] session (Fig. 3b-4). Pharmacokinetics and toxicity predictions by pkCSM for the provided small molecule are also available by clicking on the red button at the bottom-left corner of the results page.
6. The output for "Multiple Structures" jobs are shown in tabular format (Fig. 3c-1), depicting predicted affinity values, SMILES identifying the molecules and their calculated molecular properties. These results are available as a tabular file and can be downloaded (Fig. 3c-2).

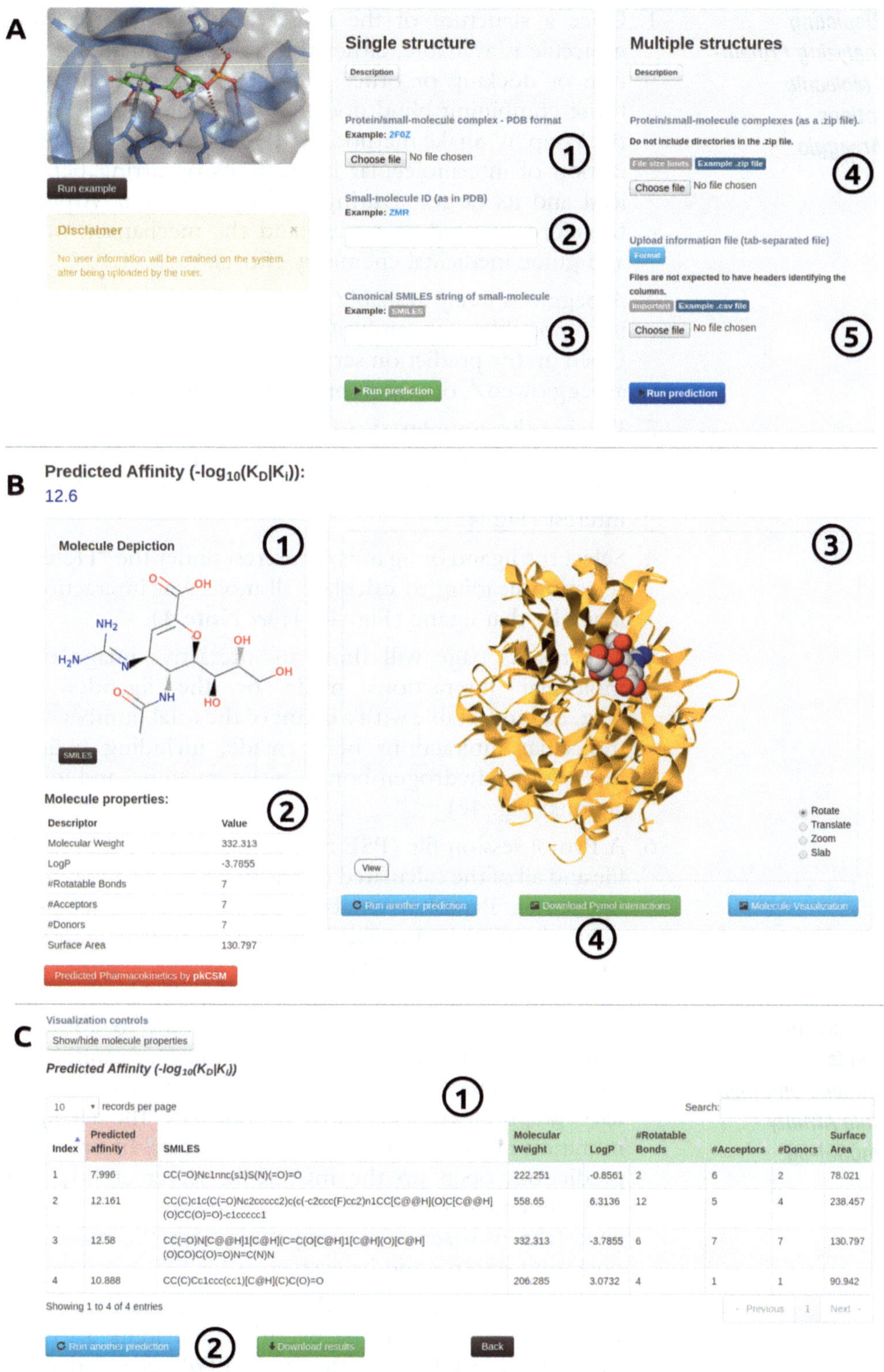

Fig. 3 CSM-lig submission and results web interface. The submission page (**a**) allows users to provide either single or multiple protein-ligand complexes for evaluation. The results page for single complex/pose assessment (**b**) provides the calculated affinity, ligand properties and depiction, as well as an interactive visualization of the complex. For multiple poses, CSM-lig provides the predicted affinities in a downloadable tabular format, together with ligand properties (**c**)

3.3 Depicting and Analyzing Protein-Small Molecule Interactions with Arpeggio

1. Once a structure of the target protein with the candidate molecule is available, either through experimental determination or docking or other alternative approach (for instance, those combining blind docking with molecular dynamics like the Wrap 'n' Shake method [46]), Arpeggio enables the visualization of intermolecular interactions occurring between the lead and its target. During lead optimization, Arpeggio can therefore be used to understand the mechanism of binding and guide medicinal chemistry efforts.
2. Arpeggio is freely available as a user-friendly web interface and is compatible with multiple operating systems and browsers. Open up the prediction server, http://biosig.unimelb.edu.au/arpeggioweb/, on a web browser of your preference.
3. Provide the complexed protein structure of interest by either uploading it as a PDB file or providing the PDB ID of the experimentally solved structure in complex with the ligand of interest (Fig. 4a-1).
4. Select the ligand or ligands of interest under the "Heteroatom" selection heading to calculate all molecular interactions being made by that ligand (Fig. 4b-1; *see* **Note 4**).
5. The results page will show an interactive image of all the molecular interactions made by the ligand(s) selected (Fig. 5a) and a table with a count of the total number of specific molecular interactions being made, including hydrophobic interactions, hydrogen bonds, pi-interactions, and ionic interactions (Fig. 4c).
6. A Pymol session file (PSE file) containing the submitted PDB file and all of the calculated interactions can be downloaded and opened in Pymol to enable visualization of the interaction network in 3D and to facilitate high-quality image generation for manuscripts (Fig. 5b).

3.4 Predicting the Effects of Mutations on Small Molecule Affinity with mCSM-lig

1. During lead optimization, it is important to consider how genetic diversity might affect the binding of candidate molecules and, in particular, if resistance is likely to arise. mCSM-lig uses graph-based signatures to calculate the change upon mutation in small molecule binding affinity. In order to run a prediction, open up the mCSM-lig server at http://biosig.unimelb.edu.au/mcsm_lig/ on a web browser of your preference (the web server is compatible with the most common operating systems and browsers).
2. Users are required to provide the protein structure in complex with the ligand of interest by either uploading a PDB file or supplying a valid four-letter code PDB accession code of a deposited experimental structure (Fig. 6a-1). Users also need to provide the mutation information, the mutation chain, the

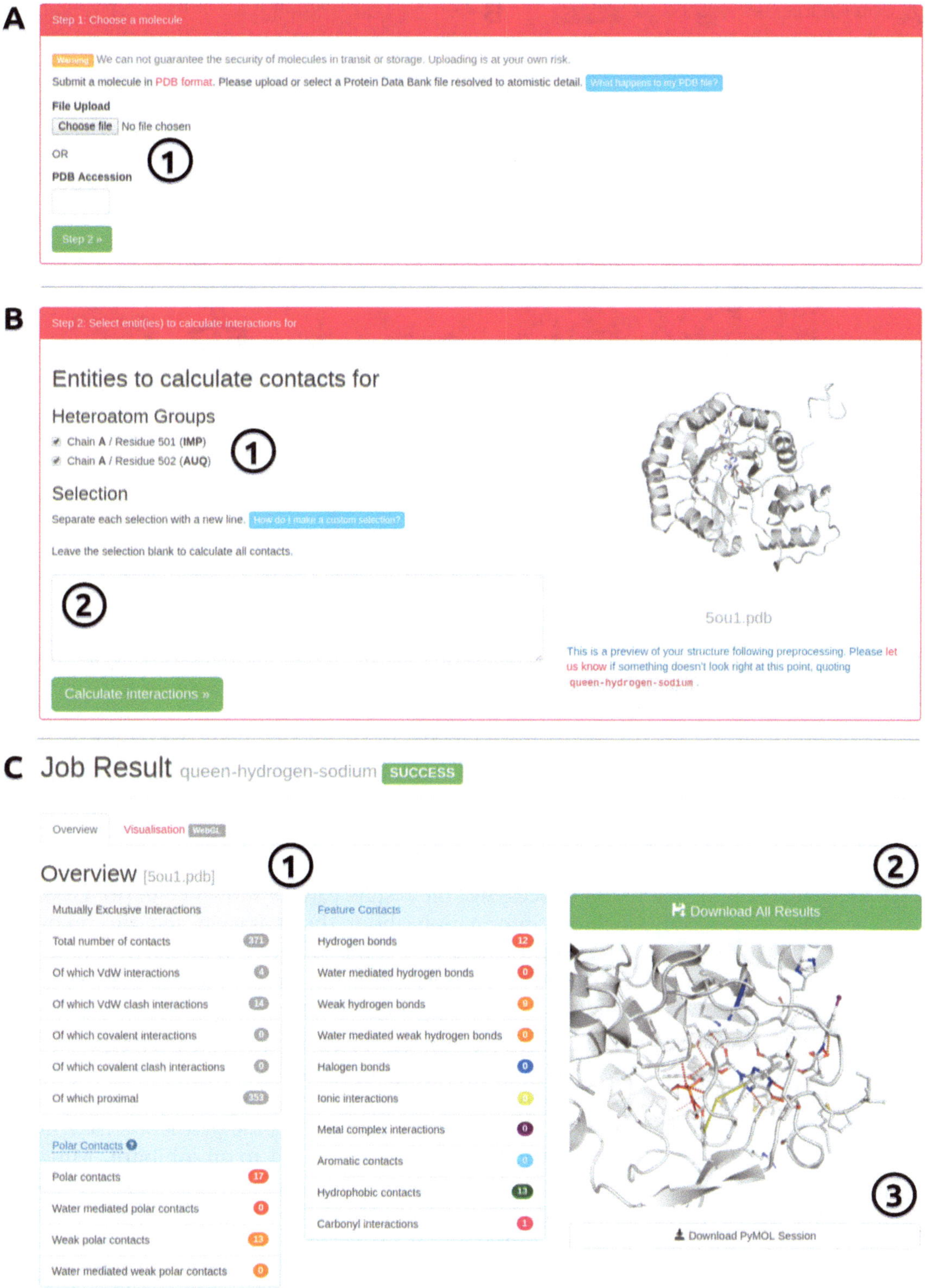

Fig. 4 Arpeggio submission and results web interface. (**a**) The submission page allows users to either provide their own PDB file or an accession code of a deposited experimental structure of the protein of interest. By selecting the molecule of interest (**b**), all molecular interactions will be calculated and displayed (**c**)

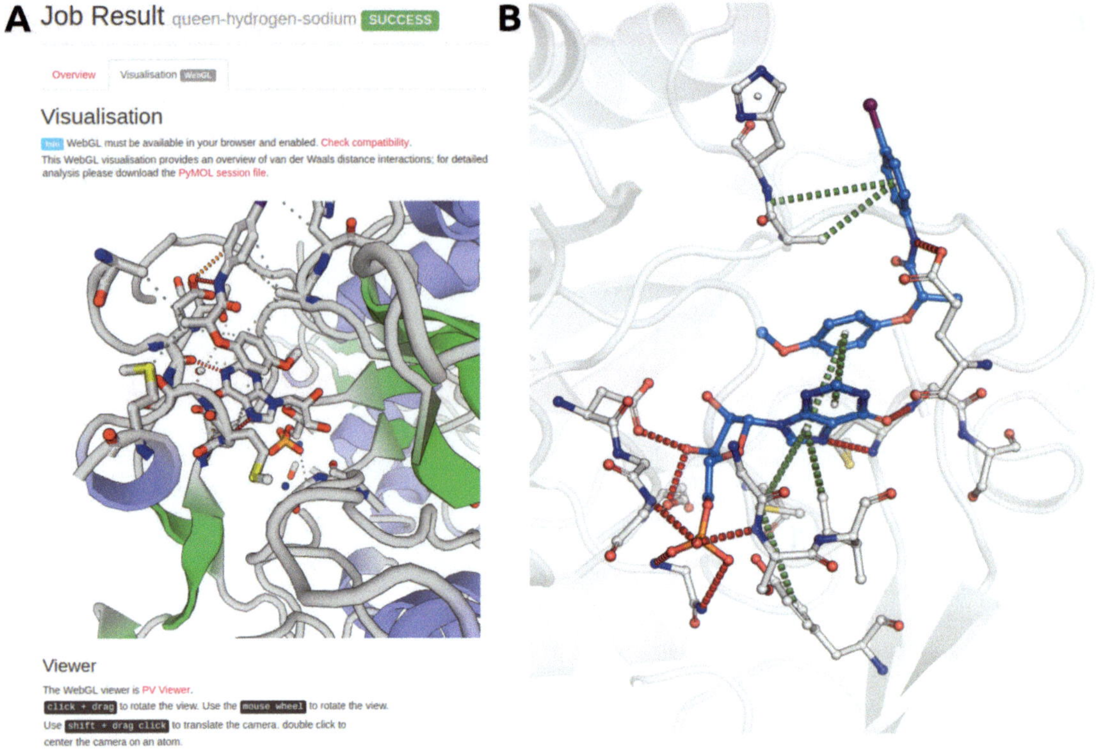

Fig. 5 Molecular interaction visualization using Arpeggio. The molecular interactions calculated by Arpeggio can be visualized either online (**a**) or by downloading the PSE file for visualization in Pymol (**b**)

three-letter code of the ligand of interest in the PDB file, and the approximate binding affinity (in nM) (Fig. 6a-2). If the binding affinity is not available, this can be approximated using CSM-lig. The mCSM-lig values do not vary significantly across most biologically relevant binding affinities.

3. After processing, the results page is shown (Fig. 6b-1), which includes information about the mutation and the predicted effects on the ligand binding affinity. An interactive molecular visualization is shown, allowing users to inspect the wild-type residue environment (Fig. 6b-2).

4. Predicted effects are outputted as the log fold change in binding affinity, in which negative values denote destabilizing mutations and positive values, stabilizing ones. The example shown in Fig. 6 and the web server depicts the prediction for a mutation on the HIV-1 protease bound to an inhibitor. Mutation from Aspartic Acid to Asparagine on residue position 30 is predicted to considerably reduce protein-ligand affinity. While users should interpret the values in the context of the protein system being studied, for competitive binding inhibitors, it is often important to consider the relative effect of a mutation on not only inhibitor binding but also the competitive ligand. This

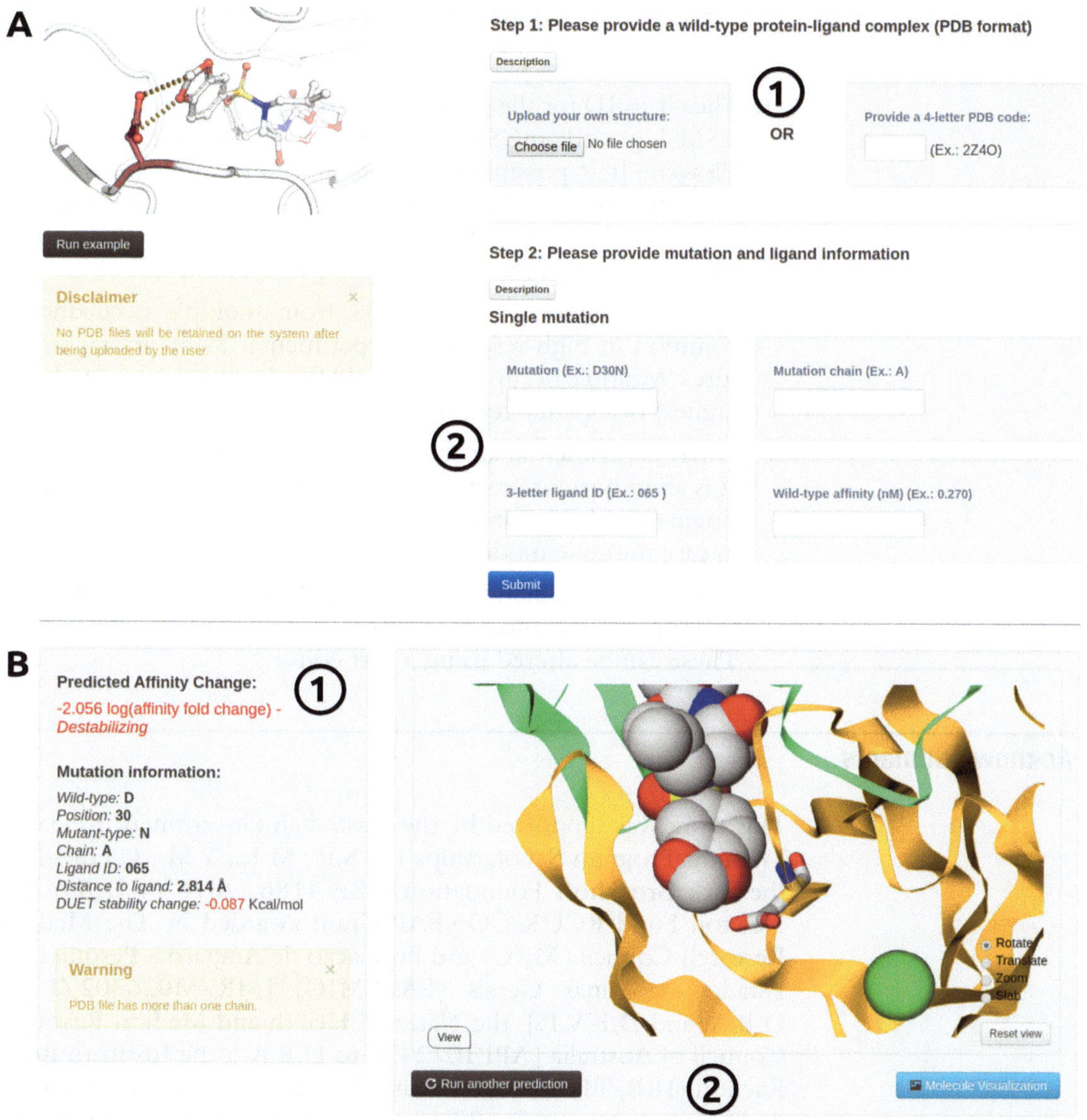

Fig. 6 mCSM-lig submission and results web interface. To predict the effects of a mutation on protein-ligand affinity, users need to provide a protein-ligand structure of interest (**a-1**) as well as mutation and ligand information (**a-2**). Once the calculations have finished, the results page will show the predicted change in ligand binding affinity (**b-1**) as well as an interactive visualization of the mutated residue within its molecular environment (**b-2**)

can be done by submitting a structure of the protein containing the ligand. Resistance mutations are more likely to affect, or have a larger effect, on inhibitor binding affinity than the natural ligand. This has been used to successfully preemptively guide detection of likely resistance variants [29–31, 47–53].

4 Notes

1. The chain ID for the provided PDB file is a mandatory field for CSM-Lig and mCSM-Lig, and blank characters are not allowed. It is possible that homology modeling tools might not automatically add a chain ID. If this is the case, the user will need to modify the PDB file prior to submission to the servers. There are several tools available to perform this task.
2. Another source of error comes from multiple occupancies, common in high-resolution experimental X-ray crystal structures. Multiple occupancies should first be filtered out, with the highest occupancy conformation normally selected.
3. NMR experimental structures often contain multiple models. It is an important practice to filter NMR structures, selecting a single model. The predictive tool will show a warning message in case multiple models are identified.
4. Arpeggio will sometimes fail if the PDB file contains an element with upper and lower case letters (e.g., Fe as opposed to FE). These can be altered using a text editor.

Acknowledgments

This work was supported by the Australian Government Research Training Program Scholarships [to S.P., M.K., Y.M., C.H.M.R.]; the Jack Brockhoff Foundation [JBF 4186, 2016 to D.B.A.]; a Newton Fund RCUK-CONFAP Grant awarded by The Medical Research Council (MRC) and Fundação de Amparo à Pesquisa do Estado de Minas Gerais (FAPEMIG) [MR/M026302/1 to D.B.A. and D.E.V.P.]; the National Health and Medical Research Council of Australia [APP1072476 to D.B.A.]; the Instituto René Rachou (IRR/FIOCRUZ Minas), Brazil, and Conselho Nacional de Desenvolvimento Científico e Tecnológico (CNPq) [to D.E.V. P., P.M.R.]; and the Department of Biochemistry and Molecular Biology, University of Melbourne [to D.B.A.].

References

1. Pires DE, Ascher DB, Blundell TL (2014) mCSM: predicting the effects of mutations in proteins using graph-based signatures. Bioinformatics 30(3):335–342. https://doi.org/10.1093/bioinformatics/btt691
2. Pires DE, Ascher DB, Blundell TL (2014) DUET: a server for predicting effects of mutations on protein stability using an integrated computational approach. Nucleic Acids Res 42 (Web Server issue):W314–W319. https://doi.org/10.1093/nar/gku411
3. Pires DE, Ascher DB (2016) CSM-lig: a web server for assessing and comparing protein-small molecule affinities. Nucleic Acids Res 44 (W1):W557–W561. https://doi.org/10.1093/nar/gkw390
4. Pires DE, Ascher DB (2016) mCSM-AB: a web server for predicting antibody-antigen affinity changes upon mutation with graph-based

signatures. Nucleic Acids Res 44(W1): W469–W473. https://doi.org/10.1093/nar/gkw458

5. Pires DE, Blundell TL, Ascher DB (2016) mCSM-lig: quantifying the effects of mutations on protein-small molecule affinity in genetic disease and emergence of drug resistance. Sci Rep 6:29575. https://doi.org/10.1038/srep29575
6. Pires DE, Chen J, Blundell TL, Ascher DB (2016) In silico functional dissection of saturation mutagenesis: interpreting the relationship between phenotypes and changes in protein stability, interactions and activity. Sci Rep 6:19848. https://doi.org/10.1038/srep19848
7. Jubb HC, Higueruelo AP, Ochoa-Montano B, Pitt WR, Ascher DB, Blundell TL (2017) Arpeggio: a web server for calculating and visualising interatomic interactions in protein structures. J Mol Biol 429(3):365–371. https://doi.org/10.1016/j.jmb.2016.12.004
8. Pandurangan AP, Ochoa-Montano B, Ascher DB, Blundell TL (2017) SDM: a server for predicting effects of mutations on protein stability. Nucleic Acids Res 45(W1): W229–W235. https://doi.org/10.1093/nar/gkx439
9. Rodrigues CH, Ascher DB, Pires DE (2018) Kinact: a computational approach for predicting activating missense mutations in protein kinases. Nucleic Acids Res 46(W1): W127–W132. https://doi.org/10.1093/nar/gky375
10. Rodrigues CH, Pires DE, Ascher DB (2018) DynaMut: predicting the impact of mutations on protein conformation, flexibility and stability. Nucleic Acids Res 46(W1):W350–W355. https://doi.org/10.1093/nar/gky300
11. Pires DE, Blundell TL, Ascher DB (2015) Platinum: a database of experimentally measured effects of mutations on structurally defined protein-ligand complexes. Nucleic Acids Res 43(Database issue):D387–D391. https://doi.org/10.1093/nar/gku966
12. Pires DEV, Ascher DB (2017) mCSM-NA: predicting the effects of mutations on protein-nucleic acids interactions. Nucleic Acids Res 45 (W1):W241–W246. https://doi.org/10.1093/nar/gkx236
13. Jafri M, Wake NC, Ascher DB, Pires DE, Gentle D, Morris MR, Rattenberry E, Simpson MA, Trembath RC, Weber A, Woodward ER, Donaldson A, Blundell TL, Latif F, Maher ER (2015) Germline mutations in the CDKN2B tumor suppressor gene predispose to renal cell carcinoma. Cancer Discov 5(7):723–729. https://doi.org/10.1158/2159-8290.CD-14-1096
14. Jubb H, Blundell TL, Ascher DB (2015) Flexibility and small pockets at protein-protein interfaces: new insights into druggability. Prog Biophys Mol Biol 119(1):2–9. https://doi.org/10.1016/j.pbiomolbio.2015.01.009
15. Usher JL, Ascher DB, Pires DE, Milan AM, Blundell TL, Ranganath LR (2015) Analysis of HGD gene mutations in patients with alkaptonuria from the United Kingdom: identification of novel mutations. JIMD Rep 24:3–11. https://doi.org/10.1007/8904_2014_380
16. Coelho MB, Ascher DB, Gooding C, Lang E, Maude H, Turner D, Llorian M, Pires DE, Attig J, Smith CW (2016) Functional interactions between polypyrimidine tract binding protein and PRI peptide ligand containing proteins. Biochem Soc Trans 44(4):1058–1065. https://doi.org/10.1042/BST20160080
17. Kano FS, Souza-Silva FA, Torres LM, Lima BA, Sousa TN, Alves JR, Rocha RS, Fontes CJ, Sanchez BA, Adams JH, Brito CF, Pires DE, Ascher DB, Sell AM, Carvalho LH (2016) The presence, persistence and functional properties of Plasmodium vivax Duffy binding protein II antibodies are influenced by HLA class II allelic variants. PLoS Negl Trop Dis 10(12):e0005177. https://doi.org/10.1371/journal.pntd.0005177
18. Nemethova M, Radvanszky J, Kadasi L, Ascher DB, Pires DE, Blundell TL, Porfirio B, Mannoni A, Santucci A, Milucci L, Sestini S, Biolcati G, Sorge F, Aurizi C, Aquaron R, Alsbou M, Lourenco CM, Ramadevi K, Ranganath LR, Gallagher JA, van Kan C, Hall AK, Olsson B, Sireau N, Ayoob H, Timmis OG, Sang KH, Genovese F, Imrich R, Rovensky J, Srinivasaraghavan R, Bharadwaj SK, Spiegel R, Zatkova A (2016) Twelve novel HGD gene variants identified in 99 alkaptonuria patients: focus on 'black bone disease' in Italy. Eur J Hum Genet 24(1):66–72. https://doi.org/10.1038/ejhg.2015.60
19. Silvino AC, Costa GL, Araujo FC, Ascher DB, Pires DE, Fontes CJ, Carvalho LH, Brito CF, Sousa TN (2016) Variation in human cytochrome P-450 drug-metabolism genes: a gateway to the understanding of Plasmodium vivax relapses. PLoS One 11(7):e0160172. https://doi.org/10.1371/journal.pone.0160172
20. White RR, Ponsford AH, Weekes MP, Rodrigues RB, Ascher DB, Mol M, Selkirk ME, Gygi SP, Sanderson CM, Artavanis-Tsakonas K (2016) Ubiquitin-dependent modification of skeletal muscle by the parasitic nematode, Trichinella spiralis. PLoS Pathog 12(11):

e1005977. https://doi.org/10.1371/journal.ppat.1005977

21. Casey RT, Ascher DB, Rattenberry E, Izatt L, Andrews KA, Simpson HL, Challis B, Park SM, Bulusu VR, Lalloo F, Pires DEV, West H, Clark GR, Smith PS, Whitworth J, Papathomas TG, Taniere P, Savisaar R, Hurst LD, Woodward ER, Maher ER (2017) SDHA related tumorigenesis: a new case series and literature review for variant interpretation and pathogenicity. Mol Genet Genomic Med 5(3):237–250. https://doi.org/10.1002/mgg3.279
22. Jubb HC, Pandurangan AP, Turner MA, Ochoa-Montano B, Blundell TL, Ascher DB (2017) Mutations at protein-protein interfaces: small changes over big surfaces have large impacts on human health. Prog Biophys Mol Biol 128:3–13. https://doi.org/10.1016/j.pbiomolbio.2016.10.002
23. Ramdzan YM, Trubetskov MM, Ormsby AR, Newcombe EA, Sui X, Tobin MJ, Bongiovanni MN, Gras SL, Dewson G, Miller JML, Finkbeiner S, Moily NS, Niclis J, Parish CL, Purcell AW, Baker MJ, Wilce JA, Waris S, Stojanovski D, Bocking T, Ang CS, Ascher DB, Reid GE, Hatters DM (2017) Huntingtin inclusions trigger cellular quiescence, deactivate apoptosis, and lead to delayed necrosis. Cell Rep 19(5):919–927. https://doi.org/10.1016/j.celrep.2017.04.029
24. Soardi FC, Machado-Silva A, Linhares ND, Zheng G, Qu Q, Pena HB, Martins TMM, Vieira HGS, Pereira NB, Melo-Minardi RC, Gomes CC, Gomez RS, Gomes DA, Pires DEV, Ascher DB, Yu H, Pena SDJ (2017) Familial STAG2 germline mutation defines a new human cohesinopathy. NPJ Genom Med 2(1):7. https://doi.org/10.1038/s41525-017-0009-4
25. Traynelis J, Silk M, Wang Q, Berkovic SF, Liu L, Ascher DB, Balding DJ, Petrovski S (2017) Optimizing genomic medicine in epilepsy through a gene-customized approach to missense variant interpretation. Genome Res 27(10):1715–1729. https://doi.org/10.1101/gr.226589.117
26. Trezza A, Bernini A, Langella A, Ascher DB, Pires DEV, Sodi A, Passerini I, Pelo E, Rizzo S, Niccolai N, Spiga O (2017) A computational approach from gene to structure analysis of the human ABCA4 transporter involved in genetic retinal diseases. Invest Ophthalmol Vis Sci 58 (12):5320–5328. https://doi.org/10.1167/iovs.17-22158
27. Andrews KA, Ascher DB, Pires DEV, Barnes DR, Vialard L, Casey RT, Bradshaw N, Adlard J, Aylwin S, Brennan P, Brewer C, Cole T, Cook JA, Davidson R, Donaldson A, Fryer A, Greenhalgh L, Hodgson SV, Irving R, Lalloo F, McConachie M, McConnell VPM, Morrison PJ, Murday V, Park SM, Simpson HL, Snape K, Stewart S, Tomkins SE, Wallis Y, Izatt L, Goudie D, Lindsay RS, Perry CG, Woodward ER, Antoniou AC, Maher ER (2018) Tumour risks and genotype-phenotype correlations associated with germline variants in succinate dehydrogenase subunit genes SDHB, SDHC and SDHD. J Med Genet 55(6):384–394. https://doi.org/10.1136/jmedgenet-2017-105127
28. Hnizda A, Fabry M, Moriyama T, Pachl P, Kugler M, Brinsa V, Ascher DB, Carroll WL, Novak P, Zaliova M, Trka J, Rezacova P, Yang JJ, Veverka V (2018) Relapsed acute lymphoblastic leukemia-specific mutations in NT5C2 cluster into hotspots driving intersubunit stimulation. Leukemia 32(6):1393–1403. https://doi.org/10.1038/s41375-018-0073-5
29. Albanaz ATS, Rodrigues CHM, Pires DEV, Ascher DB (2017) Combating mutations in genetic disease and drug resistance: understanding molecular mechanisms to guide drug design. Expert Opin Drug Discov 12 (6):553–563. https://doi.org/10.1080/17460441.2017.1322579
30. Park Y, Pacitto A, Bayliss T, Cleghorn LA, Wang Z, Hartman T, Arora K, Ioerger TR, Sacchettini J, Rizzi M, Donini S, Blundell TL, Ascher DB, Rhee K, Breda A, Zhou N, Dartois V, Jonnala SR, Via LE, Mizrahi V, Epemolu O, Stojanovski L, Simeons F, Osuna-Cabello M, Ellis L, MacKenzie CJ, Smith AR, Davis SH, Murugesan D, Buchanan KI, Turner PA, Huggett M, Zuccotto F, Rebollo-Lopez MJ, Lafuente-Monasterio MJ, Sanz O, Diaz GS, Lelievre J, Ballell L, Selenski C, Axtman M, Ghidelli-Disse S, Pflaumer H, Bosche M, Drewes G, Freiberg GM, Kurnick MD, Srikumaran M, Kempf DJ, Green SR, Ray PC, Read K, Wyatt P, Barry CE 3rd, Boshoff HI (2017) Essential but not vulnerable: indazole sulfonamides targeting inosine monophosphate dehydrogenase as potential leads against Mycobacterium tuberculosis. ACS Infect Dis 3(1):18–33. https://doi.org/10.1021/acsinfecdis.6b00103
31. Singh V, Donini S, Pacitto A, Sala C, Hartkoorn RC, Dhar N, Keri G, Ascher DB, Mondesert G, Vocat A, Lupien A, Sommer R, Vermet H, Lagrange S, Buechler J, Warner DF, McKinney JD, Pato J, Cole ST, Blundell TL, Rizzi M, Mizrahi V (2017) The inosine monophosphate dehydrogenase, GuaB2, is a vulnerable new bactericidal drug target for tuberculosis. ACS Infect Dis 3(1):5–17. https://doi.org/10.1021/acsinfecdis.6b00102

32. Trapero A, Pacitto A, Singh V, Sabbah M, Coyne AG, Mizrahi V, Blundell TL, Ascher DB, Abell C (2018) Fragment-based approach to targeting inosine-5′-monophosphate dehydrogenase (IMPDH) from Mycobacterium tuberculosis. J Med Chem 61(7):2806–2822. https://doi.org/10.1021/acs.jmedchem.7b01622
33. Pires DE, Blundell TL, Ascher DB (2015) pkCSM: predicting small-molecule pharmacokinetic and toxicity properties using graph-based signatures. J Med Chem 58 (9):4066–4072. https://doi.org/10.1021/acs.jmedchem.5b00104
34. Pires DEV, Kaminskas LM, Ascher DB (2018) Prediction and optimization of pharmacokinetic and toxicity properties of the ligand. Methods Mol Biol 1762:271–284. https://doi.org/10.1007/978-1-4939-7756-7_14
35. Sigurdardottir AG, Winter A, Sobkowicz A, Fragai M, Chirgadze D, Ascher DB, Blundell TL, Gherardi E (2015) Exploring the chemical space of the lysine-binding pocket of the first kringle domain of hepatocyte growth factor/scatter factor (HGF/SF) yields a new class of inhibitors of HGF/SF-MET binding. Chem Sci 6(11):6147–6157. https://doi.org/10.1039/c5sc02155c
36. Ascher DB, Jubb HC, Pires DE, Ochi T, Higueruelo A, Blundell TL (2015) Protein-protein interactions: structures and druggability. In: Scapin G, Patel D, Arnold E (eds) Multifaceted roles of crystallography in modern drug discovery. NATO science for peace and security series a: chemistry and biology. Springer, Netherlands, pp 141–163. https://doi.org/10.1007/978-94-017-9719-1_12
37. Pandurangan AP, Ascher DB, Thomas SE, Blundell TL (2017) Genomes, structural biology and drug discovery: combating the impacts of mutations in genetic disease and antibiotic resistance. Biochem Soc Trans 45(2):303–311. https://doi.org/10.1042/BST20160422
38. Berman HM, Westbrook J, Feng Z, Gilliland G, Bhat TN, Weissig H, Shindyalov IN, Bourne PE (2000) The Protein Data Bank. Nucleic Acids Res 28(1):235–242
39. O'Boyle NM, Banck M, James CA, Morley C, Vandermeersch T, Hutchison GR (2011) Open babel: an open chemical toolbox. J Cheminform 3:33. https://doi.org/10.1186/1758-2946-3-33
40. Hanwell MD, Curtis DE, Lonie DC, Vandermeersch T, Zurek E, Hutchison GR (2012) Avogadro: an advanced semantic chemical editor, visualization, and analysis platform. J Cheminform 4(1):17. https://doi.org/10.1186/1758-2946-4-17
41. Ascher DB, Crespi GA, Ng HL, Morton CJ, Parker MW (2008) Novel therapeutic approaches to treat Alzheimer's disease and memory disorders. J Proteomics Bioinform 1:464–476
42. Chai SY, Yeatman HR, Parker MW, Ascher DB, Thompson PE, Mulvey HT, Albiston AL (2008) Development of cognitive enhancers based on inhibition of insulin-regulated aminopeptidase. BMC Neurosci 9(Suppl 2):S14. https://doi.org/10.1186/1471-2202-9-S2-S14
43. Kawabata T (2010) Detection of multiscale pockets on protein surfaces using mathematical morphology. Proteins 78(5):1195–1211. https://doi.org/10.1002/prot.22639
44. Lipinski CA, Lombardo F, Dominy BW, Feeney PJ (2001) Experimental and computational approaches to estimate solubility and permeability in drug discovery and development settings. Adv Drug Deliv Rev 46 (1–3):3–26
45. Schrodinger, LLC (2015) The PyMOL molecular graphics system, version 1.8
46. Balint M, Jeszenoi N, Horvath I, van der Spoel D, Hetenyi C (2017) Systematic exploration of multiple drug binding sites. J Cheminform 9(1):65. https://doi.org/10.1186/s13321-017-0255-6
47. Ascher DB, Wielens J, Nero TL, Doughty L, Morton CJ, Parker MW (2014) Potent hepatitis C inhibitors bind directly to NS5A and reduce its affinity for RNA. Sci Rep 4:4765. https://doi.org/10.1038/srep04765
48. Phelan J, Coll F, McNerney R, Ascher DB, Pires DE, Furnham N, Coeck N, Hill-Cawthorne GA, Nair MB, Mallard K, Ramsay A, Campino S, Hibberd ML, Pain A, Rigouts L, Clark TG (2016) Mycobacterium tuberculosis whole genome sequencing and protein structure modelling provides insights into anti-tuberculosis drug resistance. BMC Med 14(1):31. https://doi.org/10.1186/s12916-016-0575-9
49. Hawkey J, Ascher DB, Judd LM, Wick RR, Kostoulias X, Cleland H, Spelman DW, Padiglione A, Peleg AY, Holt KE (2018) Evolution of carbapenem resistance in Acinetobacter baumannii during a prolonged infection. Microb Genom 4. https://doi.org/10.1099/mgen.0.000165
50. Holt KE, McAdam P, Thai PVK, Thuong NTT, Ha DTM, Lan NN, Lan NH, Nhu NTQ, Hai HT, Ha VTN, Thwaites G, Edwards DJ, Nath AP, Pham K, Ascher DB, Farrar J, Khor CC, Teo YY, Inouye M, Caws M, Dunstan SJ (2018) Frequent transmission of the Mycobacterium tuberculosis Beijing lineage

and positive selection for the EsxW Beijing variant in Vietnam. Nat Genet 50 (6):849–856. https://doi.org/10.1038/s41588-018-0117-9

51. Karmakar M, Globan M, Fyfe JAM, Stinear TP, Johnson PDR, Holmes NE, Denholm JT, Ascher DB (2018) Analysis of a novel pncA mutation for susceptibility to pyrazinamide therapy. Am J Respir Crit Care Med 198 (4):541–544. https://doi.org/10.1164/rccm.201712-2572LE

52. Portelli S, Phelan JE, Ascher DB, Clark TG, Furnham N (2018) Understanding molecular consequences of putative drug resistant mutations in Mycobacterium tuberculosis. Sci Rep 8 (1):15356. https://doi.org/10.1038/s41598-018-33370-6

53. Vedithi SC, Malhotra S, Das M, Daniel S, Kishore N, George A, Arumugam S, Rajan L, Ebenezer M, Ascher DB, Arnold E, Blundell TL (2018) Structural implications of mutations conferring rifampin resistance in Mycobacterium leprae. Sci Rep 8(1):5016. https://doi.org/10.1038/s41598-018-23423-1

Chapter 8

Systematic Exploration of Binding Modes of Ligands on Drug Targets

Csaba Hetényi and Mónika Bálint

Abstract

Exploration of binding sites of ligands (drug candidates) on macromolecular targets is a central question of molecular design. Although there are experimental and theoretical methods available for the determination of atomic resolution structure of drug-target complexes, they are often limited to identify only the primary binding mode (site and conformation). Systematic exploration of multiple (allosteric or prerequisite) binding modes is a challenge for present methods. The Wrapper module of our new method, Wrap 'n' Shake, answers this challenge by a fast, computational blind docking approach. Beyond the primary (orthosteric) binding mode, Wrapper systematically produces all possible binding modes of a drug scanning the entire surface of the target. In several fast blind docking cycles, the entire surface of the target molecule is systematically wrapped in a monolayer of N ligand copies. The resulted target–ligand$_N$ complex structure can be used as it is, or important ligand binding modes can be further distinguished in molecular dynamics shakers. Wrapper has been tested on important protein targets of drug design projects on cellular signaling and cancer. Here, we provide a practical description of the application of Wrapper.

Key words Pocket, Peptide, Enzyme, Interaction, Inhibitor, Receptor, Mechanism, Action, Agonist, Antagonist

1 Introduction

There is a continuous increase in the number of atomic resolution structures of biomolecules available in public repositories such as the Protein Databank (PDB [1]). This promising trend is further facilitated by emerging cutting-edge techniques such as cryo-electron microscopy [2] allowing determination of structures of large biological entities such as viruses [3]. Despite the increase in the number of solved biomolecules, and high-throughput automation of X-ray crystallography [4], the measurement of structures of biomolecular targets in complex with their ligands remains a challenge and requires considerable time and money in many cases.

Molecular docking has been introduced as a computational counterpart of experimental techniques for the determination of

Zoltán Gáspári (ed.), *Structural Bioinformatics: Methods and Protocols*, Methods in Molecular Biology, vol. 2112, https://doi.org/10.1007/978-1-0716-0270-6_8, © Springer Science+Business Media, LLC, part of Springer Nature 2020

target–ligand complex structures [5]. Thanks to its high speed, docking has been extensively applied in high-throughput screening campaigns of drug design projects [6] focusing on a known binding pocket of the target. Besides focused projects, docking has produced useful results if the search space was extended to the entire surface of the target molecule and the corresponding approach was named as blind docking [7, 8]. Blind docking has been extensively applied for finding allosteric [9–11] or multiple [12–16] binding sites. Like all other methods, docking also has numerous limitations coming from its approximations. First of all, it has been designed for focused search for drugs, and a systematic coverage of the entire target surface has not been implemented. Furthermore, starting ligand positions and steps of the search algorithm are mostly randomized which decreases reproducibility of the results. Modeling of flexibility (induced fit) and hydration of the target molecule is also oversimplified in docking programs to ensure fast results [17, 18]. Application of molecular dynamics simulations [19, 20] for blind docking is a reasonable approach to overcome the above hydration and flexibility problems of the fast methods. Nowadays, it is quite common to use realistic explicit water models with molecular dynamics, and flexibility can be obviously taken into account on both target and ligand sides. While these features of molecular dynamics considerably improve the precision of the calculated complex structure, they still cannot guarantee a systematic coverage of the entire surface of the target and correct location of the real binding pocket(s) during a single docking simulation [21].

To answer all these challenges of the blind docking problem, a new method Wrap 'n' Shake [21] was developed. The Wrapper module of Wrap 'n' Shake systematically finds all possible binding modes (sites and conformations) of a drug in several fast blind docking cycles. Wrap 'n' Shake has been tested on important protein targets of drug design projects on cellular signaling and cancer [21]. In the present paper, a detailed description of the protocol of the Wrapper module is provided to help future applications.

2 Materials

2.1 Preparation of Target and Ligand Molecules

Wrapper requires complete target and ligand molecules for proper results. Unfortunately, PDB structures of targets often have missing atoms or residues, which need to be inserted (*see* **Note 1**). In cases of missing terminal amino acids, acetyl and amide (*N*-methyl) capping groups need to be added to the N- and C-terminus, respectively. Such molecular editing and addition of hydrogen atoms can be performed by freely available modeling software such as Swiss-PdbViewer [22] or Schrödinger Maestro program package v. 9.6 [23]. Preparation of target structures is completed by energy minimization using free program packages such as GROMACS [24, 25]. For most of the protein targets, a uniform procedure with an AMBER99SB-ILDN force field [26], TIP3P explicit water model [27], and no restraints on the heavy atoms is appropriate. Ligand molecules can be built and edited by the above Maestro or other software. Protonation of the ligands (where applicable) is often helped by the pK_a plug-in in Marvin Sketch [28]. Fast energy minimization of the hydrogenated ligand structures is usually sufficient. In the first stage, molecular mechanics minimization with Maestro software is performed, using OPLS force field [29], followed by a quantum chemistry program package such as MOPAC [30] with a semiempirical parametrization such as PM6 or above.

2.2 Wrapper

The Wrapper module is available as part of a stand-alone, open source software package Wrap 'n' Shake freely downloadable from the web page of the program [31] along with full documentation. It is distributed under the terms of GNU General Public License. At present, Wrap 'n' Shake 1.1 contains software for the Wrapper module. Wrapper contains two bash scripts (pre-wrapper.sh and wrapper.sh) and a C program (wrp). After downloading the package (wns.tgz), it can be extracted using the following command:

```
$ tar -xvf wns.tgz
```

Pre-wrapper.sh and wrapper.sh can be found in wns/scripts and are readily usable under the Linux operating system. The source code of wrp can be compiled and installed into a $HOME/bin using the following commands:

```
$ cd wns/wrp/src
$ make
$ make install
```

The present version of Wrapper requires installation of external programs AutoGrid 4.2 and AutoDock 4.2 (Release 4.2.3) of the AutoDock 4.2 [32, 33] package, Python scripts of AutoDockTools [34], editconf and sasa programs of the GROMACS program package. All external programs are freely available. Organization of the components of Wrapper is shown in Fig. 1 and the programs are described as follows:

1. Script pre_wrapper.sh requires standard PDB files as input and prepares the files required by wrapper.sh. The necessary inputs for wrapper.sh are the PDBQT files of the ligand and target molecules and also grid (GPF) and docking (DPF) parameter files. The PDBQT file has the similar format to the regular PDB file, with additional columns containing the partial charges and the atom type. In wrapper, Gasteiger partial charges and the atom types of the modified AD4_parameters.dat (*see* also

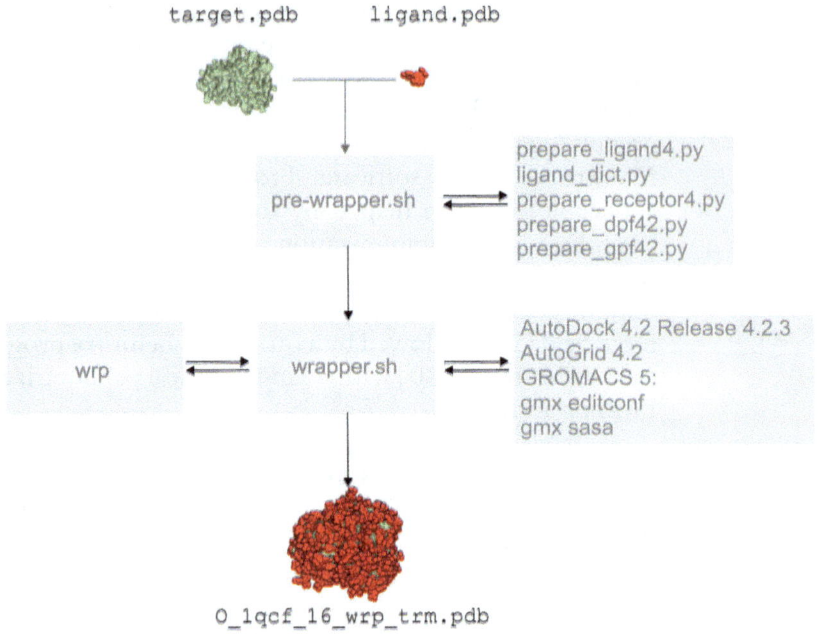

Fig. 1 Components of Wrapper and their connection with external shell scripts and programs. The figure was reproduced from the website of Wrap 'n' Shake with permission

Subheading 3.3) file are used. Notably, the original version of AD4_parameters.dat can be found in the source code folder of the AutoDock4.2 package. Both the ligand and target PDBQT apply united atom representation, which means that only the polar hydrogens are explicitly kept in the docking input file. The GPF file is the input of AutoGrid 4.2 and contains the docking (grid) box parameters. The grid box defines the search space where the docking calculations are performed. The GPF file also lists the names of target and ligand files and their atom types. The DPF file is the input file of AutoDock 4.2 and contains the parameters of the search algorithm and docking runs. The DPF also contains the names of map files generated by AutoGrid 4.2 for each atom type.

2. Wrapper.sh is the director of the Wrapper module. In several blind docking cycles, it covers the entire surface of the target with a monolayer of numerous ligand copies. Wrapper.sh works in symbiosis with program wrp of the present package detailed in the next point. The blind docking cycles are performed by external programs of the AutoDock 4.2 package and performed in separate working directories. After each cycle, free surface area of the target is calculated by external programs of the GROMACS package. Wrapper.sh reads PDBQT files of the ligand and target molecules and supplies the results as a single PDB file. For the ligand, a template file (ligand_templ.pdbqt) is also required for post-processing the wrapped target and used

in the trimming mode of Wrapper. During Wrapper, all ligand copies are renamed as "LIG" by the wrp program, and after ligand minimization, all atom names are renamed by MOPAC. Thus, the ligand template file is used for renumbering and renaming the ligand atoms and residue name after Wrapper. This ensures an exact match of the ligand atom names and ligand residue name with the molecular dynamics topology, which is required if the user merges the target ligand complex to use in a Shaker step. The atoms of the template file must have exactly the same order and number of heavy atoms as the input ligand.pdbqt file. The template file can be prepared by following the same input preparatory steps as for the ligand.pdb, except MOPAC minimization. Note that all hydrogen atoms must be added (Subheading 2.1) and the MOPAC energy minimization step is not required. After adding all hydrogen atoms, the PDB template file can be converted to a PDBQT file, using the command line of the python script below or the graphical interface of ADT program:

```
$pythonsh $PATH_TO/prepare_ligand4.py -l ligand_templ.pdb -o ligand_templ.pdbqt -v
-d $PATH_TO/ligand_dict.py -F
```

In this way, the same number and order of atoms is obtained in the template file as in the input PDBQT of the ligand. Wrapper.sh also produces log files containing reports on finished cycles with interaction energy and accessible surface values.

3. Wrp is an open source C program and serves as the background engine of the Wrapper module. It is called by wrapper.sh and performs clustering and ranking of the docked ligand conformations and subsequent assignation of excluded atoms. In wrapping mode, wrp results in a PDBQT file including the target, and all ligand copies accumulated up to the actual cycle and also a statistical file with ranking and intermolecular energy results (E_{inter}), calculated by the AutoDock 4.2 scoring function [35]. Wrp can also work in trimming mode where excess ligand copies not interacting with the target are removed after the final cycle and the results are written into a single PDB file identical with that one mentioned at wrapper.sh. This step is also initiated by script wrapper.sh. Repeated use of wrp in wrapping mode provides the target structure systematically covered in a monolayer of ligand copies. The work of wrp is adjusted by distance tolerance values as described in Subheading 3.4.
4. External python scrips (Table 1) of AutoDock Tools (ADT) are required by pre-wrapper.sh. The scripts are freely available [32, 34]. After ADT installation, these scripts can be found in

Table 1
Python scripts of ADT

Python script name	Input	Output
prepare_ligand4.py	PDB	PDBQT
ligand_dict.py	PDB	PDBQT
prepare_receptor4.py	PDB	PDBQT
prepare_dpf42.py	PDBQT	DPF
prepare_gpf4.py	PDBQT	GPF

a separate directory of the user: `$USER_HOME/MGLTools-1.5.6/MGLToolsPckgs/AutoDockTools/Utilities24`

The pythonsh binary is also installed, and insertion of an alias line in the .bashrc system file is advised, for easy access: `alias pythonsh=$USER_HOME/MGLTools-1.5.6/bin/pythonsh`

The python scripts generate PDBQT, DPF, and GPF files required by AutoDock 4.2 using the parameters described in Table 2. Based on the generated PDBQT files, ADT scripts also prepare grid and docking parameter files as required by AutoDock 4.2 [32].

We recommend the use of flexible ligand structures with torsional restriction on the aromatic and amide bonds only. Accordingly, branching of the torsion tree in the DPF files is generated with all default torsions of the ligand molecules as automatically assigned by ADT.

5. Blind docking runs of wrapper cycles are performed by external program package AutoDock 4.2. including program AutoGrid 4.2 for calculation of grid maps of the target molecule with pre-calculated energy values and the docking engine AutoDock 4.2 with a Lamarckian genetic algorithm. Docking parameters were used as described in a previous study [8]. The source code of the package was modified in order to be able to produce all the necessary map files in case of multiple target files. Original source code limits the number grid map generation to 14 atom types. Therefore, to produce grid map for all atom types, in autocomm.h file, line number 93 needs to be changed as follows.

   Original source code:

```
#define MAX_ATOM_TYPES (14 - NUM_NON_VDW_MAPS)
```

   Replaced by:

```
#define MAX_ATOM_TYPES (34 - NUM_NON_VDW_MAPS)
```

Table 2
Blind docking parameters

Parameter	Value
Grid parameters	
Grid spacing	0.375 Å
Number of grid points (*x*,*y*,*z*)	200,200,200
Docking parameters	
Search method	Lamarckian genetic algorithm
Population size	250
Maximum number of energy evaluations	20 million
Maximum number of generations	2000 million
Number of top individuals to survive to next generation	1
Rate of gene mutation	0.02
Rate of crossover	0.8
Alpha parameter of Cauchy distribution	0.0
Beta parameter of Cauchy distribution	1.0
Number of iterations of Solis and Wets local search	300
Consecutive successes before changing rho	4
Consecutive failures before changing rho	4
Size of local search space to sample	1
Lower bound on rho	0.01
Probability of performing local search on individual	0.06
Number of hybrid GA-LS runs	100

6. External GROMACS programs editconf and sasa [25] are called for calculation of accessible surface area of the target–ligand complex using a PDB file as input. The editconf command transforms the input pdb file intro gromacs .gro file, and the sasa program performs the calculations. GROMACS sasa calculates the ASA for the entire target–ligand complex, but wrapper.sh will eliminate the surface calculated for the ligand, by deleting rows, with residue name "LIG" from the total_atomarea_lig.xvg file obtained from GROMACS. Wrapper.sh also produces a log file containing the free target surface not covered by ligand copies.

3 Methods

3.1 Overview

Wrapper builds a monolayer of ligand copies covering the entire target molecule. Wrapper performs a series of automated, fast blind docking cycles. The algorithm ensures a complete and systematic coverage of the surface of the target with ligand copies. Wrapper uses a modified docking force field and clustering allowing maximal ligand–target and minimal ligand–ligand interactions. The popular docking program package AutoDock 4.2 is piped into Wrapper and performs consecutive fast blind docking cycles without the need of initial ligand positions or any other interventions of the user. The outcome of Wrapper is a single PDB file including the structure of the target wrapped in a monolayer of ligand copies, i.e., the structure of a target–ligand$_N$ complex. The application of Wrapper is described using an example (Fig. 2) of the complex of hematopoietic cell kinase (HCK, target, in green) and 1-ter-butyl-3-p-tolyl-1H-pyrazolo[3,4-D]pyrimidin-4-ylamine (PP1, ligand, in red). The complex structure was published under PDB code 1qcf, and this code will be used in the names of input and output files of the example also provided for download on the web page of the program [31].

3.2 Input Files

Wrapper requires complete, energy-minimized structures of the ligand (1qcf_ligand.pdb, red) and target (1qcf_target.pdb, green) molecules in Protein Databank (*.pdb) format. Preparation of target and ligand molecules is described in Subheading 2.

3.3 Pre-wrapper.sh

From both target and ligand structures, pre-wrapper.sh produces PDBQT input files (1qcf_target.pdbqt, 1qcf_ligand.pdbqt) and parameter files (1qcf_target.gpf, 1qcf_target.dpf) as required by AutoDock 4.2 called by wrapper.sh. The docking box is set to cover the entire surface of the target molecule. For this, the center of the box is set to that of the target molecule (default option), and grid maps of 200 grid points in all three spatial directions are generated. Notably, if the size of the target exceeds ca. 450 amino acids corresponding to the largest proteins of our test set (Fig. 3), the number of grid points of 200 should be increased in the following command of the pre-wrapper.sh script calling prepare_gpf4.py in order to cover the whole target in one BD cycle.

```
$SCRIPTPATH/pythonsh $SCRIPTPATH/prepare_gpf4.py
-l $ligand_name.pdbqt -r $target_name.pdbqt -p spacing=0.375
-p npts='200,200,200' -p ligand_types='A,..,YY,LL' -v
```

With this, the numbers of grid points are specified in GPF for all three directions of space. The user must also consider the shape of the target and change the box dimensions in one or all directions

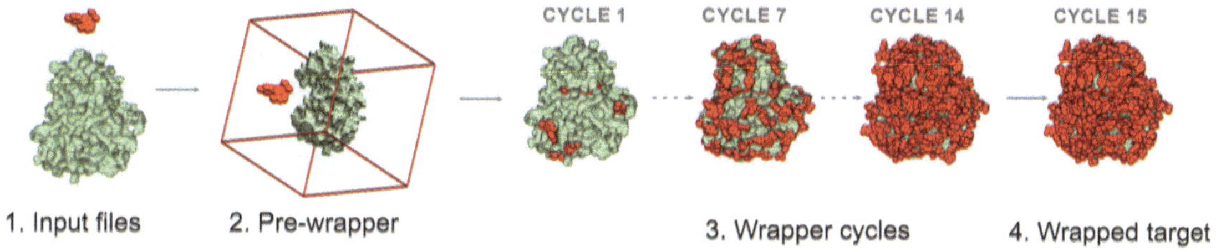

Fig. 2 Main stages of Wrapper. The target (hematopoietic cell kinase, green) is wrapped in numerous copies of the ligand (1-ter-butyl-3-p-tolyl-1H-pyrazolo[3,4-D]pyrimidin-4-ylamine, red) molecule in several blind docking cycles. The docking box (red lines) covers the entire surface of the target molecule. The figure was reproduced from the website of Wrap 'n' Shake with permission

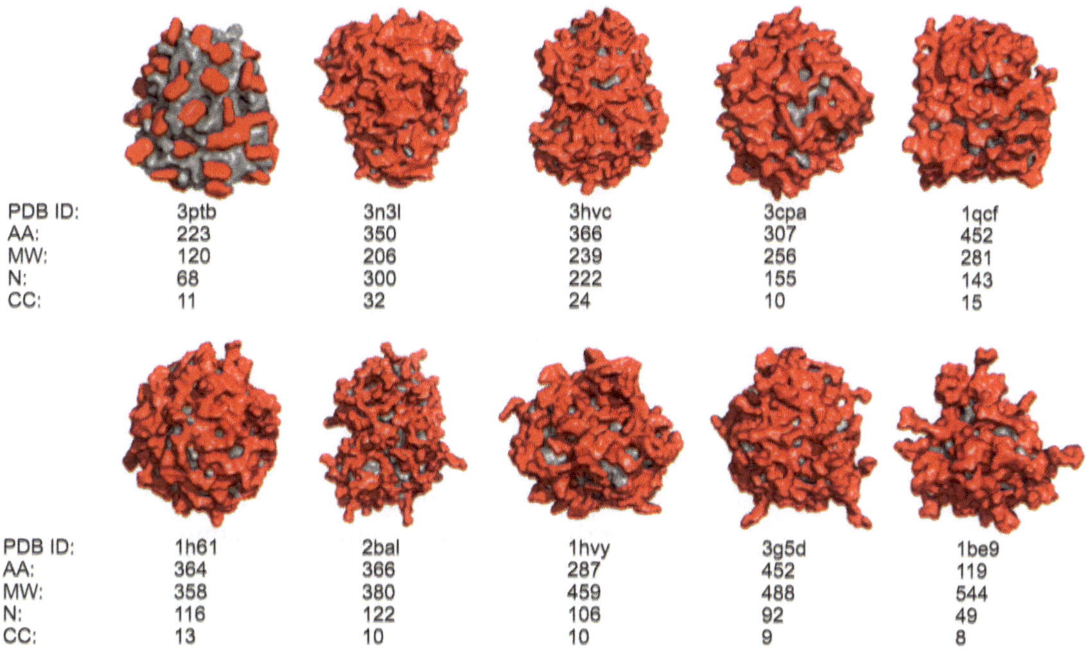

Fig. 3 Targets (grey) wrapped in a monolayer of ligand (red) copies. *AA* count of amino acids of the target, *MW* molecular weight of the ligand, *N* number of ligand copies, *CC* count of cycles. (The figure was reproduced from the website of Wrap 'n' Shake with permission)

accordingly (*see* **Note 2**). An edge of the box can be calculated in Ångström as the product number of grid points and grid spacing (a value of 0.375 Å was used; Table 2). Pre-wrapper.sh also adds new entries of excluded atom types LL and YY (commonly marked as X in our original publication [21]) to the DPF and GPF files. This step is performed only once, as the same parameter files can be used in all wrapping cycles later. This step is necessary for generation of the map files of the new atom types. Gasteiger partial charges are added to both the ligand and target. Addition of hydrogen atoms to the ligand or target is skipped as the minimized PDB files already have all atoms. The nonpolar hydrogens are merged (Subheading 2.2). All default active torsions are kept for the ligand, but the target is treated rigidly, without active torsions. Parameter files have the settings as described in Subheading 2.2 (Table 2).

Pre-wrapper also performs three important administrative modifications on text files.

1. The first row of the parameter files (both the DPF and GPF) is updated to the actual path of the modified AD4_parameters.dat.

   Default:

```
autodock_parameter_version 4.2
```

   Modified:

```
parameter_file $USER_DEFINED_PATH/AD4_parameters.dat
```

2. New lines of atom types LL and YY are inserted after the last line of standard atom type maps.

```
map 1qcf_target.YY.map
map 1qcf_target.LL.map
```

3. Two lines of atom types LL and YY are inserted to the end of AD4_parameters.dat file (the modified file can be also downloaded from our web page [31]).

```
atom par YY 3.60 1E-04 00.0000 0.00000 0.0 0.0 0 0 0 0
atom par LL 3.60 1E-04 00.0000 0.00000 0.0 0.0 0 0 0 0
```

   The user may decide to prepare the input PDBQT, DPF, and GPF using the graphical interface of ADT instead of pre-wrapper.sh. In this case, after generating the DPF and GPF, the above detailed three changes should be also done by manual editing of the files. Whereas the use of pre-wrapper.sh is not mandatory as file preparations can be arranged as described above; however, the use of pre-wrapper.sh is recommended to avoid human mistakes especially if multiple target files or a library of ligand structures are handled.

3.4 Wrapper.sh and wrp

Wrapper.sh performs the coverage of target surface with a monolayer of N ligand copies ending up in a target–ligand$_N$ complex. Several fast BD cycles are performed all of them resulting in 100 docked ligand copies. The count of necessary BD cycles (CC) depends on the size and shape of the target molecule as indicated in Fig. 3. Ligand copies and interacting target surface elements are excluded from successive BD cycles via assignation of a new "excluded" atom type to the atoms involved. In this way, unbound target sites can be distinguished from those covered with ligand copies, ligand-ligand interactions are minimized, and target–ligand interactions are maximized for the largest possible coverage of the target surface. Further details on structural and physical chemistry of the Wrapper algorithm can be found in the original publication of Wrap 'n' Shake [21].

The BD cycles follow a uniform protocol. Grid map files (1qcf_target_∗.map) of chemical and excluded (YY, LL) atom types are calculated by Autogrid 4.2 along with a log file. The corresponding ∗.YY.map and ∗.LL.map files are generated before the docking runs. One hundred BD runs are performed in each cycle, and the docked ligand structures are collected in a log file (1qcf_1.dlg for the first cycle) by AutoDock 4.2. The log file is evaluated by the wrp program, which first ranks and clusters the docked ligand conformations.

Docked ligand conformations of the DLG file are clustered and ranked based on their interaction energy (E_{inter}, the first energy component of estimated free energy of binding in the DLG file) values with the target and the closest distance between each heavy atom of the ligand copies (dmin). In the initial clustering phase, wrp (wrapper mode) sorts the 100 docked ligand conformations according to E_{inter}. Ligand conformation of the lowest E_{inter} from among the 100 docked ligand copies is selected as the representative of Cluster 1. Ligand conformation of the second lowest E_{inter} is selected as a representative of a new Cluster 2 if dmin>drnk, where drnk is a ranking tolerance, a measure of separation of clusters from each other. If dmin≤drnk, then ligand conformation of the second lowest E_{inter} is placed into Cluster 1. In this way, all 100 ligand conformations are clustered, and the representatives are evenly spread over the target surface without clashing each other. In our protocol, drnk was set to 2 Å, which is approximately a covalent bond distance (1.5 Å) plus a 0.5 Å added. The results of clustering are summarized in .sta file type (O_1qcf_1_wrp.sta) after each wrapper cycle.

Wrp in wrapper mode assigns the new atom type (YY, LL) of the abovementioned excluded atoms in the target file (YY) and the docked ligand copies (cluster representatives LL). Excluded atoms are assigned using a target–ligand interface tolerance and an assignation tolerance. Both of these tolerance values were set to 3.5 Å in our default settings. Merging of the modified target and ligand copies results in a target–ligand complex O_1qcf_1_wrp.pdbqt file. This file is moved from the working directory of the current cycle into the directory of the next cycle and used as target input for programs AutoGrid 4.2 and AutoDock 4.2 if none of the exit criteria described below are achieved. After each cycle, the free (unliganded) accessible surface area (ASA) is calculated by external GROMACS program sasa, as described in Subheading 2, Point 6 (Msroll in the 1.0 version). Wrapping ends if ASA $\leq$ 1% or the interaction energy E_{inter} value of any cluster representative in the cycle is $\geq$0 kcal/mol. Otherwise, the resulted PDBQT file is forwarded to the next cycle as described above. ASA and E_{inter}

evaluations are calculated for each wrapper cycle and stored in two separate files (O_1qcf_1_surface_percentage.log and O_1qcf_1_lowest_energy.log). These files are generated in the working directory of each cycle and moved to "stats" folder where statistical evaluation of Wrapper takes place.

For our test system 1qcf, wrapping finished in 16 cycles and 1qcf_16_wrp.pdbqt is the result after the last cycle. All files of the complete Wrapping process of 16 wrapping cycles can be downloaded as a single compressed package (O_1qcf_wrp.tgz).

After the last (16th) wrapping cycle, a trimming mode of wrp is involved to remove ligand copies positioned far from the target surface. This is necessary, as some ligand copies may dock to distant regions of the docking box depending on the actual target. The trimming step also performs formal post-processing of the 1qcf_16_wrp.pdbqt file using a template file (1qcf_ligand_templ.pdbqt) described in Subheading 2.2. The resulted O_1qcf_16_wrp_trm.pdb file has all atoms renamed according to the standards of PDB file format allowing the use of this file of the molecular dynamics steps of a Shaker process (*see* **Note 4**).

3.5 Output, Benchmark

In our example, the target structure was wrapped in a monolayer of $N = 143$ ligand copies in 15 cycles (Fig. 3). The CPU time of a cycle of 100 docking runs took 11 h for this system on an Intel Xeon E5520. In general, CPU times of a cycle varied between some hours and 1–2 days for the test systems listed in Fig. 3 depending on the size of the target molecule and the size and number of rotatable torsion of the ligand (*see* **Note 3**). The count of cycles (CC in Fig. 3) necessary for complete wrapping depends both on the size and geometry of the partners. The largest ligand (system 1be9) fully covered its relatively small target in less than ten cycles. The largest CC of 32 was found for system 3n3l, where the ligand is relatively small and the target is large. The special geometry of ligand benzamidine is probably a reason for the unique wrapping pattern corresponding to unexpectedly low N and CC values obtained in the case of system 3ptb.

4 Notes

1. During pre-wrapper.sh, it is useful to check the net charge (sum of partial charges of all atoms in the PDBQT file) of the target and ligand molecules. The value of the net charge of a PDBQT file should be close to an integer. For example, a net charge of 3.5 indicates that the structure of the molecule is erroneous (missing/extra atoms), or partial charges could not be assigned correctly by ADT. In this way, checking of net charge helps the detection of error occurring during the preparation of target or ligand structures. Special attention must also be given to the

charge assigned on systems with coordinating ions (e.g., Fe^{3+}, Ni^{2+}, etc.) as the partial charges assigned for such atoms by ADT are not always correct [36].

2. The user should check if the grid box covers the whole target; otherwise, parts of the target surface excluded from the box will not be analyzed for possible binding sites. The grid box can be visualized by a python script called gbox.py downloadable from the website of Wrap 'n' Shake [31].
3. We suggest running pre-wrapper.sh on a simple workstation (personal computer, PC) as it requires only some seconds to finish. Wrapper.sh can also be run on a simple PC under Linux. However, as complete wrapping of a target usually takes several hours or days of CPU time, its frequent application may require a dedicated PC or a server node.
4. The Shaker protocol of Wrap 'n' Shake [21] can be used for distinction of important binding modes and structural refinements on hydration and induced fit effects in successive molecular dynamics steps. The wrapped target is placed in a simulation box and hydrated with explicit water molecules. The hydrated complex is subjected to a series of simulations and filtering steps between the MD runs, where loosely bound ligand copies are removed. Refinement of bound ligand structure can be performed with all target atoms released.

Acknowledgments

We acknowledge a grant of computer time from CSCS Swiss National Supercomputing Centre and the Governmental Information Technology Development Agency, Hungary. We acknowledge that the results of this research have been achieved using the DECI resource Archer based in the UK at the National Supercomputing Service with support from the PRACE aisbl. The work was supported by the K123836 grant from the National Research, Development, and Innovation Office, Hungary. The University of Pécs is acknowledged for a support by the 17886-4/23018/FEKUTSTRAT excellence grant. The work was supported by EFOP-3.6.2-16-2017-00008 and the ÚNKP-19-4 New National Excellence Program of the Ministry for Innovation and Technology. The work was supported by the János Bolyai Research Scholarship of the Hungarian Academy of Sciences.

References

1. Rose PW, Beran B, Bi C, Bluhm WF, Dimitropoulos D, Goodsell DS et al (2011) The RCSB Protein Data Bank: redesigned web site and web services. Nucleic Acids Res 39(Suppl 1):392–401
2. Nogales E (2018) Profile of Joachim Frank, Richard Henderson, and Jacques Dubochet,

2017 Nobel laureates in chemistry. Proc Natl Acad Sci U S A 115(3):441–444

3. Cheng Y, Glaeser RM, Nogales E (2017) How Cryo-EM became so hot. Cell 171 (6):1229–1231
4. Hui R, Edwards A (2003) High-throughput protein crystallization. J Struct Biol 142 (1):154–161
5. Kitchen DB, Decornez H, Furr JR, Bajorath J (2004) Docking and scoring in virtual screening for drug discovery: methods and applications. Nat Rev Drug Discov 3(11):935–949
6. Doman TN, McGovern SL, Witherbee BJ, Kasten TP, Kurumbail R, Stallings WC et al (2002) Molecular docking and high-throughput screening for novel inhibitors of protein tyrosine phosphatase-1B. J Med Chem 45(11):2213–2221
7. Hetényi C, van der Spoel D (2002) Efficient docking of peptides to proteins without prior knowledge of the binding site. Protein Sci 11 (7):1729–1737
8. Hetényi C, Van Der Spoel D (2006) Blind docking of drug-sized compounds to proteins with up to a thousand residues. FEBS Lett 580:1447–1450
9. Hocker H, Rambahal N, Gorfe AA (2014) LIBSA - a method for the determination of ligand-binding preference to allosteric sites on receptor ensembles. J Chem Inform Model 54 (2):530–538
10. Schindler CEM, De Vries SJ, Zacharias M (2015) Fully blind peptide-protein docking with pepATTRACT. Structure 23 (8):1507–1515
11. Whalen KL, Tussey KB, Blanke SR, Spies MA (2011) Nature of allosteric inhibition in glutamate racemase: discovery and characterization of a cryptic inhibitory pocket using atomistic MD simulations and pKa calculations. J Phys Chem B 115(13):3416–3424
12. García-Sosa AT, Sild S, Maran U (2008) Design of multi-binding-site inhibitors, ligand efficiency, and consensus screening of avian influenza H5N1 wild-type neuraminidase and of the oseltamivir-resistant H274Y variant. J Chem Inf Model 48(10):2074–2080
13. Roumenina L, Bureeva S, Kantardjiev A, Karlinsky D, Andia-Pravdivy JE, Sim R et al (2007) Complement C1q-target proteins recognition is inhibited by electric moment effectors. J Mol Recognit 20(5):405–415
14. Bugatti A, Giagulli C, Urbinati C, Caccuris F, Chiodelli P, Oreste P et al (2011) Molecular interaction studies of HIV-1 matrix protein p17 and heparin: identification of the heparin-binding motif of p17 as a target for the development of multitarget antagonists. J Biol Chem 288(2):1150–1161
15. Kovács M, Tóth J, Hetényi C, Málnási-Csizmadia A, Seller JR (2004) Mechanism of blebbistatin inhibition of myosin II. J Biol Chem 279(34):35557–35563
16. Agarwal T, Annamalai N, Khursheed A, Kumar T, Bin H, Haris M (2015) Molecular docking and dynamic simulation evaluation of Rohinitib—Cantharidin based novel HSF1 inhibitors for cancer therapy. J Mol Graph Model 61:141–149
17. Rastelli G, Ferrari AM, Costantino L, Gamberini MC (2002) Discovery of new inhibitors of aldose reductase from molecular docking and database screening. Bioorg Med Chem 10 (5):1437–1450
18. García-Sosa AT, Mancera RL (2010) Free energy calculations of mutations involving a tightly bound water molecule and ligand substitutions in a ligand-protein complex. Mol Inform 29(8–9):589–600
19. Shan Y, Kim ET, Eastwood MP, Dror RO, Seeliger MA, Shaw DE (2011) How does a drug molecule find its target binding site? J Am Chem Soc 133(24):9181–9183
20. Buch I, Giorgino T, De Fabritiis G (2011) Complete reconstruction of an enzyme-inhibitor binding process by molecular dynamics simulations. Proc Natl Acad Sci U S A 108 (25):10184–10189
21. Bálint M, Jeszenői N, Horváth I, van der Spoel D, Hetényi C (2017) Systematic exploration of multiple drug binding sites. J Cheminform 9(1):65
22. Guex N, Peitsch MC, Schwede T (2009) Automated comparative protein structure modeling with SWISS-MODEL and Swiss-PdbViewer: a historical perspective. Electrophoresis 30 (Suppl 1):162–173
23. Maestro, Schrödinger, LLC (2017) New York, NY
24. Abraham MJ, Murtola T, Schulz R, Páll S, Smith JC, Hess B et al (2015) Gromacs: high performance molecular simulations through multi-level parallelism from laptops to supercomputers. SoftwareX 1–2:19–25
25. GROMACS (2018). Available from: http://manual.gromacs.org/current. Accessed 01 Oct 2018
26. Lindorff-Larsen K, Piana S, Palmo K, Maragakis P, Klepeis JL, Dror RO et al (2010) Improved side-chain torsion potentials for the Amber ff99SB protein force field. Proteins 78(8):1950–1958
27. Jorgensen WL, Chandrasekhar J, Madura JD, Impey RW, Klein ML (1983) Comparison of

simple potential functions for simulating liquid water. J Chem Phys 79(2):926–935

28. Chemaxon (2014) Marvin Sketch, v. 6.3.0. Chemaxon, Budapest
29. Jorgensen WL, Tirado-Rives J (2005) Potential energy functions for atomic-level simulations of water and organic and biomolecular systems. Proc Natl Acad Sci U S A 102 (19):6665–6670
30. MOPAC (2012) MOPAC. Stewart JJP, computational chemistry
31. Wrap'n'Shake (2017). Available from: http://www.wnsdock.xyz. Accessed 01 Oct 2018
32. Morris GM, Huey R, Lindstrom W, Sanner MF, Belew RK, Goodsell DS et al (2009) AutoDock4 and AutoDockTools4: automated docking with selective receptor flexibility. J Comput Chem 28(1):73–86
33. Autodock 4.2 (2009). Available from: http://www.autodock.scripps.edu. Accessed 28 Sept 2018
34. AutoDock Tools 1.5.6 (2009). Available from: http://mgltools.scripps.edu/downloads. Accessed 28 Sept 2018
35. Morris GM, Goodsell DS, Halliday RS, Huey R, Hart WE, Belew RK et al (1998) Automated docking using a Lamarckian genetic algorithm and an empirical binding free energy function. J Comput Chem 19 (14):1639–1662
36. Bikadi Z, Hazai E (2009) Application of the PM6 semi-empirical method to modeling proteins enhances docking accuracy of AutoDock. J Cheminform 1(1):1–16

Chapter 9

Using MemBlob to Analyze Transmembrane Regions Based on Cryo-EM Maps

Georgina Csizmadia, Bianka Farkas, Eszter Katona, Gábor E. Tusnády, and Tamás Hegedűs

Abstract

Transmembrane proteins include membrane channels, pores, and receptors and, as such, comprise an important part of the proteome, yet our knowledge about them is much less complete than about soluble, globular proteins. An important aspect of transmembrane protein structure is their exact position within the lipid bilayer, a feature hard to investigate experimentally at the atomic level. Here we describe MemBlob, a novel approach utilizing difference electron density maps obtained by cryo-EM studies of transmembrane proteins. The idea behind is that the nonprotein part of such maps carries information on the exact localization of the membrane mimetics used in the experiment and can be used to extract the positional information of the protein within the membrane. MemBlob uses a structural model of the protein and an experimental electron density map to provide an estimation of the surface residues interacting with the membrane.

Key words Transmembrane region, Cryo-EM map, Lipid interface

1 Introduction

Transmembrane (TM) proteins represent around ~30% of eukaryotic proteomes. Their structure determination is notoriously difficult compared to that of soluble proteins, although there is significant progress in this area in the last decades. Recent developments, including those in cryo-EM techniques, allow for the determination of a larger number of novel structures at high resolution [1]. Even so, the structural description of a transmembrane protein cannot be complete without information on its localization within the lipid membrane. This issue is usually addressed using computational tools, where various modeling approaches are applied to provide information about the membrane plane relative to the

Georgina Csizmadia and Bianka Farkas contributed equally to this work.

Zoltán Gáspári (ed.), *Structural Bioinformatics: Methods and Protocols*, Methods in Molecular Biology, vol. 2112,
https://doi.org/10.1007/978-1-0716-0270-6_9,

© Springer Science+Business Media, LLC, part of Springer Nature 2020

protein. Methods like TMDET [2] determine the best localization of two planes corresponding to the membrane outer and inner surfaces by optimizing an objective function. These methods, however, contain necessary simplifications like assumptions on the thickness of the membrane and lack the potential to incorporate experimental information about the position of the bulk membrane phase at sufficient resolution. MEMPROTMD [3] applies molecular dynamics calculations to build a lipid bilayer around the molecule. Molecular dynamics can also be applied to identify binding sites of specific lipid molecules within the membrane [4].

In this chapter we describe MemBlob, a recently developed method that utilizes cryo-EM maps to extract information about the localization of lipid membranes around TM proteins. The basic idea is that in many cases the full cryo-EM map contains electron density data of the protein structure and the surrounding lipid molecules. By subtracting the reconstructed electron density of the protein itself from the full map, the resulting "blob" can be used to specify the sites on the protein that are in contact with the membrane. We introduce the MemBlob server and database and discuss the usage of the server through the structure of a potassium channel (PDB ID 5U70) as an example.

2 Methods

2.1 Input Files for MemBlob

The MemBlob server needs three input files, two of which, a PDB format structure file and an MRC format electron density file (gzipped), are mandatory (*see* **Note 1**). The server also utilizes the output of TMDET for the given structure. Such a file can be obtained by the user either from the PDBTM database or by submitting the structure to the TMDET web server at http://tmdet.enzim.hu. If the user does not provide this file, the MemBlob server will try to fetch it from the PDBTM database, assuming that the first four characters of the uploaded PDB file correspond to a PDB ID. If this step is unsuccessful, MemBlob submits the uploaded PDB to TMDET to get the necessary information.

2.2 Overview of the MemBlob Method

The MemBlob pipeline (Fig. 1) is described in [5]. As a first step, the electron density map calculated for the PDB structure is subtracted from the submitted experimental cryo-EM map (provided in the MRC format file) (*see* **Note 2**). Using the output of TMDET, the z-axis and the origin of the membrane bilayer is predicted, and different layers of x–y planes at an interval of 2 Å are retrieved. The values from these planes are then projected to a cylindrical mantle around the protein at intervals corresponding to a 10° rotation along the z-axis. Values are smoothened using a Savitzky-Golay filtering step, and the first minima along the z-axis are used to define the boundaries of the membrane. The values are then projected to

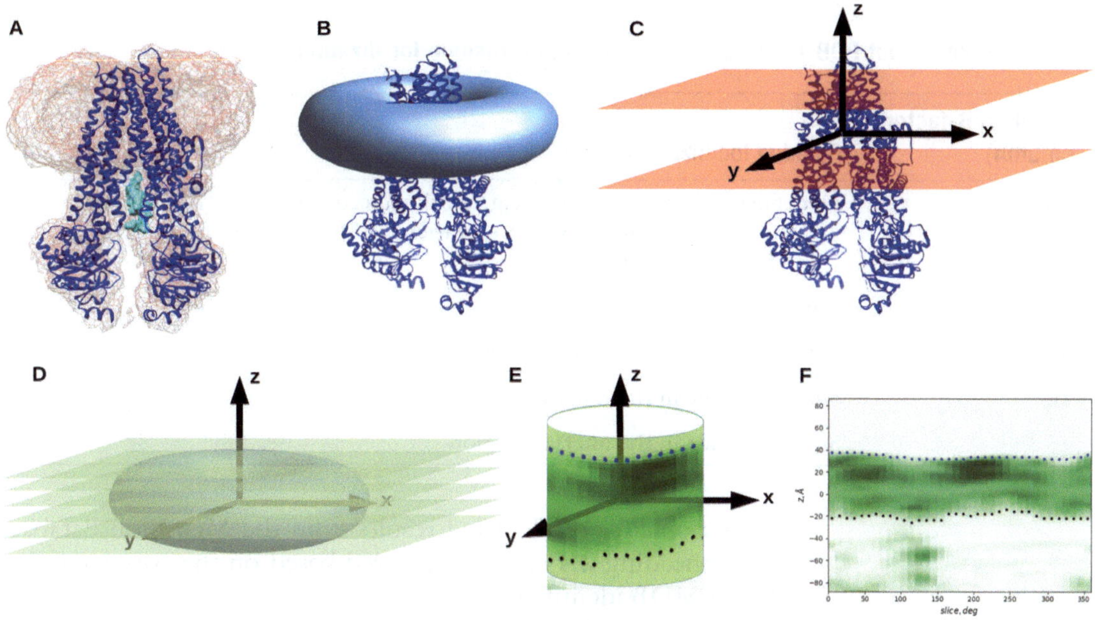

Fig. 1 Schematic representation of the main steps of the MemBlob pipeline. (**A**) Structure of a transmembrane protein fitted to the electron density map. The "blob" is clearly visible at the top of the structure. (**B**) Schematic representation of the "blob" obtained from the difference map and its relation to the protein structure. (**C**) The membrane position is predicted by TMDET and is used to define a coordinate system. (**D**) Electron density values are extracted from *x*–*y* planes spaced 2 Å from each other along the *z*-axis. (**E**) The values are projected with 10° intervals to a cylindrical mantle. (**F**) The mantle is displayed as a plane

the atoms of the protein, assigning localization information to each residue. An interface region from the boundary of the membrane toward the origin is defined at both sides using a user-adjustable cutoff that is 8 Å by default. Residues on the protein surface and interior are identified using DSSP. The main output of the method is a PDB file with labels for the position of the residues (Table 1).

2.3 The MemBlob Web Server

The MemBlob web server is accessible at http://memblob.hegelab.org. To initiate a calculation, the user should submit at least two input files, the PDB and the corresponding (and aligned) cryo-EM map file. The input form offers the option to submit a user-calculated difference electron density map; in this case, the "Difference map" checkbox should be checked, and the server will use this map instead of calculating the difference map itself. The user can also submit an XML format file containing the output of the TMDET algorithm.

The sole option for the calculation that can be set by the user is the thickness of the interface region, defined as the region starting from the membrane boundary toward the origin (in the "middle" of the membrane).

Table 1
Labels in the output PDB file and the corresponding residue localizations

Label (in B-factor column)	Residue localization
−10	Undefined/unknown residue (e.g., small molecule for which ASA cannot be calculated and residues with no sufficient residues to identify and labeled by UNK in the PDB file)
0	Buried residue
5	Surface residue in the hydrophobic core region (lipid phase)
10	Surface residue in the water phase
15	Surface residue in the interface region

Optionally, precalculated results from the MemBlob database (*see* Subheading 2.4) can be retrieved based on the corresponding PDB/EMDB identifier.

Clicking on "Submit query" will initiate the calculation which takes several minutes when the pipeline runs from scratch but provides immediate results for precalculated entries. The results page shows the slices (y–z and x–z planes) of smoothed densities at 0°, 90°, 180°, and 270° as well as a "mantle" map with densities from 0° to 350° along the protein surface. The average distances of the boundaries from the origin (determined by TMDET) are shown in a table. The page also includes an interactive structure viewer where the different regions of the molecule are color-coded. It is highly recommended that the user visually checks the output in all cases (*see* **Note 2**).

2.4 The MemBlob Database

The MemBlob database, available from the MemBlob website, contains currently 92 entries corresponding to MemBlob runs for transmembrane protein structures determined by cryo-EM at a resolution of 4 Å or better. For each entry, the corresponding PDB code and EMDB map along with the UniProt ID of the protein as well as the membrane mimetics used for the structure determination are listed. For 34 of these structures, the MemBlob pipeline could not be applied successfully because of various reasons including the absence of a visible membrane "blob" or that their cryo-EM maps exhibited a very low signal to noise ratio. The reason for unsuccessful processing or other relevant remarks are shown in the last column of the list. For each entry, the success or failure of MemBlob analysis is indicated by a green or red icon, clicking on which the results can be viewed.

3 Using MemBlob to Identify the TM Region in the Slo2.2 K^+-Channel

As an example, we selected the Na^+-dependent K^+-channel Slo2.2 [6], PDB ID 5U70. This structure represents an open structure of the channel, obtained in high (300 mM) NaCl concentration and micelles as membrane mimetics. The biological unit contains four identical chains with fourfold symmetry, and this is reflected by the electron density map.

To analyze the structure, we recommend first to download the EM map from the MemBlob database page for the entry 5U70 (http://memblob.hegelab.org/results?sid=5u70). The required EM data file is named "em_8515.gz". After downloading, the file should be unzipped and renamed to have the extension .mrc to be easily accessible by chimera.

Next, go to the corresponding page on RCSB PDB (https://www.rcsb.org/structure/5u70), and on the "Download files" tab on the right, choose "Biological assembly 1." After unzipping, the user should have a file named "5u70.pdb1."

In chimera, the cryo-EM map can be opened by clicking File → Open and then selecting file type "MRC density map." After selecting and opening the file, select Actions → Surface → Transparency, and set to 50% or another desired value to make the map transparent. As a next step, open the structure by clicking File → Open and selecting file type "PDB." The structure represented in the file 5u70.pdb1 should align well with the map that can be explored by visual inspection while rotating the structure. The fourfold symmetry can be best explored by coloring the chains differently in chimera (Fig. 2A).

Next we will explore the precomputed results on the MemBlob results page for the 5U70 entry (http://memblob.hegelab.org/results?sid=5u70). Four cross sections along the *z*-axis are shown, showing the electron density of the "blob" (i.e., the difference map obtained by subtracting the one calculated from the protein structure from the full experimental one) along with a simplified representation of the protein structure in the plane of the cross section. It is clear that at the lower part, which is the transmembrane region of the structure, there is extra electron density corresponding to nonprotein molecules, which form the membrane-mimetic micelle in this case. The membrane boundaries predicted by TMDET are also shown in Fig. 2B, which depict the cross section at an angle of 90°. In this case, all other cross sections shown on the results page (taken at angles 0°, 180°, and 270°) are similar because of the fourfold symmetry of the structure. Figure 2C and D show the membrane boundaries determined from the electron density at 10° intervals. The fourfold symmetry of the structure is clearly visible in these maps, including the width of the membrane-interacting region. On Fig. 2D, the smoothed boundaries along with those

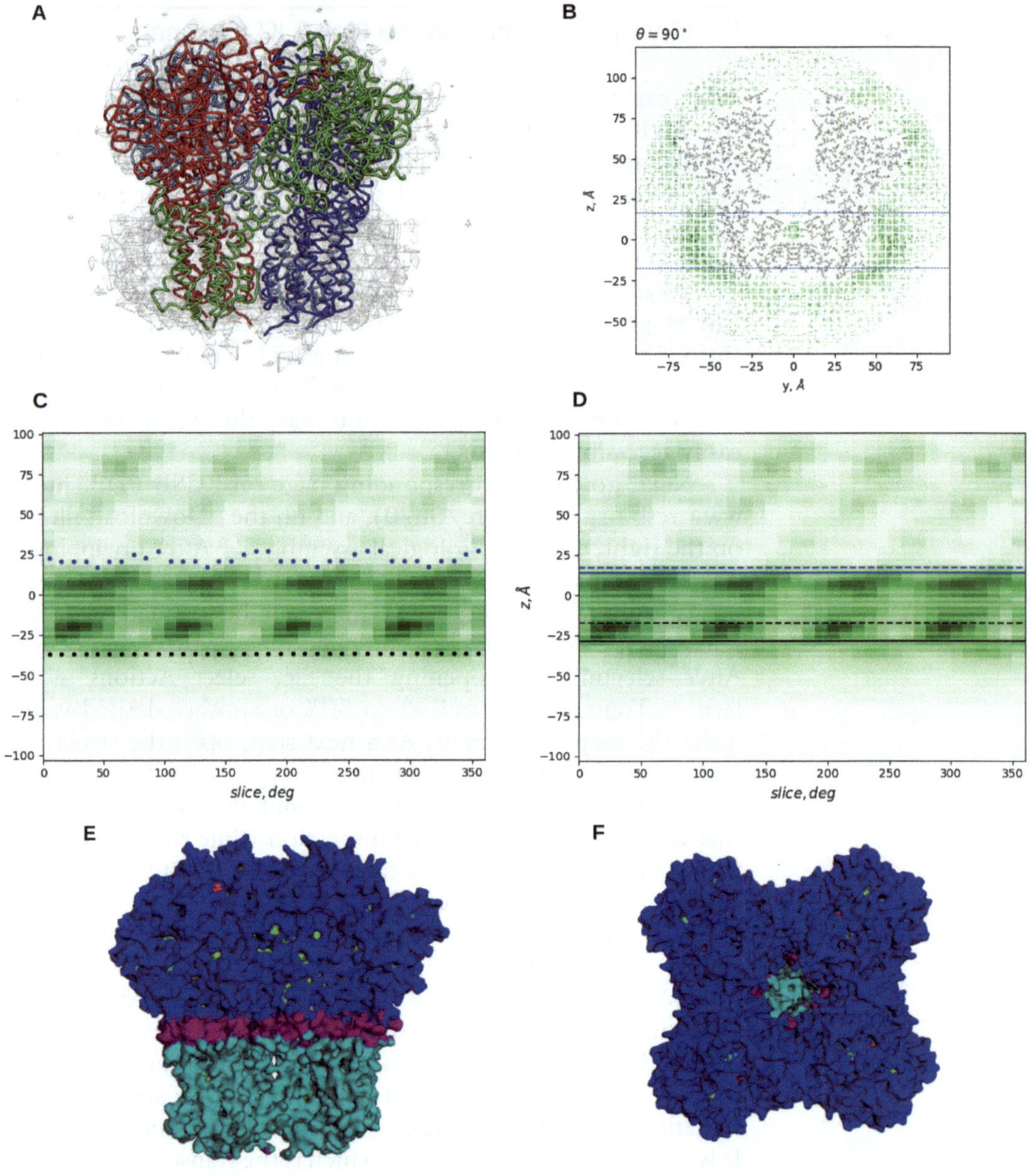

Fig. 2 Analysis of the Slo 2.2. Na+ channel with MemBlob. (**A**) Structure of the biological assembly with the electron density map. Note the "blob" at the bottom of the structure corresponding to electron density coming from the nonprotein parts, which are the membrane-mimetic micelles in this case. (Figure prepared with Chimera.) (**B**) Cross section of the electron density at 90° with the membrane boundaries predicted by TMDET marked. (**C**) Boundaries of the membrane for each slice as determined from the electron density and (**D**) compared to that predicted by TMDET. (**E**, **F**) Side and top view of the channel structure with color-coding: cyan, membrane-interacting residues; blue, solvent-exposed residues; magenta, residues at the membrane interface

predicted by TMDET are shown together. Figure 2E and F show a color-coded structural representation indicating the residues interacting with the membrane and the aqueous phase and those in the membrane interface region. Analysis of the obtained results shows that the TMDET and MemBlob-predicted boundaries do not match closely; in fact, 5U70 is one of the structures where the largest difference has been observed between prediction and MemBlob calculation. In addition, the width of the membrane-interacting region of the protein fluctuates along the structure. These results stress the importance of the inclusion of experimental data in analyzing the exact mode of membrane insertion of TM proteins.

4 Notes

1. The PDB file and the provided cryo-EM map should be aligned. The user might visually check their correspondence with the program Chimera [7] by opening both the MRC and the PDB files (make sure to select the appropriate file format, and the MRC file should be unzipped first) and selecting Actions → Surface → mesh from the menu. For the MemBlob server, the maximum sizes of the gzipped cryo-EM map and the PDB file for upload are 460 MB and 10 MB, respectively.
2. The electron density map for the submitted PDB file is calculated for a resolution of 6 Å using the VMD MDFF package [8]. The two maps are scaled together before subtraction, and points below 10% of the maximum intensity are treated as zero. The difference map is then converted to a set of 3D points with density values. If there is no visible difference, "blob," then MemBlob will not be able to determine the position of the membrane. This can be checked by visual inspection of the results of the method. *See* also the comments for the unsuccessfully processed records in the MemBlob database for more details.

Acknowledgments

This work has been supported by grants from the National Research, Development and Innovation Office NKFIH K111678, NKFIH K119287, NKFIH K125607, and NKFIH K127961, the Cystic Fibrosis Foundation (CFF HEGEDU18I0), and the Semmelweis Science and Innovation Fund.

References

1. Allen JP (2019) Recent innovations in membrane-protein structural biology. F1000Res 8:211
2. Tusnády G, Dosztányi Z, Simon I (2005) TMDET: web server for detecting transmembrane regions of proteins by using their 3D coordinates. Bioinformatics 21:1276–1277
3. Stansfeld PJ, Goose JE, Caffrey M, Carpenter EP, Parker JL, Newstead S, Sansom MP (2015) MemProtMD: automated insertion of membrane protein structures into explicit lipid membranes. Structure 23:1350–1361
4. Hedger G, Samson MSP (2016) Lipid interaction sites on channels, transporters and receptors: recent insights from molecular dynamics simulations. Biochim Biophys Acta 1858:2391–2400
5. Farkas B, Csizmadia G, Katona E, Tusnády GE, Hegedűs T (2019) MemBlob database and server for identifying transmembrane regions using cryo-EM maps. Bioinformatics. https://doi.org/10.1093/bioinformatics/btz539
6. Hite RK, MacKinnon R (2017) Structural titration of Slo2.2, a Na^+-dependent K^+ channel. Cell 168:390–399
7. Pettersen EF, Goddard TD, Huang CC, Couch GS, Greenblatt DM, Meng EC, Ferrin TE (2004) UCSF Chimera—a visualization system for exploratory research and analysis. J Comput Chem 25:1605–1612
8. Trabuco LG, Villa E, Schreiner E, Harrison CB, Schulten K (2009) Molecular dynamics flexible fitting: a practical guide to combine cryo-electron microscopy and X-ray crystallography. Methods 49:174–180

Chapter 10

Structural Characterization of Protein–Protein Interactions with pyDockSAXS

Brian Jiménez-García, Pau Bernadó, and Juan Fernández-Recio

Abstract

Structural characterization of protein–protein interactions can provide essential details to understand biological functions at the molecular level and to facilitate their manipulation for biotechnological and biomedical purposes. Unfortunately, the 3D structure is available for only a small fraction of all possible protein–protein interactions, due to the technical limitations of high-resolution structural determination methods. In this context, low-resolution structural techniques, such as small-angle X-ray scattering (SAXS), can be combined with computational docking to provide structural models of protein–protein interactions at large scale. In this chapter, we describe the pyDockSAXS web server (https://life.bsc.es/pid/pydocksaxs), which uses pyDock docking and scoring to provide structural models that optimally satisfy the input SAXS data. This server, which is freely available to the scientific community, provides an automatic pipeline to model the structure of a protein–protein complex from SAXS data.

Key words Protein–protein interactions, Structural modeling, Small-angle X-ray scattering (SAXS), Computational docking, FTDock, CRYSOL, pyDock

1 Introduction

Protein–protein interactions are essential for the majority of biological processes in the cell. The high-resolution description of the 3D structure of these specific protein complexes can improve our understanding of the biological functions and facilitate rational intervention for biotechnological and biomedical purposes. Unfortunately, high-resolution structural data is available for only a tiny fraction of such complexes, due to the limitations of current structural biology methods. In this context, small-angle X-ray scattering (SAXS) has emerged as a powerful low-resolution method for the characterization of biomolecules and macromolecular assemblies [1–5]. The structural information coded in a SAXS curve can be fully exploited when combined with computational approaches [6]. This combination is especially adequate for providing detailed models of protein–protein complexes [7]. One of the first methods

Zoltán Gáspári (ed.), *Structural Bioinformatics: Methods and Protocols*, Methods in Molecular Biology, vol. 2112,
https://doi.org/10.1007/978-1-0716-0270-6_10, © Springer Science+Business Media, LLC, part of Springer Nature 2020

for rigid-body modeling of SAXS data is SASREF [8], which uses simulated annealing to simultaneously fit structural models for protein–protein complexes to multiple SAXS/SANS profiles. However, relying on SAXS data alone is not sufficient to resolve degeneracy in the resulting models, since multiple docking orientations provide similar overall shapes and, therefore, show similar descriptions of the experimental data. This limitation can be overcome by using approaches with the capacity to discriminate between different arrangements. In that context, computational protein–protein docking strategies, which rely on the chemical nature of the interacting surface, turn out to be powerful tools to be combined with SAXS data and improve the quality of the resulting models. Computational docking methods can generate a large number of poses that are geometrically and energetically coherent [9]. The incorporation of SAXS experimental data can narrow the set of docking solutions that are consistent with such experimental data. The first reported method to implement and validate this strategy was pyDockSAXS [10], and since then other methodologies appeared [11–16]. pyDockSAXS, when systematically tested on a standard protein–protein docking benchmark, showed twofold increase in the predictive success rates with respect to the individual approaches, energy-based docking, or SAXS fitting alone. This method was implemented in a web server that provided an automatic pipeline for the structural characterization of protein–protein complexes with SAXS data [17]. In this chapter, we will review the pyDockSAXS methodology, with detailed running instructions, example cases, and advises for efficient application.

2 Materials

Our method is available as a web service, freely accessible at https://life.bsc.es/pid/pydocksaxs. The web front-end acts as a proxy to the user, removing any complexity aroused from a local installation of the software. Via a user-friendly interface, the user is capable of uploading molecular structural information, in PDB format, and experimental SAXS data, compatible with the CRYSOL [18] software, to obtain a set of complex predictions consistent with the experimental data provided (if possible). Our protocol pyDockSAXS, described in [10], samples the rigid-body translational and rotational space in search for the best 10,000 protein complex conformations by means of FTDock [19] software and then rescores them by the pyDock scoring energy [20]. After this step, the capacity to describe the experimental SAXS curve is evaluated with the program CRYSOL [18]. A final score combining the agreement with the experimental SAXS curve and the protein docking scoring energy is calculated to filter out the best predictions.

The server runs on a cluster with reserved resources for this service. Allocated resources consist in a computation node composed of 16 cores (4 Intel Xeon E5620 Quad Core at 2.4 GHz) with 11 TB of total available disk space and 256 GB of physical memory. This configuration allows the user to compute docking predictions in a high-performance computing (HPC) environment.

3 Methods

3.1 Input Files

There are two possible modes of using pyDockSAXS server: default mode (in which docking models are generated for two given interacting proteins) and advanced mode (in which the user can provide a previously generated docking set).

3.1.1 Default Mode

In default mode, pyDockSAXS server requires three different files from the user:

(a) **Receptor structure file**. This file is in plain-text PDB format and contains the information of the receptor protein structure. In order to avoid inaccurate results or software failure, the PDB structure must contain all the atomic information for every backbone and side-chain atom. Residues with incomplete backbone information are removed in the protocol, while incomplete side-chain atoms are rebuilt using SCWRL version 3.0 software [21]. Multichain PDB files are totally supported, while multi-model files are trimmed to the first model. Alternative atom positions are not considered in the protocol. An example of this file can be found online in the *Help* page, section *Sample Data*, file *1PPE_rec.pdb*, which are the coordinates of bovine β-trypsin extracted from the X-ray structure of its complex with CMTI-I (pdb code 1PPE). The use of the bound form in this example is only for the purpose of clarity, but obviously in a real case, the coordinates (structure or model) of the individual input proteins will be used most of the times.

(b) **Ligand structure file**. It is the same as the receptor structure file, but now containing the structure for the protein ligand. An example of this file can be found online in the *Help* page, section *Sample Data*, file *1PPE_lig.pdb*, corresponding to the coordinates of CMTI-I in complex with bovine β-trypsin (pdb code 1PPE). It is a common practice to define the receptor as the largest molecular partner in the complex, in terms of number of atoms and/or the maximum diameter of the minimum ellipsoid containing the protein, because it is usually faster for FFT-based methods to sample smaller mobile partners.

Table 1
Example of SAXS data file

Randomized data, RELERR = 2.00%, file 1ppe_ref00.iMon Dec 18 14:30:51 2006		
0.5000E−02	0.1420E+08	0.2909E+06
0.7500E−02	0.1418E+08	0.2899E+06
0.1000E−01	0.1467E+08	0.2886E+06
0.1250E−01	0.1458E+08	0.2869E+06
0.1500E−01	0.1459E+08	0.2848E+06

Only the header and the first five rows are shown. The complete sample data of this table can be found at https://life.bsc.es/pid/pydocksaxs/static/data/1PPE_curve.dat

(c) **SAXS experimental data**. A file containing SAXS experimental data compatible with CRYSOL software version 2.8 (Table 1). This data represents an experimental scattering curve. The first line is always treated as a title. The following lines should contain momentum transfer, nonzero intensity, and standard deviation in a free format (separated by blank spaces or commas). If standard deviations are not present, the errors are estimated automatically with the help of a polynomial smoothing procedure. An example of this file can be found online in the *Help* page, section *Sample Data*, file *1PPE_curve.dat*.

3.1.2 Advanced Mode

In expert mode (option stated as *advanced users*), an extra file is required (*see* **Note 1**):

(a) **Rigid-body docking set**. This file represents a previously computed rigid-body docking by pyDock [20] software or pyDock-WEB [22] web server. It is identified by the extension *.rot* and contains a set of 13 numerical columns separated by spaces (Table 2). The first nine columns represent the Euler angles of the rotation matrix applied to the ligand structure to obtain the final complex pose. Columns 10–12 represent the translation vector of the ligand structure in respect of the origin of coordinates (0, 0, 0). The last column (13) is a numerical identifier of the docking pose. This docking set file is calculated by pyDock or pyDockWEB, and although it could be calculated by any other external docking program, this is strongly discouraged to avoid inconsistences (unless strictly checking that the format is correct). If this file has been generated in a previous protein–protein docking run or by the pyDockSAXS protocol, you can upload it to speed up the calculations of the protocol. An example of this file can be found online in the *Help* page, section *Sample Data*, file *1PPE.rot*.

Table 2
Example of .rot file describing a set of docking models

−0.956	0.004	−0.294	−0.277	0.325	0.905	0.099	0.946	−0.309	−2.306	−5.272	22.608	1
0.223	0.712	−0.665	−0.063	−0.671	−0.739	−0.973	0.207	−0.105	−7.227	10.897	24.014	2
−0.384	0.510	−0.769	0.791	−0.247	−0.559	−0.476	−0.824	−0.309	4.021	−8.084	28.232	3
0.207	−0.743	−0.636	−0.230	−0.669	0.707	−0.951	0.000	−0.309	30.032	−1.054	34.559	4
−0.780	0.504	−0.372	−0.175	−0.745	−0.644	−0.601	−0.437	0.669	18.081	−2.460	−10.433	5
−1.000	0.000	0.000	−0.000	−1.000	0.000	0.000	−0.000	1.000	3.318	1.055	38.074	6
0.420	−0.788	−0.450	−0.889	−0.258	−0.378	0.182	0.559	−0.809	8.239	−8.787	28.232	7
−0.763	0.336	0.552	−0.513	−0.834	−0.201	0.393	−0.437	0.809	33.547	−7.381	16.281	8
−0.541	0.393	−0.743	−0.588	−0.809	−0.000	−0.601	0.437	0.669	26.517	−12.302	6.439	9
−0.856	−0.143	−0.497	0.507	−0.038	−0.861	0.104	−0.989	0.105	2.615	17.224	7.142	10

Only the first 10 rows are shown. The complete sample data for this file can be found at https://life.bsc.es/pid/pydocksaxs/static/data/1PPE.rot

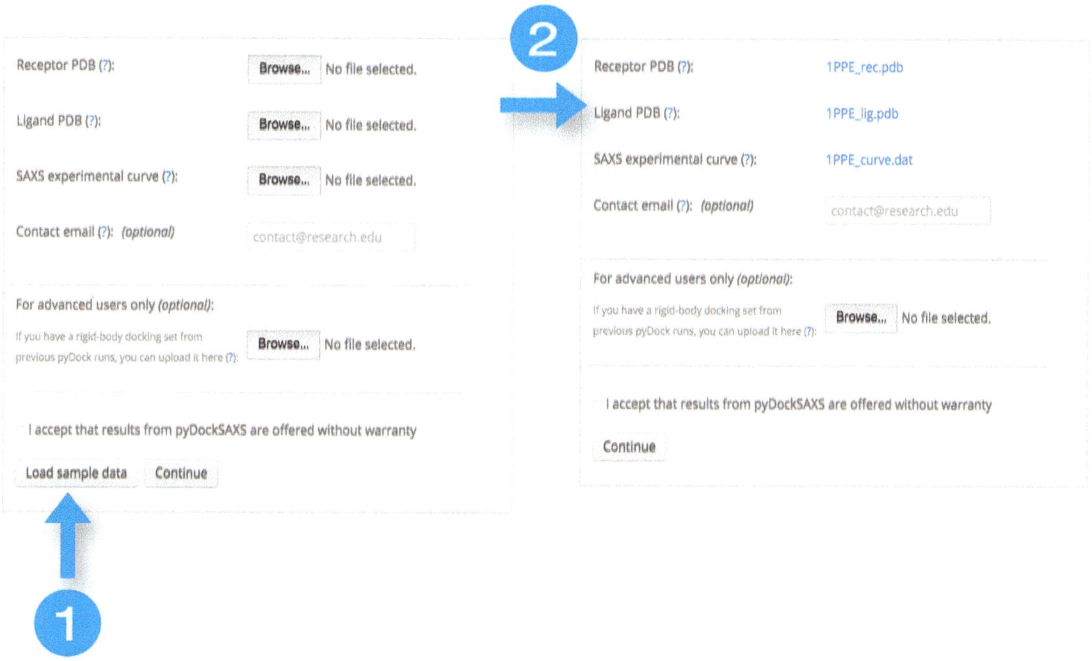

Fig. 1 Automatic loading of the example data of a protein–protein complex (PDB code 1PPE) in the pyDockSAXS webserver. The user has to click on the *Load sample data* button (1), and the sample files will appear in the same page on the right side (2)

There is also the possibility to load sample data for the complex between *bovine β-trypsin* and *CMTI-I* (PDB code 1PPE). The user has only to click on the *Load sample data* button, and the sample files will appear in the same view (Fig. 1).

Fig. 2 Chain selection page. The user is asked to select the chains for both receptor and ligand involved in the protein–protein docking prediction step

Once the input files have been uploaded, then click on *Continue* button.

3.2 Using the pyDockSAXS Protocol Web Server

3.2.1 Chain Selection

Once the user has provided the input files (*see* **Note 2**), a new view asks the user to select the chains for both receptor and ligand involved in the protein–protein docking prediction step (Fig. 2). If no chain is selected for receptor or for the ligand, an error will appear asking the user to select at least one chain per subunit. Figure 2 shows the available chains for the receptor and ligand input files previously extracted from the structure of the complex between *bovine β-trypsin* and *CMTI-I* (PDB code 1PPE), used as example.

3.2.2 CRYSOL Parameters Selection

After the selection, a new view for configuring additional parameters for CRYSOL [18] is displayed (Fig. 3). Only after selecting the option ***For advanced users only:*** *use custom CRYSOL parameters*, the *Constant subtraction* and the *Angular units* options will be enabled. Here we detail these two options:

(a) **Constant subtraction**. This operation accounts for possible systematic errors due to mismatched buffers in the experimental data. This is a free parameter that is added to all intensities of the scattering profile to improve CRYSOL [18] fitting.

(b) **Angular units**. By default, an attempt is made to estimate the unit scale of the SAXS curve. If angular units are explicitly selected, they will be used by the CRYSOL software and may incur in prediction failure. There are five available options:

- 1/Å, $s = 4\pi\sin(\theta)/\lambda$
- 1/nm, $s = 4\pi\sin(\theta)/\lambda$
- 1/Å, $s = 2\sin(\theta)/\lambda$
- 1/nm, $s = 2\sin(\theta)/\lambda$
- Automatic (by default)

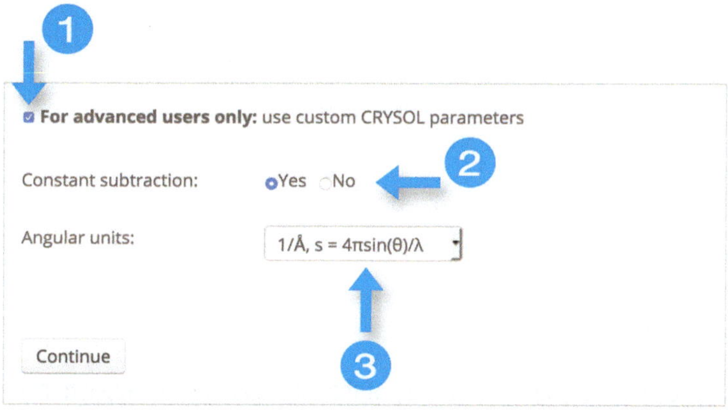

Fig. 3 Advanced mode for selecting CRYSOL-specific parameters. After selecting the option **For advanced users only**: *use custom CRYSOL parameters* (1), the *Constant subtraction* (2) and the *Angular units* (3) options will be enabled

(c) Where s is the momentum transfer, 2θ is the scattering angle, and λ is the X-ray wavelength.

3.2.3 Data Submission

In the final step, a summary of the data provided by the user is displayed (Fig. 4). This data is:

- Contact email (if provided). It would be used to notify the user after completion of the computation in the server.
- Receptor PDB structure file name and selected chains.
- Ligand PDB structure file name and selected chains.
- A plot of the scattering curve of the experimental data provided (*see* **Note 3**).

When clicking on *Submit job* button, the user will be redirected to the job results page, which displays the current status of the job and it is automatically refreshed every 5 min.

3.3 Results Page

The job has finished when the status in the job results page is set to *calculated*. At this point, the page shows four basic blocks of data (Fig. 5):

(a) **Results files**. A compressed file in TAR-GZIP format containing all the results predicted is provided for downloading. Please refer to Subheading 3.4 for more details.

(b) **Table of predicted energies**. This table, which can be also downloaded in *PDF* format, shows the top 100 predictions as sorted by the pyDockSAXS score, with the different energies calculated by the protocol (Table 3). The order of each docking model is identified in the *RANK* column (from 1, the best one, to 100). For each conformation (*Conf* column), values for electrostatics (*Ele* column), desolvation (*Desolv* column),

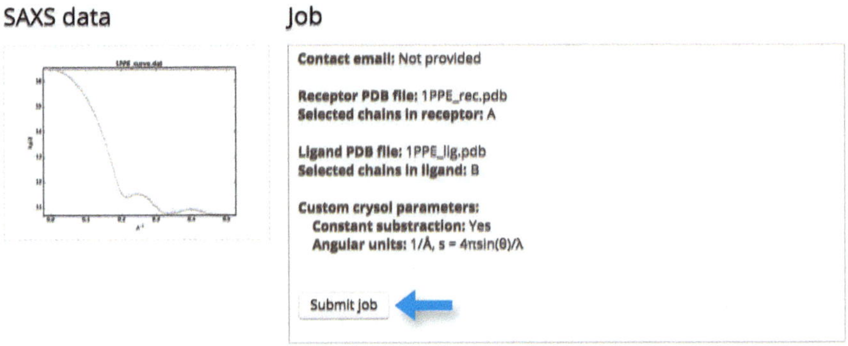

Fig. 4 Submission page. A summary of the main parameters appears before submitting the job to the server

and Van der Waals (*VDW* column) energies calculated by pyDock are displayed. The column "pyDock" represents the pyDock energy, which is calculated as the sum of *Ele*, *Desolv*, and 10% of the *VDW* column. *Crysol* column indicates the value of χ defining the goodness of fit to the SAXS data computed with CRYSOL 2.8 (*see* **Note 4**). Finally, the *pyDock-SAXS* column indicates the final score calculated by the protocol.

(c) **SAXS curves for the top ten predicted models compared to the input experimental curve**. An interactive plot where every fitted SAXS curve is in the same color as the model represented in the 3D visualization section (Fig. 5, panel 3). The number on the right of the color identifies the conformation (*Conf* column in the predicted energy table).

(d) **Top 10 predicted models in a 3D interactive visualization**. Receptor protein is fixed and displayed using van der Waals spheres and white color (Fig. 5, panel 4). Ligand models are displayed in backbone-only mode and in different colors. Models can be activated or deactivated using the checkboxes below the 3D representation.

3.4 Output Files

Output files are organized in four different folders: *input_data*, *pydock*, *top100*, and *fit_top10_SAXS*. The tag *xxx* is a numerical identifier of the job in the pyDockSAXS web server:

(a) ***input_data***. This folder contains the original data provided by the user.

(b) ***pydock***. This folder contains the files generated by the pyDock software:

- *setup.log*: a description of how pyDock reads and parses the original PDB structures provided by the user.
- *project_[xxx]_rec.pdb*: receptor PDB structure after parsing in pyDock.

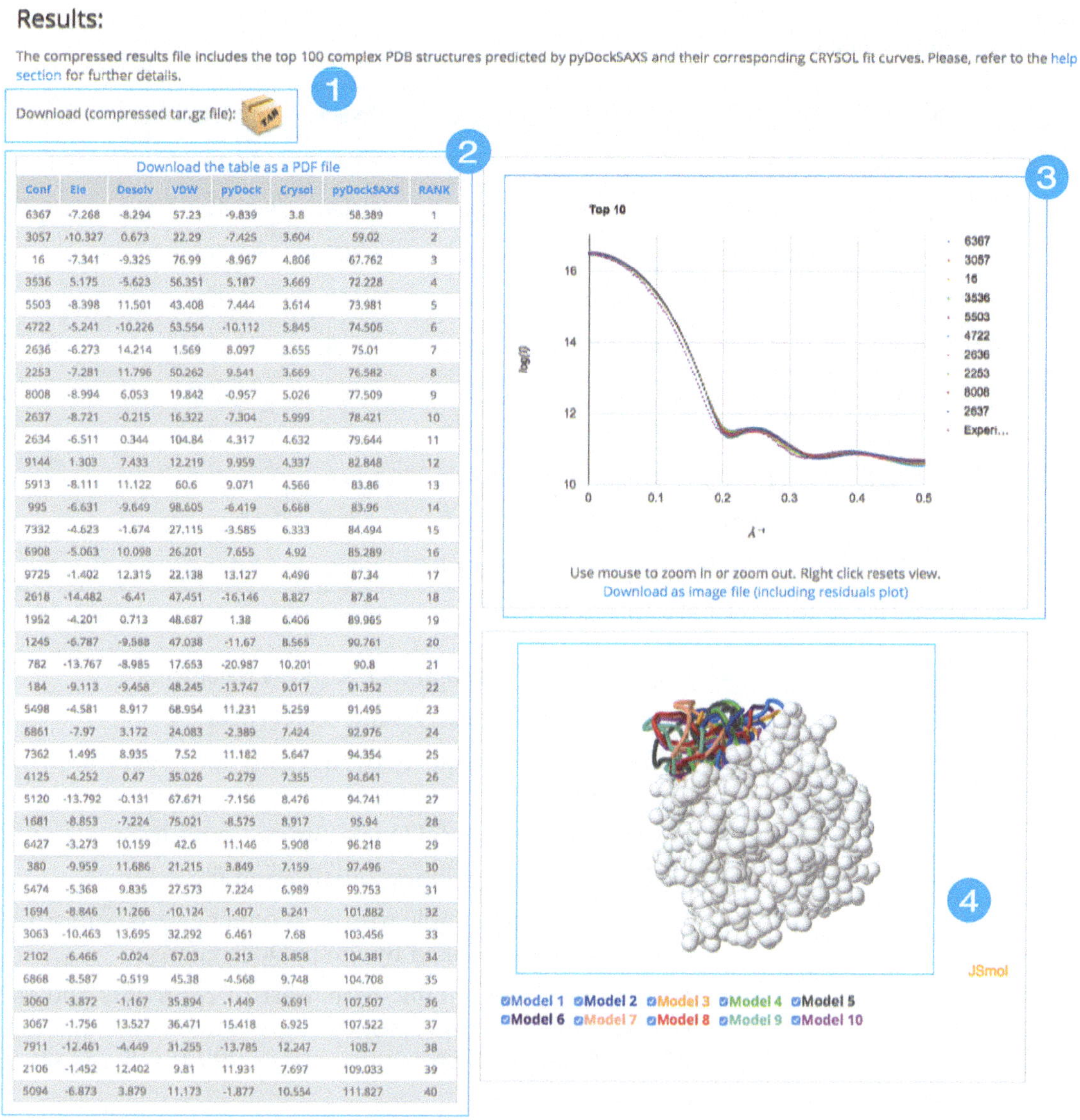

Results:

The compressed results file includes the top 100 complex PDB structures predicted by pyDockSAXS and their corresponding CRYSOL fit curves. Please, refer to the help section for further details.

Download (compressed tar.gz file):

Download the table as a PDF file

Conf	Ele	Desolv	VDW	pyDock	Crysol	pyDockSAXS	RANK
6367	-7.268	-8.294	57.23	-9.839	3.8	58.389	1
3057	-10.327	0.673	22.29	-7.425	3.604	59.02	2
16	-7.341	-9.325	76.99	-8.967	4.806	67.762	3
3536	5.175	-5.623	56.351	5.187	3.669	72.228	4
5503	-8.398	11.501	43.408	7.444	3.614	73.981	5
4722	-5.241	-10.226	53.554	-10.112	5.845	74.506	6
2636	-6.273	14.214	1.569	8.097	3.655	75.01	7
2253	-7.281	11.796	50.262	9.541	3.669	76.582	8
8008	-8.994	6.053	19.842	-0.957	5.026	77.509	9
2637	-8.721	-0.215	16.322	-7.304	5.999	78.421	10
2634	-6.511	0.344	104.84	4.317	4.632	79.644	11
9144	1.303	7.433	12.219	9.959	4.337	82.848	12
5913	-8.111	11.122	60.6	9.071	4.566	83.86	13
995	-6.631	-9.649	98.605	-6.419	6.668	83.96	14
7332	-4.623	-1.674	27.115	-3.585	6.333	84.494	15
6908	-5.063	10.098	26.201	7.655	4.92	85.289	16
9725	-1.402	12.315	22.138	13.127	4.496	87.34	17
2618	-14.482	-6.41	47.451	-16.146	8.827	87.84	18
1952	-4.201	0.713	48.687	1.38	6.406	89.965	19
1245	-6.787	-9.588	47.038	-11.67	8.565	90.761	20
782	-13.767	-8.985	17.653	-20.987	10.201	90.8	21
184	-9.113	-9.458	48.245	-13.747	9.017	91.352	22
5498	-4.581	8.917	68.954	11.231	5.259	91.495	23
6861	-7.97	3.172	24.083	-2.389	7.424	92.976	24
7362	1.495	8.935	7.52	11.182	5.647	94.354	25
4125	-4.252	0.47	35.026	-0.279	7.355	94.641	26
5120	-13.792	-0.131	67.671	-7.156	8.476	94.741	27
1681	-8.853	-7.224	75.021	-8.575	8.917	95.94	28
6427	-3.273	10.159	42.6	11.146	5.908	96.218	29
380	-9.959	11.686	21.215	3.849	7.159	97.496	30
5474	-5.368	9.835	27.573	7.224	6.989	99.753	31
1694	-8.846	11.266	-10.124	1.407	8.241	101.882	32
3063	-10.463	13.695	32.292	6.461	7.68	103.456	33
2102	-6.466	-0.024	67.03	0.213	8.858	104.381	34
6868	-8.587	-0.519	45.38	-4.568	9.748	104.708	35
3060	-3.872	-1.167	35.894	-1.449	9.691	107.507	36
3067	-1.756	13.527	36.471	15.418	6.925	107.522	37
7911	-12.461	-4.449	31.255	-13.785	12.247	108.7	38
2106	-1.452	12.402	9.81	11.931	7.697	109.033	39
5094	-6.873	3.879	11.173	-1.877	10.554	111.827	40

Fig. 5 Results section. The page shows four basic blocks of information provided to the user by the pyDockSAXS webserver: results files ready to be downloaded (1), a table with the predicted energies for the top 100 docking poses (2), SAXS curves for the top ten predicted models compared to the input experimental curve (3), and a 3D graphics visualization of the top ten predicted models (4)

- *project_[xxx]_rec.pdb.amber*: AMBER94 [23] force field values for atoms in receptor PDB structure used by pyDock.
- *project_[xxx]_rec.pdb.H*: receptor PDB structure parsed by pyDock, including hydrogens.
- *project_[xxx]_lig.pdb*: ligand PDB structure parsed by pyDock.

Table 3
Example of results table

Conf	Ele	Desolv	VDW	pyDock	Crysol	pyDockSAXS	RANK
6367	−7.268	−8.294	57.23	−9.839	3.8	58.389	1
3057	−10.327	0.673	22.29	−7.425	3.604	59.02	2
16	−7.341	−9.325	76.99	−8.967	4.806	67.762	3
3536	5.175	−5.623	56.351	5.187	3.669	72.228	4
5503	−8.398	11.501	43.408	7.444	3.614	73.981	5
4722	−5.241	−10.226	53.554	−10.112	5.845	74.506	6
2636	−6.273	14.214	1.569	8.097	3.655	75.01	7
2253	−7.281	11.796	50.262	9.541	3.669	76.582	8
8008	−8.994	6.053	19.842	−0.957	5.026	77.509	9
2637	−8.721	−0.215	16.322	−7.304	5.999	78.421	10

Only the header and the top ten conformations and their respective energies are shown

- *project_[xxx]_lig.pdb.amber*: same as receptor.
- *project_[xxx]_lig.pdb.H*: same as receptor.
- *project_[xxx].rot*: rigid-body docking set generated from the FTDock [19] output.
- *project_[xxx].ftdock*: FTDock software output.
- *project_[xxx].ini*: pyDock initialization file.
- *project_[xxx].ene*: a table with a list of generated conformations scored and ranked by the pyDock energy.
- *project_[xxx].saxs*: a table with a list of *chi-square* and the *radius of gyration* values for each generated conformation. When *chi-square* is larger than 10, a 999.0 value is set in the file.
- *project_[xxx].ene.saxs*: a table with a list of generated conformations scored and ranked by the pyDockSAXS energy.

(c) **top100**. This folder contains the top 100 structures scored by pyDockSAXS and their corresponding CRYSOL fitted curve. File names follow the pattern *RankNumber_project_ProjectID_ConformationNumber.extension*, where *RankNumber* corresponds to the *RANK* column and *ConformationNumber* to the *Conf* column in the predicted energy table.

(d) **fit_top10_SAXS**. This folder contains the top ten fitted curves calculated by CRYSOL and in line with the *Chi-square* value. The file name format follows the pattern *RankNumber_ConformationNumber.fit*, where *RankNumber* corresponds to the *RANK* column and *ConformationNumber* to the *Conf* column in the predicted energy table.

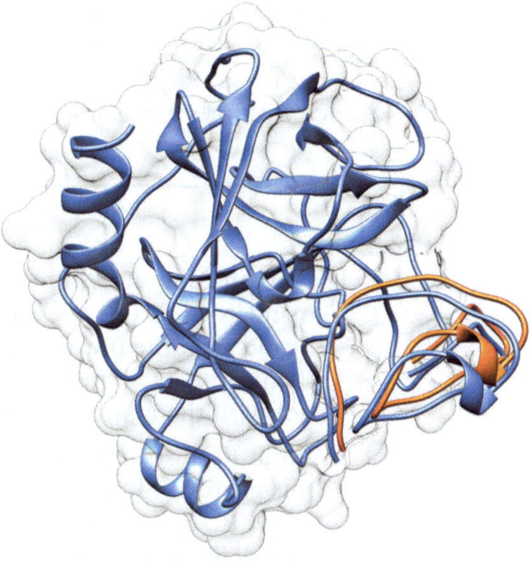

Fig. 6 Predicted models for the example case (bovine β-trypsin and CMTI-I). The position of the ligand in the docking model ranked 1 by pyDockSAXS is shown in orange ribbon. This can be compared with the complex reference (PDB code 1PPE) shown in blue ribbon (receptor in blue ribbon and white surface). This model has ligand RMSD 2.4 Å with respect to the reference, after superimposing the corresponding receptor molecules

4 Discussion

As a test exercise, we can compare the docking models provided by the pyDockSAXS server in the example case (bovine β-trypsin and CMTI-I) with the X-ray structure of the reference complex (PDB 1PPE). In order to evaluate the results, we can calculate standard measures in docking prediction assessment. One of the most popular measures is the ligand RMSD, which is the root mean standard deviation of the ligand atoms in the model with respect to those in the reference, after optimally superimposing the receptor molecules. Figure 6 shows in orange the model ranked 1 by pyDockSAXS for the example case in comparison with the reference complex in blue (receptor molecules from model and reference are superimposed). The ligand RMSD for this model is 2.4 Å, which indicates that the method has worked well in this case. Interestingly, the models with the best fitting to the SAXS curve and with best pyDock energy are further from the reference structure, which shows that the combination of energy-based and SAXS-based scoring is improving the predictive results as compared with the two individual approaches (*see* **Note 5**).

Of course, in a real case, one does not have the reference complex to compare. The method provides a series of models that can be used to interpret or guide experimental results (*see* **Note 5**).

There are some hints that can indicate reliability of the predictions, such as convergence of the best-scoring models toward the same structure, good pyDock and/or CRYSOL scoring values, consistency with other available experimental data, etc.

5 Notes

1. If a precomputed rigid-body docking set is provided, it is mandatory to use the same input structures for receptor and ligand as in the previous docking prediction. This is crucial, as the protocol will not check if the docking set is compatible or not with the input structures.
2. Input PDB structures are one of the major sources of failure of this protocol. One of the reasons is the existing heterogeneity in the PDB file format. Despite the syntax of the PDB format is well defined, third-party software that users can apply to analyze, visualize, or generate PDB files is very diverse and may interpret or modify the file format. In our protocol, only lines starting with the keyword "ATOM," containing only protein information, i.e., atom coordinates, type, and information related to the standard 20 residues, are parsed. No water, cofactors, small molecules, DNA, or RNA data are accepted by the protocol.
3. The second major source of failure of the protocol is the diverse set of errors and mistakes on the experimental SAXS data provided. Wrong identification of the units, wrong use of the constant subtraction, or several initial lines in the data file (only the first line is identified by the CRYSOL software as a header or title) are typical examples. Despite many of these errors are controlled in the web server, unorthodox input formats can escape this sanitization step.
4. Predictions might not be compatible with the scattering data provided. In that case, the protocol is not capable of providing a good model for the assembly, and thus the top predictions displayed in the results page (or the top predictions in the energy table) will show a wrong *CRYSOL* score (9999.0) to penalize them.
5. The user should note that SAXS contribution to the identification of correct docked models will strongly depend on the shape of the interacting proteins. In cases with anisotropic proteins (i.e., elongated or flattened), SAXS data will be more discriminative. However, in cases with spherical proteins, the different docking models will yield similar SAXS curves, and therefore the scoring will rely mostly on the pyDock energy.

Acknowledgments

This work was supported by the Spanish Ministry of Science (grant BIO2016-79930-R), the European Union H2020 programme (grant MuG 676566), and the Labex EpiGenMed, an "Investissements d'avenir" program (ANR-10-LABX-12-01). The CBS is a member of France-BioImaging (FBI) and the French Infrastructure for Integrated Structural Biology (FRISBI), two national infrastructures supported by the French National Research Agency (ANR-10-INSB-04-01 and ANR-10-INSB-05, respectively).

References

1. Koch MH, Vachette P, Svergun DI (2003) Small-angle scattering: a view on the properties, structures and structural changes of biological macromolecules in solution. Q Rev Biophys 36:147–227
2. Putnam CD, Hammel M, Hura GL, Tainer JA (2007) X-ray solution scattering (SAXS) combined with crystallography and computation: defining accurate macromolecular structures, conformations and assemblies in solution. Q Rev Biophys 40:191–285
3. Jacques DA, Trewhella J (2010) Small-angle scattering for structural biology—expanding the frontier while avoiding the pitfalls. Protein Sci 19:642–657
4. Cordeiro TN, Herranz-Trillo F, Urbanek A, Estaña A, Cortés J, Sibille N, Bernadó P (2017) Small-angle scattering studies of intrinsically disordered proteins and their complexes. Curr Opin Struct Biol 42:15–23
5. Bernadó P, Shimizu N, Zaccai G, Kamikubo H, Sugiyama M (2018) Solution scattering approaches to dynamical ordering in biomolecular systems. Biochim Biophys Acta Gen Subj 1862:253–274
6. Hub JS (2018) Interpreting solution X-ray scattering data using molecular simulations. Curr Opin Struct Biol 49:18–26
7. Yang S (2014) Methods for SAXS-based structure determination of biomolecular complexes. Adv Mater 26:7902–7910
8. Petoukhov MV, Svergun DI (2005) Global rigid body modeling of macromolecular complexes against small-angle scattering data. Biophys J 89:1237–1250
9. Ritchie DW (2008) Recent progress and future directions in protein-protein docking. Curr Protein Pept Sci 9:1–15
10. Pons C, D'Abramo M, Svergun DI, Orozco M, Bernado P, Fernandez-Recio J (2010) Structural characterization of protein-protein complexes by integrating computational docking with small-angle scattering data. J Mol Biol 403:217–230
11. Schneidman-Duhovny D, Hammel M, Sali A (2011) Macromolecular docking restrained by a small angle X-ray scattering profile. J Struct Biol 173:461–471
12. Schneidman-Duhovny D, Hammel M, Tainer JA, Sali A (2016) FoXS, FoXSDock and MultiFoXS: single-state and multi-state structural modeling of proteins and their complexes based on SAXS profiles. Nucleic Acids Res 44 (W1):W424–W429
13. Sønderby P, Rinnan Å, Madsen JJ, Harris P, Bukrinski JT, Peters GHJ (2017) Small-angle X-ray scattering data in combination with RosettaDock improves the docking energy landscape. J Chem Inf Model 57:2463–2475
14. Schindler CEM, de Vries SJ, Sasse A, Zacharias M (2016) SAXS data alone can generate high-quality models of protein-protein complexes. Structure 24:1387–1397
15. Schneidman-Duhovny D, Hammel M (2018) Modeling structure and dynamics of protein complexes with SAXS profiles. Methods Mol Biol 1764:449–473
16. Bonvin AMJJ, Karaca E (2013) On the usefulness of ion-mobility mass spectrometry and SAXS data in scoring docking decoys. Acta Crystallogr D Biol Crystallogr 69:683–694
17. Jiménez-García B, Pons C, Svergun DI, Bernadó P, Fernández-Recio J (2015) pyDockSAXS: protein–protein complex structure by SAXS and computational docking. Nucleic Acids Res 43(W1):W356–W356
18. Svergun DI, Barberato C, Koch MHJ (1995) CRYSOL – a program to evaluate X-ray solution scattering of biological macromolecules from atomic coordinates. J Appl Crystallogr 28:768–773

19. Gabb HA, Jackson RM, Sternberg MJ (1997) Modelling protein docking using shape complementarity, electrostatics and biochemical information. J Mol Biol 272:106–120
20. Cheng TM, Blundell TL, Fernandez-Recio J (2007) pyDock: electrostatics and desolvation for effective scoring of rigid-body protein-protein docking. Proteins 68:503–515
21. Wang Q, Canutescu AA, Dunbrack RL Jr (2008) SCWRL and MolIDE: computer programs for side-chain conformation prediction and homology modeling. Nat Protoc 3:1832–1847
22. Jiménez-García B, Pons C, Fernández-Recio J (2013) pyDockWEB: a web server for rigid-body protein-protein docking using electrostatics and desolvation scoring. Bioinformatics 29:1698–1699
23. Cornell WD, Cieplak P, Bayly CI, Gould IR, Merz KM, Ferguson DM, Spellmeyer DC, Fox T, Caldwell JW, Kollman PA (1995) A second generation force field for the simulation of proteins, nucleic acids, and organic molecules. J Am Chem Soc 117:5179–5197

Chapter 11

Protein–Protein Modeling Using Cryo-EM Restraints

Mikael Trellet, Gydo van Zundert, and Alexandre M. J. J. Bonvin

Abstract

Recent improvements in cryo-electron microscopy (cryo-EM) in the past few years are now allowing to observe molecular complexes at atomic resolution. As a consequence, numerous structures derived from cryo-EM are now available in the Protein Data Bank. However, if for some complexes atomic resolution is reached, this is not true for all. This is also the case in cryo-electron tomography where the achievable resolution is still limited. Furthermore the resolution in a cryo-EM map is not a constant, with often outer regions being of lower resolution, possibly linked to conformational variability. Although those low- to medium-resolution EM maps (or regions thereof) cannot directly provide atomic structure of large molecular complexes, they provide valuable information to model the individual components and their assembly into them. Most approaches for this kind of modeling are performing rigid fitting of the individual components into the EM density map. While this would appear an obvious option, they ignore key aspects of molecular recognition, the energetics and flexibility of the interfaces. Moreover, this often restricts the modeling to a unique source of data, the EM density map.

In this chapter, we describe a protocol where an EM map is used as restraint in HADDOCK to guide the modeling process. In the first step, rigid-body fitting is performed with PowerFit in order to identify the most likely locations of the molecules into the map. These are then used as centroids to which distance restraints are defined from the center of mass of the components of the complex for the initial rigid-body docking. The EM density is then directly used as an additional restraint energy term, which can be combined with all the other types of data supported by HADDOCK. This protocol relies on the new version 2.4 of both the HADDOCK webserver and software. Preparation steps consisting of cropping the EM map and rigid-body fitting of the atomic structure are explained. Then, the EM-driven docking protocol using HADDOCK is illustrated.

Key words Biomolecular interactions, Information-driven docking, Cryo-EM data, Flexibility, HADDOCK, Molecular modeling

1 Introduction

To drive all essential functions of the cells, biomolecules interact with each other forming complexes of different scales and stabilities. Deciphering the three-dimensional (3D) structure of such

Electronic supplementary material: The online version of this chapter (https://doi.org/10.1007/978-1-0716-0270-6_11) contains supplementary material, which is available to authorized users.

Zoltán Gáspári (ed.), *Structural Bioinformatics: Methods and Protocols*, Methods in Molecular Biology, vol. 2112, https://doi.org/10.1007/978-1-0716-0270-6_11, © Springer Science+Business Media, LLC, part of Springer Nature 2020

molecular complexes provides insights into the molecular determinants of these interactions and opens the route to tuning them in order to prevent or promote functions linked, for example, to diseases. Several experimental techniques exist to solve the 3D structure of molecules. Depending on the flexibility, mobility, and environment of those proteins, some techniques will be more efficient than others. They might also picture the system at different resolutions. X-ray crystallography and NMR have been for a long time the sole providers of high-resolution atomic structures stored in the Protein Data Bank (PDB). However, the past few years have seen the rise in the number of high-resolution structures solved by cryo-electron microscopy (cryo-EM). Cryo-EM has undergone a revolution in terms of the achievable resolution, thanks to both technical (e.g., the direct electron detectors) and software advances [1, 2].

Despite those advances, there will still be plenty of cases where cryo-EM will not achieve atomistic resolution (also typically difficult to reach in cryo-electron tomography). The resolution within one large macromolecular complex is also not a constant, meaning that parts of the complexes, often on the periphery or the more flexible parts, might only be seen at lower resolution. In those cases, one has to rely to fitting structures or models of the components of a complex into the density. This can be done via different ways: Manual fitting using specialized tools [3, 4], exhaustive search and rigid-body fitting [5], or flexible fitting, using different strategies to account for the atomic structures flexibility [6]. Often this modeling does not take into account flexibility (or only to a limited degree) and usually ignores the energetics at the interface of the fitted components, with the result that the interfaces in those complexes often have a poor quality with many clashes.

We have previously published a protocol that makes use of cryo-EM densities in flexible docking based on our information-driven, integrative modeling platform HADDOCK [7]. In this chapter, we illustrate the use of cryo-EM data as restraints to drive the modeling of a protein–protein complex using the new HADDOCK2.4 web portal, which now supports such kind of data. The protocol illustrates various steps, from the preparation/cropping of the original cryo-EM map to rigid-body fitting into the cryo-EM density to extract centroids position and finally to the setup of HADDOCK-EM run using its web portal version.

2 Overview

This section describes the different steps and their background in order to perform a protein–protein docking run in HADDOCK using an EM density map as restraint.

HADDOCK makes use of a variety of restraints (often expressed in terms of ambiguous or unambiguous distance restraints) throughout the entire docking process to drive and score the complex formation. These restraints can be derived from various experimental information sources such as NMR chemical shift perturbations, hydrogen/deuterium exchange, chemical cross-linking detected by mass spectrometry, mutagenesis, etc. [8–11].

When using cryo-EM data, however, HADDOCK needs to first convert the information provided by the EM map into distance restraints in order to drive the molecules to their potential location. This can be done by extracting centroids from the EM map as described in [12]. The centroids are provided as 3D coordinates to HADDOCK and are automatically converted to unambiguous (or ambiguous in cases where circular symmetry is present or the identity between subunits is uncertain) distance restraints between the centroids and the center of mass of the subunits, as illustrated in Fig. 1. These restraints draw, during the initial rigid-body step of HADDOCK, the molecules toward their location within the EM map. Once the rigid complex is formed and oriented correctly in the density, the cryo-EM density-based restraint energy term in HADDOCK is applied, and the refinement protocol proceeds through the various steps of HADDOCK. For details, *see* Subheading 2.3 and the original HADDOCK-EM publication [7].

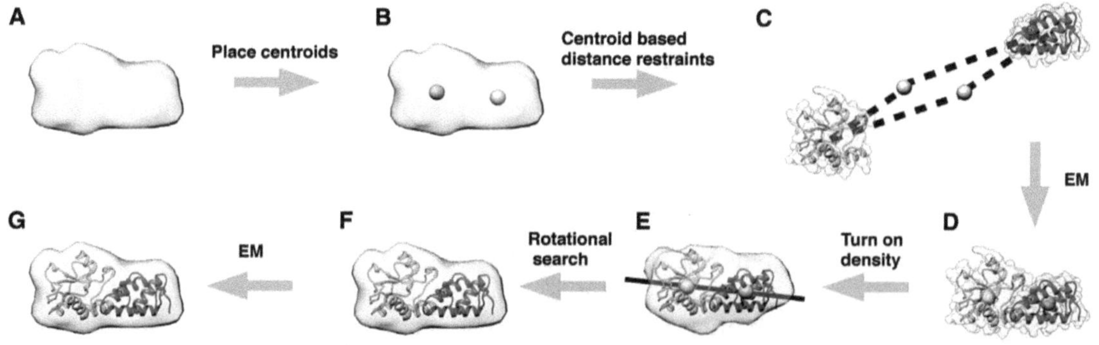

Fig. 1 Representation of the Rigid-Body Docking Protocol in HADDOCK-EM as illustrated in [7]. (**a**) Simulated cryo-EM data of colicin E7/IM7 complex (PDBid 7CEI). (**b**) Centers of mass of each subunit represented with gray spheres within the EM map. (**c**) Distance restraints in HADDOCK it0 step are defined between the COM of chain A (light gray) and B (dark gray) and their corresponding centroids. (**d**) Example of a complex obtained after the first rigid-body minimization (it0). (**e**) After the position, the relative orientation of each subunit should be determined. (**f**) A line drawn between the two centroids is used as axis to perform a rotational search. The complex with the highest cross-correlation value is chosen. (**g**) Excluding the centroid-based restraints, a final rigid-body minimization is performed against the cryo-EM data and assessed, thanks to a cross-correlation-based potential

2.1 High-Resolution Atomic Structure Rigid-Body Fitting into Cryo-EM Densities

The rigid-body fitting into the cryo-EM map will be performed using PowerFit [12], making use of our web server [13]. PowerFit fits atomic structures into density maps by performing a full-exhaustive six-dimensional cross-correlation search between the atomic structure and the density. It takes as input an atomic structure in PDB or mmCIF format and a cryo-EM density with its resolution, and outputs positions and rotations of the atomic structure corresponding to high correlation values and the top ten best scoring rigid poses. PowerFit uses the local cross-correlation function as its base score. The score is by default enhanced with an optional Laplace prefilter and a core-weighted version that minimizes the effect overlapping densities from neighboring subunits.

From the fitted structure, one can extract the 3D coordinates of the centroids (their center of mass position into the map), an information required by HADDOCK-EM.

2.2 Cryo-EM Density Map Cropping

In order to reduce data noise and save computational time, we strongly advise to crop the cryo-EM map to the region of interest. Cropping can be straightforwardly performed using UCSF Chimera [3]. A step-by-step protocol to extract a subregion of a density map is available at https://www.cgl.ucsf.edu/chimera/docs/UsersGuide/midas/mask.html. In this protocol, we will use fitting results from PowerFit to crop the map with respect to the predicted molecular subunits' location.

2.3 Protein–Protein HADDOCKing with EM Restraints

2.3.1 Docking Protocol

The docking protocol in HADDOCK consists of three successive steps:

- *it0*: Rigid-body energy minimization (RBEM),
- *it1*: Semiflexible simulated annealing (SA) in torsion angle space (TAD/SA),
- *water*: Final restrained molecular dynamics in explicit solvent.

Pre- and post-processing steps are performed: (1) to build missing atoms in the preliminary step and (2) to launch a variety of analyses and clustering of solutions in the final step. For further details, please refer to [14, 15].

The HADDOCK-EM protocol requires as input an EM density map and its resolution together with the centroid coordinates of each of the subunits to be docked. Some changes have been made to the default HADDOCK docking protocol to account for the cryo-EM data parameters, mainly in it0, where centroids, approximate location of the subunits' COMs in the density map obtained during the fitting step (*see* Subheading 2.2), are used to place the subunits. As for the center of mass docking protocol of HADDOCK [16], additional distance restraints are generated between the COMs of the subunits. The main difference here lies in the fact that distance restraints are not created between the subunits themselves but between each subunit and one or several (in case of ambiguity) centroid coordinates.

Other cryo-EM-related required parameters for HADDOCK are either directly extracted from the map or have optimized default values. Some of these can be controlled through the web portal interface, for expert tuning of results.

Rigid-Body Energy Minimization (RBEM, it0)

In the initial docking stage, the interacting partners are considered rigid and separated in space and placed on a sphere centered on the midpoint of the centroids. For each docking trial, each subunit is randomly rotated around its center of mass and translated within a 10 Å box of to ensure unbiased starting configurations. In the case of unambiguous centroid-based restraints, HADDOCK will fit the subunits' COMs on the centroids to which they are associated. In the case of ambiguous restraints, each subunit would be ambiguously linked to any of the centroid given as input. Then, selection of the best conformation will solely rely on the HADDOCK score.

The centroid-based distance restraint is described by a soft square potential between two pseudo-atoms, one of which corresponds to the centroid and the other to the COM of the subunit.

Optimization steps have been performed to derive the best values for (1) the force constant of the centroid-based distance restraints that drives the COMs to the centroids, (2) the weight for the cross-correlation energy term, and (3) the weight of the LCC term in the HADDOCK score for *it0*. The default values in our protocol stand, respectively, at 50, 15,000, and 100. Those three values can be changed in the submission interface of HADDOCK2.4.

Binary systems will undergo a supplementary optimization step that aims at optimizing their orientation within the EM map. For this, an exhaustive 4° rotation search along the axis joining the centroids is performed, and at each step, the cross-correlation value is calculated to assess the pose. The orientation with the maximal cross-correlation value is kept. Finally, a rigid-body minimization is performed against the map using a combination of the cross-correlation, van der Waals, and electrostatic energy terms. Models are then scored by the traditional HADDOCK score plus a LCC term that reports on the overall quality of the fitting within the EM map. Typically, 2000 models are generated and scored from which typically the 400 models with the best HADDOCK score (*see* Subheading "Scoring") will go to the semiflexible simulated annealing stage of HADDOCK.

Semiflexible Simulated Annealing in Torsion Angle Space (TAD/SA, *it1*)

After a first rigid-body simulated annealing stage, the semiflexible simulated annealing stage, which starts with a short rigid-body molecular dynamics phase, optimizes the side chain conformations at the interface and then both backbone and side chains. The flexible regions are automatically defined for each docking model as the residues within 5 Å from a partner molecule. The parameters for *it1* are the same as in a typical docking run with HADDOCK, with the exception of adding the cross-correlation energy term used both during the simulated annealing protocol and in the scoring.

Restrained Molecular Dynamics in Explicit Solvent (Water)

The structures obtained after simulated annealing are finally refined in an explicit solvent layer to further improve the scoring. This is done by a short molecular dynamics simulation in water, solvating the complex in an 8 Å shell of TIP3P water molecules [17].

Scoring

The EM protocol introduces a new term to the HADDOCK score, namely, the local cross-correlation value (LCC) computed for a given model which is added to the equation defining the score, with an optimal weight for the three stages:

$$HS_{EM-it0} = 0.01 * E_{vdw} + 1.0 * E_{elec} + 0.01 * E_{AIR} + 1.0 * E_{desolv} - 0.01 * BSA - 400 * LCC.$$

$$HS_{EM-it1} = 1.0 * E_{vdw} + 1.0 * E_{elec} + 0.1 * E_{AIR} + 1.0 * E_{desolv} - 0.01 * BSA - 10{,}000 * LCC$$

$$HS_{EM-itw} = 1.0 * E_{vdw} + 0.2 * E_{elec} + 0.1 * E_{AIR} + 1.0 * E_{desolv} - 10{,}000 * LCC$$

The other terms of the scoring function are the intermolecular van der Waals (E_{vdw}) and electrostatic (E_{elec}) energies calculated with the OPLS force field and an 8.5 Å nonbonded cutoff [18], an empirical desolvation potential (E_{desolv}) [19], the ambiguous interaction restraint energy (E_{AIR}), and the buried surface area (BSA).

2.3.2 Clustering of Final Solutions

All models generated by HADDOCK are clustered either based on their fraction of common contacts [20] (FCC, default) or on their interface-ligand-RMSD (i-l-RMSD) depending on the user's choice.

3 Methods

The HADDOCK-EM protocol requires some preliminary steps outside the traditional HADDOCK pipeline and independent from the web server. As explained in the previous sections, atomic structures will first be fit into an EM map region, then the EM map will be cropped, followed by a final fitting step.

To follow our protocol in its entirety, the 3D viewer program UCSF Chimera (https://www.cgl.ucsf.edu/chimera/) is needed. The protocol described here is based on version 1.12.0. Python2 or Python3 should also be installed (we recommend the latest stable versions Python 2.7.15 or Python 3.6). All other steps will simply make use of a standard web browser with JavaScript enabled. A registration to the CSB portal is required to use both PowerFit and HADDOCK (*see* **Note 1**). Complementary to the HADDOCK registration, users must request GURU access via their profile page to get access to the EM restraint parameters.

In the following sections, we illustrate our protocol on a test case taken from the use cases illustrated in [7]. The complex studied describes the interaction between two proteins of the 30S subunit of the ribosome (chains F and R). An atomic model of the entire complex is available (PDBid: 2YKR) as well as the 9.8 Å resolution cryo-EM map from which it was derived (EMDBid: 1884). The necessary files are provided in a tar archive available in the Supplementary Material. This protocol describes a two-body docking example. The same recipe can be extended to more components by repeating the PowerFit steps as many times as there are components and providing all the independent structures and centroid positions to the HADDOCK submission portal. The protocol should be able to run on any operating system since it mainly relies on a web browser, Chimera, and some Python scripts.

3.1 Preprocessing of the Cryo-EM Map

In this section, we will crop the cryo-EM map to only keep the part that is relevant for our docking. This step is optional and significantly depends on (1) the size of the map and (2) the preliminary information we have about the structure localization within this map. In our example, we already know the location of the subunits we want to dock in the EM map. Without this information, a very first step would have been to perform a fitting of the subunits within the EM map we plan to use to identify their possible location (as described in Subheading 3.2, for instance). Such fitting should always start from the largest components since these are easier to identify in the EM map.

To crop the map, we will use UCSF Chimera. Chimera has a very complete support for density maps and allows to quickly observe, analyze, and manipulate such maps via a customised user interface. We will follow the instructions given in the Chimera documentation [21] with little modifications accounting for most recent versions of Chimera. For an online version of the documentation, please refer to https://www.cgl.ucsf.edu/chimera/docs/UsersGuide/midas/mask.html.

1. Open Chimera.
2. Load the cryo-EM map (`emd_1884.map`).
3. Load the crystal structure of the 30S subunit bound to RsgA (`2ykr_FR.pdb`).

   Warning: At this stage, be careful to not move the complex independently from the EM map during the session. This could lead to erroneous results during the next steps.
4. If the *Model Panel* window is not displayed, go to *Tools* > - *General Controls* > *Model Panel.*
5. Select "2ykr_FR.pdb."
6. Click on *Action* > *Surface* > *Show* to generate the surface of the protein.
7. A new line should appear in the *Model Panel* window with the name "MSMS main surface of 2ykr_FR.pdb."
8. Click on *Tools* > *General Controls* > *Command Line.*

9. In the new dialog window that opened at the bottom of the viewer main window type:

   mask #0 #1

   where #0 represents the identifier of your EM map and #1 the identifier of your protein–protein surface.

10. A new volume representation should appear (*see* Fig. 2) together with a new line in the *Model Panel* window named "emd_1884.map masked."

11. Save the new masked map, in the *Volume Viewer* window: *File* > *Save map as*... (`1884_masked.map`).

3.2 Getting Centroid Coordinates by Fitting the Atomic Structure into the New Cryo-EM Map

In this section, we will use the PowerFit web server to obtain the centroid coordinates of the two subunits of the complex. PowerFit performs an exhaustive search to identify the best fit of our crystal structure within the new masked cryo-EM map obtained in Subheading 3.1. The best solutions are ranked according to a cross-correlation score (similar to the one used in the HADDOCK protocol).

Note that access to the PowerFit web server requires registration (https://wenmr.science.uu.nl/auth/register/—select **PowerFit** as registered service).

1. Go to http://milou.science.uu.nl/services/POWERFIT.
2. Add the cryo-EM map file (`1884_masked.map`) to the "Cryo-EM map" field.
3. Add the atomic structure of the complex (`2ykr_F.pdb`) to the "Atomic structure" field.
4. Put `9.8` as "Map resolution" (in Angstroms).

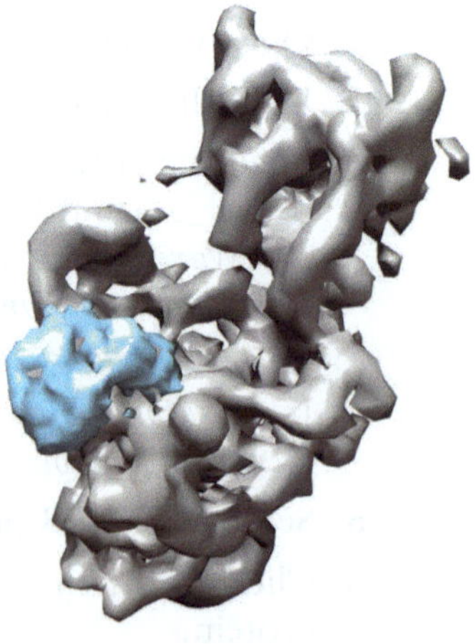

Fig. 2 Chimera snapshot illustrating the EM map of 30S ribosomal subunit with RsgA bound in the presence of GMPPNP, EMDBid 1884 [24], in white and, in blue, subpart of the EMDBid 1884 EM map masked by the subunits F and R of atomic structure 2ykr

5. By default the server will redirect the computation to GPGPU grid resources provided by the federated sites of EGI. To run locally on our server, you might choose to uncheck "Redirect submission to grid (GPU) resources."
6. Enter your credentials (email + password) and click on "Submit."
7. The run should take about 5 min (*see* **Note 2**). The status of your job will be updated every 30 s. Once the job is finished, you will get an email, and, if you have left the page open, you will be redirected to the results page, similar to the one shown in Fig. 3.
8. On this page, reach the "Solutions" section. The table presented here reports the 15 best nonredundant solutions ranked by correlation score. We will focus on the best solution.
9. Click on the first link of the page corresponding to "Archive of the complete run." This will download the output of PowerFit under an archive file.
10. Untar the archive.
11. Redo **steps 1–10** by only changing the PDB file provided in **step 3**. But give this time the other protein, `2ykr_R.pdb`, as atomic structure input.
12. *Optional*: For each run (and then each set of output files extracted from the archives), open files `lcc.mrc` and `fit_1.pdb` with Chimera. Check that the atomic structure is well fitted within the density map file.
13. Using a terminal (or the windows command-prompt), run the python script `em_tools/centroid-from-structure.py` providing the best fit chains PDB files (`fit_1.pdb`) in each run archive as unique arguments.

```
> python centroid-from-structure.py fit_1.pdb
Parsed file: fit_1.pdb
Corresponding centroid (x, y, z):
11.90 -2.48 75.54
> python centroid-from-structure.py fit_1.pdb
Parsed file: fit_1.pdb
Corresponding centroid (x, y, z):
17.40 5.80 58.00
```

14. Save the centroid coordinates for later.

3.3 Preparation of Input Files

Each PDB provided to HADDOCK has to respect the PDB format with proper syntax and clear chain identifiers (*see* **Note 3**). The two input chains for the docking run are the chains F and R of 2YKR and are, respectively, provided in files `2ykr_F.pdb` and `2ykr_R.pdb`.

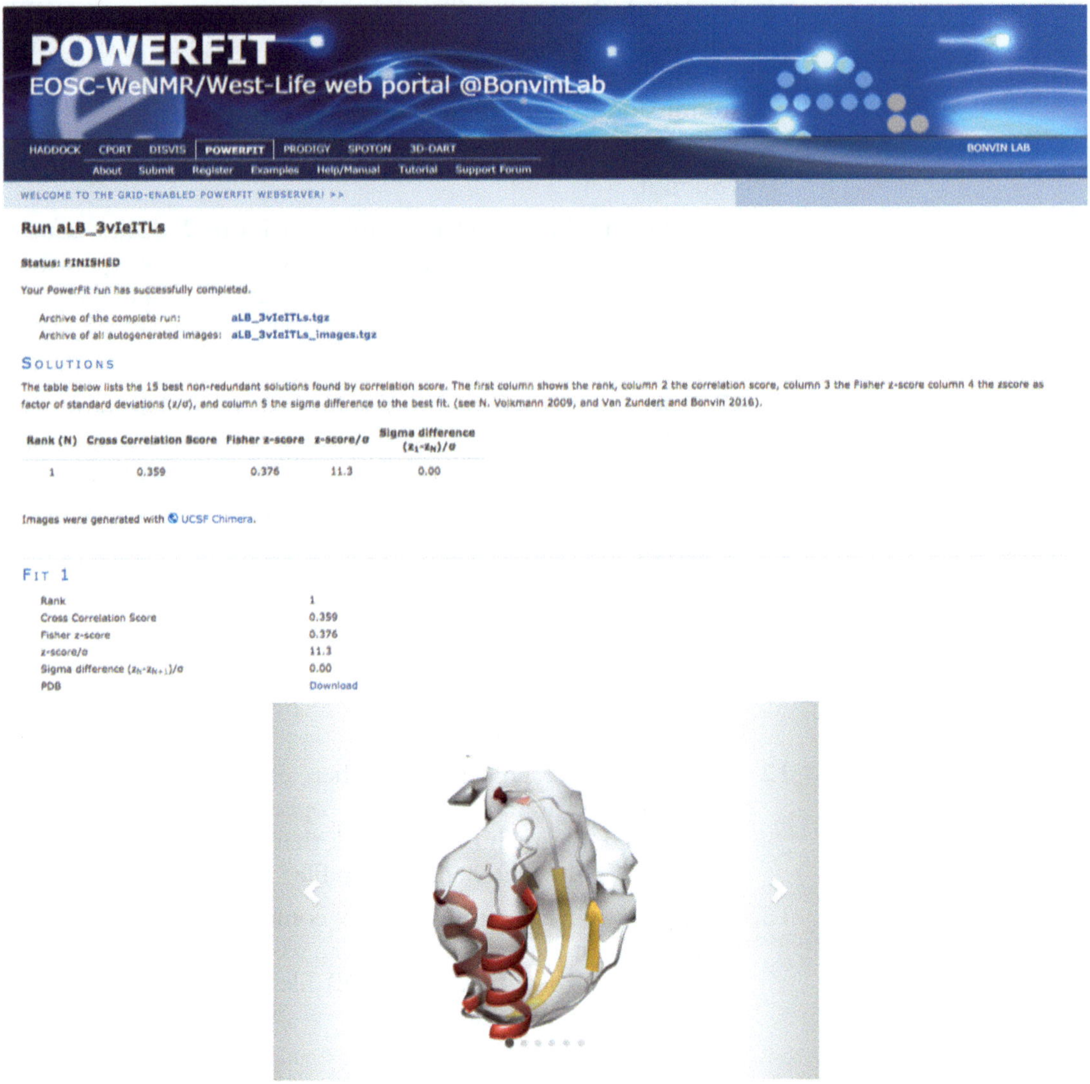

Fig. 3 Screen capture of PowerFit results page after fitting of chain F of 2ykr in the masked map obtained from EMDBid 1884

The PDB file of the protein must be checked to avoid any double occupancies or residue insertion codes. If present, these can be removed by manual editing of the file or automatically by using the `pdb_delocc.py` script provided as part of the PDB-tools repository maintained by the HADDOCK team (https://github.com/haddocking/pdb-tools).

The EM map obtained after the previous step of cropping can be submitted as it is. The HADDOCK2.4 new web server processes and converts automatically any map under MRC or CCP4 format to XPLOR format, the latter being the only one read by CNS

(Crystallography and NMR System) [22], the computational engine used by HADDOCK.

3.4 Docking Two Subunits of the 30S Ribosome with the HADDOCK2.4 Web Server

For this docking, we will make use of the new HADDOCK web server available in its beta version at (https://wenmr.science.uu.nl/haddock2.4/). Registration is required to make use of the new interface and can be accessed through the corresponding submenu in the portal. Following the activation by the HADDOCK support team, users must request GURU access to be able to use EM restraints. This can be done in their own user profile page.

1. Open an Internet browser and go to https://wenmr.science.uu.nl/haddock2.4/. Click on the Submit subsection. You will find the page illustrated in Fig. 4.
2. We advise to give a name to your docking run. Be aware that no space or special characters other than "-" or "_" are allowed. We propose here to name the run "2ykr_em_modelling."
3. Define the number of molecules to dock (in this case, the default value of **2**).
4. There is no precise order for the molecules, and either of the PDB files can be provided first, but we do advise as a general rule to provide the largest component as first molecule (*see* **Note 4**). By default, we will use chain F as first molecule. In the section "First molecule," at the entry *"Where is the structure provided?"* Leave option `I am submitting it`. Leave "Which chain of the structure must be used?" to `All` (*see* **Note 3**). Next to "PDB structure to submit," press the `Choose file` button and move to the location where the tutorial data were unpacked. Go to the `pdbs/`directory and select the *2ykr_F.pdb* file. Keep both Nter and Cter to `False`.
5. In the section "Second molecule," at the entry "Where is the structure provided?" Leave option `I am submitting it`. Leave "Which chain of the structure must be used?" to `All` (*see* **Note 3**). Next to "PDB structure to submit," press the "Choose file" button and move to the location where the tutorial data were unpacked. Go to the `pdbs/`directory and select the `2ykr_R.pdb` file. Keep both Nter and Cter to False.
6. Click "Next" and wait for the second step interface to load (should not take more than a few seconds).
7. Leave the Molecule 1 and 2 parameters empty. Go to section "EM restraints (optional)" and unfold it as illustrated in Fig. 5.
8. Check `Use density/XREF restraints?` (set to True)
9. Next to "EM map," press the "Choose file" button and move to the location where the tutorial data were unpacked. Go to the `em_maps/` directory and select the `1884_masked.map` file (or select the one you generated at Subheading 3.2).

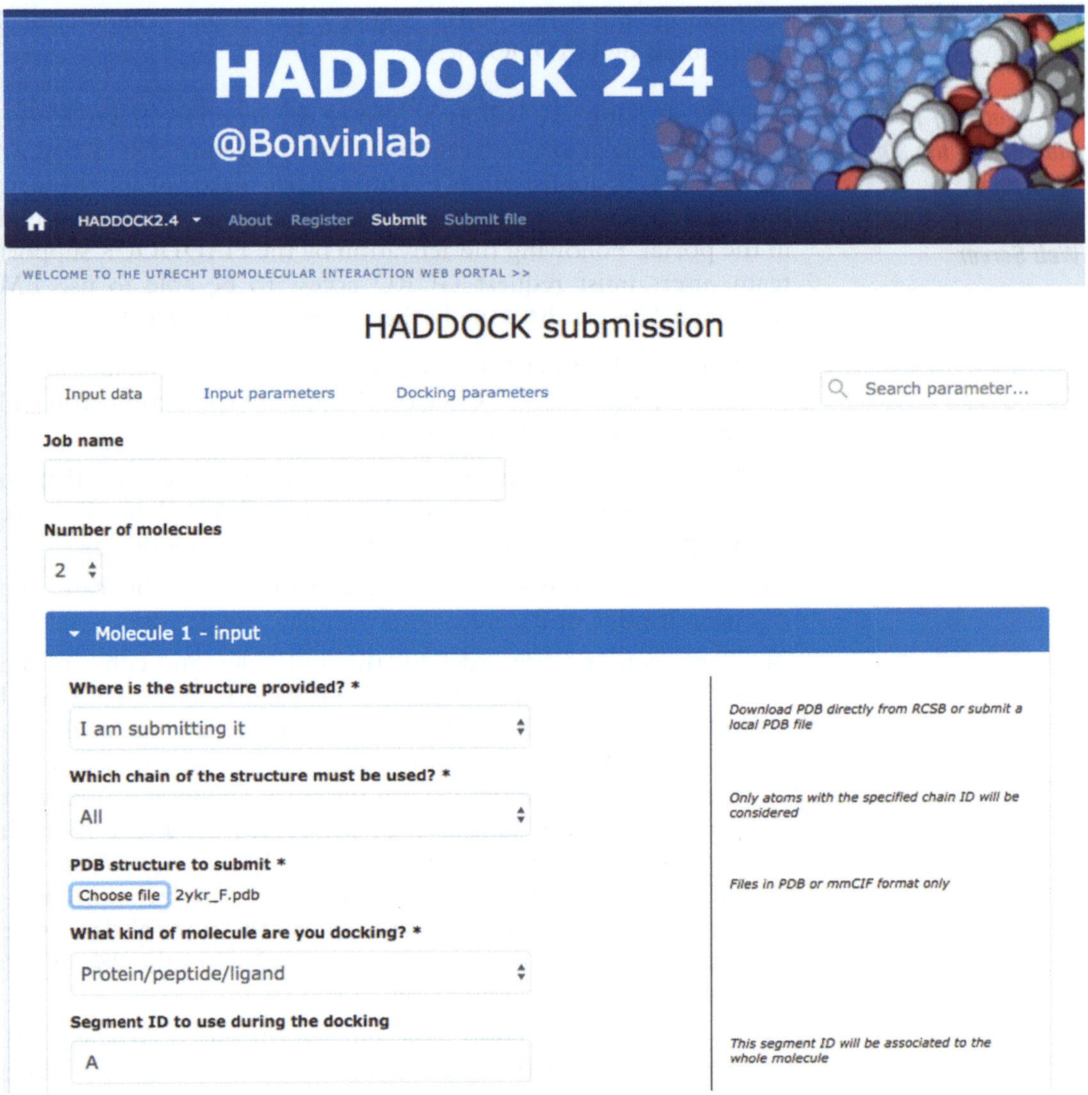

Fig. 4 Illustration of HADDOCK 2.4 submission page at the *Input data* first step

10. Set **9.8** in “Resolution of data in angstrom” field.
11. If this is not the case, check `Use centroid restraints?` (set to True).
12. In “MOLECULE 1 > Centroid position in absolute coordinates”, enter the coordinates you saved from Subheading 3.3 for chain A.
13. In “MOLECULE 2 > Centroid position in absolute coordinates”, enter the coordinates you saved from Subheading 3.3 for chain B.
14. Click `Next` and wait for the third step interface to load (should not take more than a few seconds).

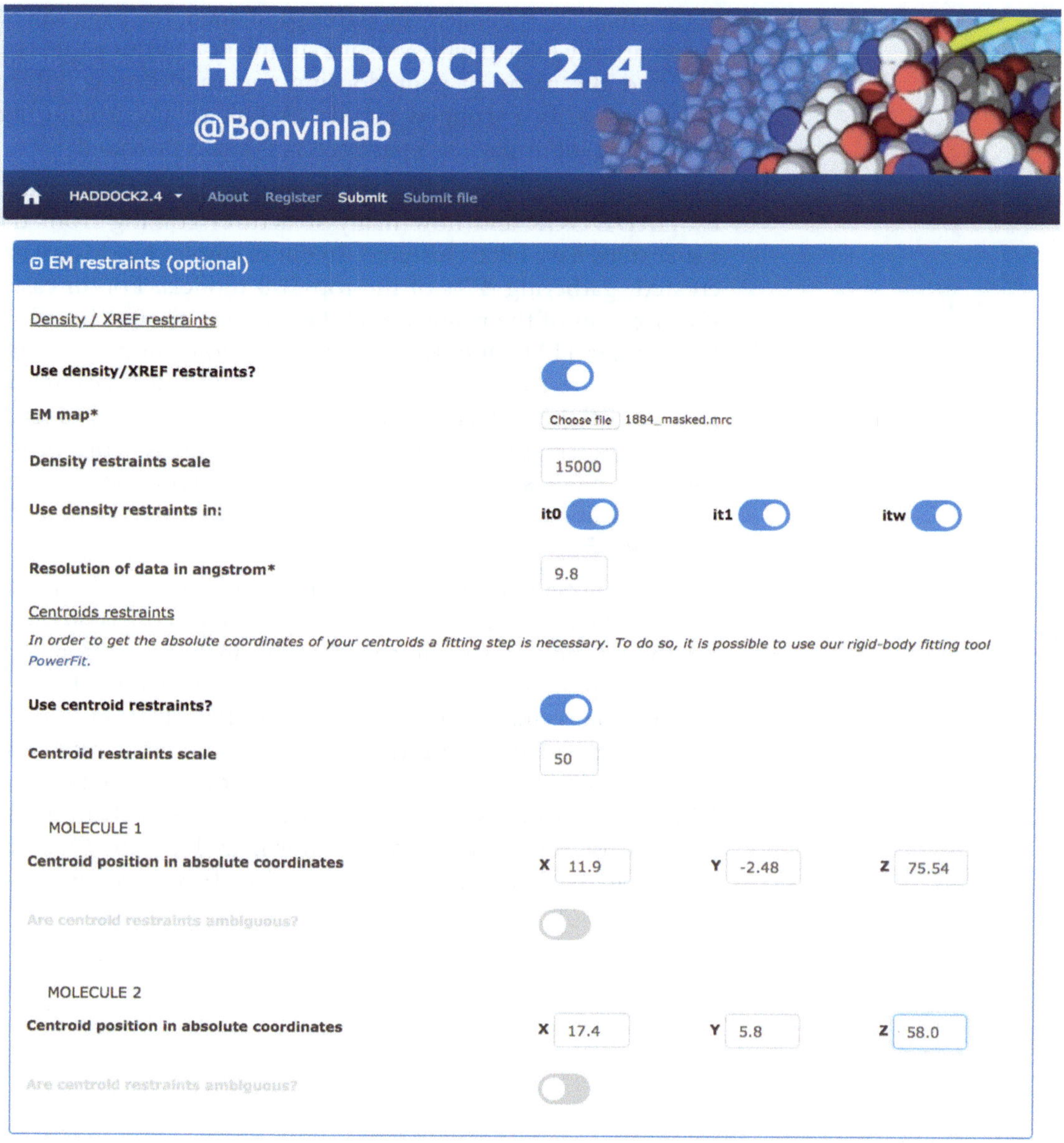

Fig. 5 Illustration of HADDOCK 2.4 submission page at the *Input parameters* second step

15. Leave default parameters and click `Submit` at the bottom of the page.
16. After a few seconds, you will be redirected to a page reporting the status of your job, a short summary of the docking input, and a progression report. This page will be updated every 30 s to report the progression of your job.
17. Within typically a few hours, depending on the web server load, you will receive another email reporting the final status of your job. If successful, a result page will be available at the link given

in the email, or, if you left the status page open, the page will be automatically loaded with a results summary. On this page, you will find the name of your docking run as well as a link to download it as a gzipped tar file. A link to the unique file containing input data and parameters is again provided.

18. The results page also indicates the number of clusters created by HADDOCK and how many structures coming from the *water* steps have been clustered. In our example, **12** clusters are created, gathering **47%** of the top 200 models. For an easier visualization of the results, only the ten best clusters based on the average HADDOCK score of its top four models are displayed in the summary page. You can find information and analyses of the last cluster in the gzipped tar file. For each cluster, information relative to the HADDOCK score of the top four models, the cluster size, and different statistics and energy values are reported as we can see for cluster 1 in Fig. 6a (*see* **Note 5**).

19. At last, an interactive representation of different CAPRI assessment criteria with respect to the HADDOCK score is provided for the ten best clusters in the "Results analysis" section. The first three plots show the HADDOCK score versus the fraction of common contacts (FCC—*see* **Note 6**), the i-RMSD, and the l-RMSD calculated using the top-ranked model as reference, respectively (*see* **Note 7**). The last three plots show the van der Waals, electrostatics, and AIR energy versus i-RMSD. An example of one of the plots is shown in Fig. 6b. One can note that the Eair values are all equal to 0 because no other

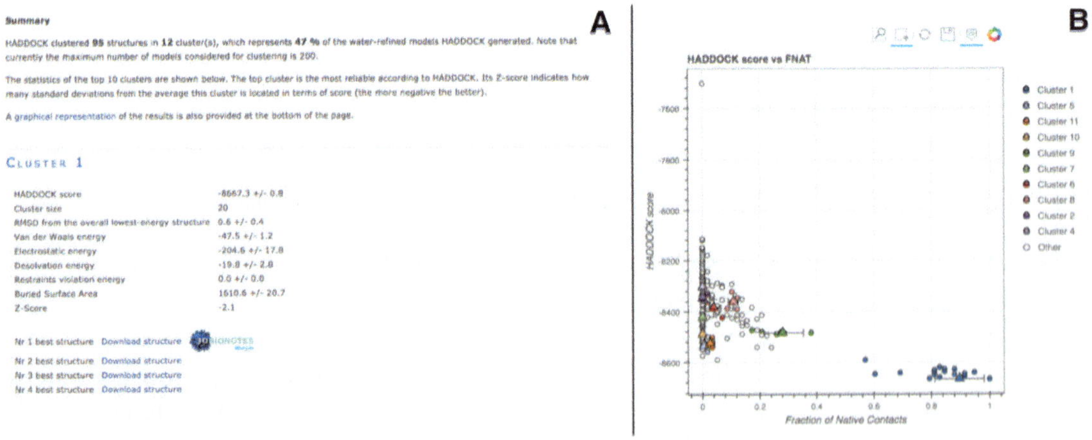

Fig. 6 Illustration of HADDOCK 2.4 results page after docking subunits F and R from 2ykr using as sole restraints the EM map information. (**a**) Extract of the cluster analysis for cluster 1. (**b**) Snapshot of one of the interactive plots provided in the Model Analysis section. In this plot, the HADDOCK score is plotted against the Fraction of Native Contacts

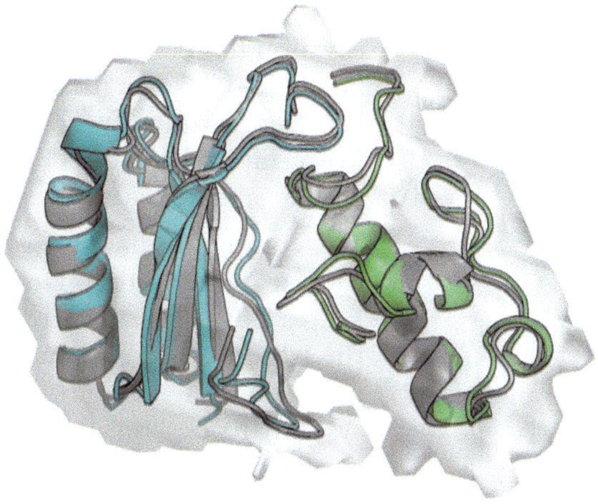

Fig. 7 Comparison of the best scoring models generated by HADDOCK, in blue (chain F) and green (chain R), and the reference structure (PDBid 2ykr) in dark grey. The EM map used to fit the two subunits and drive the docking run is shown as a transparent surface

restraints than the EM map-derived ones have been used to drive this docking.

20. It is possible to manually compare a reference structure with the best models of each cluster generated by HADDOCK. The 3D structures of these models can be directly downloaded from the results page. They are also located in the root of the docking run you downloaded as a gzipped tar file. Their name follows the following syntax: `cluster2_1.pdb`. This file is, for instance, the best model according to its HADDOCK score in the second cluster given by HADDOCK. The clusters are reported on the result page in the order of their HADDOCK score (from best to worst) (*see* **Note 8**).

    You can use fitting software such as ProFit [23] to get precise values of RMSD. PyMol is also useful since it has its own fitting algorithm and will give you a RMSD value as well as a visual feedback of the differences between the clustered models and the reference structure. Keep in mind that your reference structure has to be formatted in the same way that the PDB models generated by HADDOCK. ProFit considers only structures with an identical number of atoms. A superposition between the best HADDOCK model and the reference structure is reported in Fig. 7.

4 Notes

1. Registration to the CSB portal is mandatory to make use of both PowerFit and HADDOCK and can be done following

this link: https://wenmr.science.uu.nl/register. Once the registration has been done, check your mailbox for a confirmation link and click on this link (or copy/paste it in your web browser) to give us the possibility to activate your account. PowerFit and HADDOCK will only work for users logged in with a validated account.

2. The computational time of PowerFit scales almost linearly with the size of the system. However, using GPU resources allows to keep a PowerFit run with default parameters under 30 min for the largest systems. A hard limit of 200 MB for the size of the files that can be uploaded on the server prevents too large systems to be considered without previous trimming and/or cleaning.
3. Defining the largest molecule as first molecule for docking can be important for the final clustering because, in case of RMSD clustering, the structures are first fitted on the interface residues of the first molecule and then the RMSD is calculated on the interface residues of the second molecule. The interface residues are defined from an analysis of contacts in the generated models (at it1 and water, respectively). Defining the largest molecule first should thus result in a better fitting and clustering. However, one should note that the default clustering method is FCC and the order of the molecules does not impact the FCC calculation algorithm.
4. The PDB files provided to HADDOCK have to be correctly formatted to avoid any issues during the simulation process. There should be no overlap in residue numbering between different chains of a PDB. One can check the proper format of its PDB file using the pdb_format.py script provided as part of the PDB-tools repository maintained by the HADDOCK team (https://github.com/haddocking/pdb-tools). Missing atoms in the PDB files are not problematic since HADDOCK will rebuild them automatically.
5. The *Z*-score indicates how many standard deviations from the average a cluster is located in terms of its HADDOCK score. So the more negative, the better.
6. The FCC stands for fraction of common contacts and is calculated by comparing the lists of contacts at the interface between the components of a complex for two different structures. A contact is defined when two residues from different chains of the complex are closer than 5 Å from each other. The FCC is calculated as the fraction of common residue pairs shared between the two structures.
7. All reported RMSDs are calculated with respect to the lowest scoring model (the best model according to the HADDOCK score). The i-l-RMSD, which is used for clustering, is calculated on the interface backbone atoms of all chains except the

first one after fitting on the backbone atom of the interface of the first molecule. The i-RMSD is calculated by fitting on the backbone atoms of all the residues involved in intermolecular contacts within a cutoff of 10 Å. The l-RMSD is obtained by first fitting on the backbone atoms of the first molecule and then calculating the RMSD on the backbone atoms of the remaining chains.

8. The naming of clusters in HADDOCK is linked to their size and not their score. This originates from the clustering software. By definition, the largest cluster is always called cluster1, followed by cluster2 and so on. The cluster size however does not correlate per se with the HADDOCK score. Refer to the result page (or open in a web browser the `index.html` file provided in the tar archive) to see the cluster order based on the HADDOCK score.

Acknowledgments

This work is supported by the European H2020 e-Infrastructure grants (West-Life grant no. 675858 and BioExcel grants no. 675728 and 823830).

References

1. Bai X, McMullan G, Scheres SH (2015) How cryo-EM is revolutionizing structural biology. Trends Biochem Sci 40:49–57
2. Kimanius D, Forsberg BO, Scheres SH et al (2016) Accelerated cryo-EM structure determination with parallelisation using GPUs in RELION-2. Elife 5. https://doi.org/10.7554/eLife.18722
3. Pettersen EF, Goddard TD, Huang CC et al (2004) UCSF Chimera – a visualization system for exploratory research and analysis. J Comput Chem 25:1605–1612
4. Baker TS, Johnson JE (1996) Low resolution meets high: towards a resolution continuum from cells to atoms. Curr Opin Struct Biol 6:585–594
5. Esquivel-Rodríguez J, Kihara D (2013) Computational methods for constructing protein structure models from 3D electron microscopy maps. J Struct Biol 184:93–102
6. McGreevy R, Teo I, Singharoy A et al (2016) Advances in the molecular dynamics flexible fitting method for cryo-EM modeling. Methods 100:50–60
7. van Zundert GCP, Melquiond ASJ, Bonvin AMJJ (2015) Integrative modeling of biomolecular complexes: HADDOCKing with cryo-electron microscopy data. Structure 23:949–960
8. van Dijk AD, Boelens R, Bonvin AMJJ (2005) Data-driven docking for the study of biomolecular complexes. FEBS J 272:293–312
9. Melquiond ASJ, Bonvin AMJJ (2010) Data-driven docking: using external information to spark the biomolecular rendez-vous. In: Zacharias M (ed) Protein-protein complexes. Imperial College Press, London, pp 182–208
10. Karaca E, Bonvin AMJJ (2013) Advances in integrative modeling of biomolecular complexes. Methods 59:372–381
11. Rodrigues JPGLM, Bonvin AMJJ (2014) Integrative computational modelling of protein interactions. FEBS J 281:1988–2003
12. van Zundert GCP, Bonvin AMJJ (2015) Fast and sensitive rigid-body fitting into cryo-EM density maps with PowerFit. AIMS Biophys 2:73–87
13. van Zundert GCP, Trellet M, Schaarschmidt J et al (2017) The DisVis and PowerFit web servers: explorative and integrative modeling of biomolecular complexes. J Mol Biol 429:399–407

14. Dominguez C, Boelens R, Bonvin AM (2003) HADDOCK: a protein-protein docking approach based on biochemical or biophysical information. J Am Chem Soc 125:1731–1737
15. de Vries SJ, van Dijk AD, Krzeminski M et al (2007) HADDOCK versus HADDOCK: new features and performance of HADDOCK2.0 on the CAPRI targets. Proteins 69:726–733
16. de Vries SJ, Melquiond ASJ, Kastritis PL et al (2010) Strengths and weaknesses of data-driven docking in critical assessment of prediction of interactions. Proteins 78:3242–3249
17. Jorgensen WL, Chandrasekhar J, Madura JD et al (1983) Comparison of simple potential functions for simulating liquid water. J Chem Phys 79:926–935
18. Jorgensen WL, Tirado-Rives J (1988) The OPLS [optimized potentials for liquid simulations] potential functions for proteins, energy minimizations for crystals of cyclic peptides and crambin. J Am Chem Soc 110:1657–1666
19. Fernández-Recio J, Totrov M, Abagyan R (2004) Identification of protein–protein interaction sites from docking energy landscapes. J Mol Biol 335:843–865
20. Rodrigues JPGLM, Trellet M, Schmitz C et al (2012) Clustering biomolecular complexes by residue contacts similarity. Proteins 80:1810–1817
21. Goddard TD, Huang CC, Ferrin TE (2005) Chimera documentation – subregions. https://www.cgl.ucsf.edu/chimera/docs/UsersGuide/midas/mask.html
22. Brünger AT, Adams PD, Clore GM et al (1998) Crystallography & NMR system: a new software suite for macromolecular structure determination. Acta Crystallogr D Biol Crystallogr 54:905–921
23. Martin ACR, Porter C (2010) ProFit. http://www.bioinf.org.uk/software/profit/
24. Guo Q, Yuan Y, Xu Y et al (2011) Structural basis for the function of a small GTPase RsgA on the 30S ribosomal subunit maturation revealed by cryoelectron microscopy. Proc Natl Acad Sci U S A 108:13100–13105

Chapter 12

Modeling of Multimolecular Complexes

Dina Schneidman-Duhovny and Haim J. Wolfson

Abstract

Macromolecular complexes play a key role in cellular function. Predicting the structure and dynamics of these complexes is one of the key challenges in structural biology. Docking applications have traditionally been used to predict pairwise interactions between proteins. However, few methods exist for modeling multi-protein assemblies. Here we present two methods, CombDock and DockStar, that can predict multi-protein assemblies starting from subunit structural models. CombDock can assemble subunits without any assumptions about the pairwise interactions between subunits, while DockStar relies on the interaction graph or, alternatively, a homology model or a cryo-electron microscopy (EM) density map of the entire complex. We demonstrate the two methods using RNA polymerase II with 12 subunits and TRiC/CCT chaperonin with 16 subunits.

Key words Macromolecular assembly, Subunit assembly, Protein complexes, Cross-linking by mass spectrometry, Protein-protein docking, Integer linear programming

1 Introduction

The majority of proteins function when associated in multimolecular assemblies [1]. It is estimated that a protein interacts with nine other proteins on average [2, 3]. While there has been a significant progress in predicting complexes of pairwise protein interactions [4], prediction of the structures of multimolecular complexes remains a challenge [5, 6]. Methods for predicting the structure of symmetric oligomers have been developed previously [7–11]. The multimolecular version of HADDOCK (*see* previous chapter and [12–14]) is driven by experimental and bioinformatics data but limits the number of subunits to six, apparently due to computational complexity constraints. Multi-LZerD [15] builds the multimolecular assembly applying a stochastic search driven by a genetic algorithm. Kuzu et al. [16] construct the multimolecular complex iteratively, where in each iteration the subassembly is grown by one subunit. Experimental MS-based data is translated into spatial restraints and integrated into the scoring function using

Zoltán Gáspári (ed.), *Structural Bioinformatics: Methods and Protocols*, Methods in Molecular Biology, vol. 2112, https://doi.org/10.1007/978-1-0716-0270-6_12, © Springer Science+Business Media, LLC, part of Springer Nature 2020

the IMP platform [17]. The generation of candidate models is done by an exhaustive Monte Carlo search of the conformational space.

We have previously developed one of the first methods for the assembly task, **CombDock** [18], which formulates the multimolecular complex detection task as a search for an optimally scoring spanning tree of the assembly interaction graph. In this "assembly interaction graph," the vertices are the individual subunits, and the edges represent the docking solutions (rigid transformations) between the pairs of interacting subunits. CombDock applies a heuristic branch and bound technique to build an optimally scoring spanning tree, which represents the subunit interaction graph and the spatial pose of the individual subunits. Recently, we have developed **DockStar** [19], a global assembly method, which requires prior knowledge of the interaction graph and uses cross-linking data to deduce it. The optimization of the multimolecular assembly is formulated as an integer linear programming (ILP) task. While CombDock is general and can be applied without any prior interaction knowledge, DockStar is a significantly faster alternative for cases where the interaction graph or, alternatively, a homology model or EM density map of the entire complex is available.

2 Materials

2.1 Software

The following software packages are used in the protocols described below:

1. CombDock: a program for multimolecular assembly based on pairwise docking hypotheses.
2. DockStar: a program for multimolecular assembly based on pairwise docking hypotheses to an anchor subunit or placements within a homology model or EM density map scaffold.

   Example files and scripts for CombDock can be downloaded from http://bioinfo3d.cs.tau.ac.il/CombDock/CombDock-Download.zip.
   Example files for DockStar are available at the webserver http://bioinfo3d.cs.tau.ac.il/DockStar/help.

3 Methods

3.1 Macromolecular Assembly with CombDock

The input to CombDock consists of a set of protein structural models. The goal is to predict the native complex formed by the interactions between the proteins. The algorithm consists of three main steps [18]. In the first step, pairwise docking is applied on each pair of input structures to generate a set of docked configurations. In the second step, combinatorial optimization is used to combine different subsets of the configurations from pairwise docking to generate consistent clash-free complex models. In the third step, the generated complexes are scored and clustered in order to discard redundant models.

Optionally, CombDock supports distance constraints and restraints. Constraints require that all the output models satisfy the constrained distances, while restraints only require satisfaction of a fraction of distances. For example, constraints can be used to enforce sequence connectivity by specifying limits on distances between consecutive domains of the same protein. Restraints are useful for information coming from cross-linking mass spectrometry datasets or coevolutionary variation, where false positives are possible. Based on the dataset quality, the user can specify which fraction of the restraints has to be satisfied in the output models.

3.1.1 Inputs

The input to CombDock is three or more subunit structure files in the PDB format and transformations (three rotational and three translational parameters) between pairs of subunits generated by a suitable protein-protein docking software (Subheading 3.1.2). In principle, transformations between all pairs of input subunits should be given. However, it is possible to provide a transformation list only for pairs that are known to be in contact. For example, based on the cross-link dataset, CombDock can assemble the complex based on pairs with cross-links between them only, as long as there are enough cross-links to connect all the subunits.

Here we run CombDock on RNA polymerase II subunits with cross-linking data [20, 21]. RNA polymerase II is a eukaryotic complex that catalyzes DNA transcription to synthesize mRNA strands. Eukaryotic RNA polymerase II contains 12 subunits, Rpb1 to Rpb12. Rpb1 and Rpb2 are the largest subunits that cover over half of the complex with 1733 and 1224 residues, respectively. We have divided these two proteins into domain subunits, since in real-life scenario we are unlikely to have the structure of the whole protein chain for such long proteins. Rpb1 was divided into five subunits and Rpb2 was divided into four subunits (Fig. 1a). We also used Rpb3, Rpb10, and Rpb11 as additional subunits (Fig. 1a). Together, these 12 subunits contain 83% of the RNA polymerase II core. Subunits Rpb4–9 and Rpb12 were excluded due to the lack of cross-links that can connect them to the core.

To run CombDock, we prepare a file with a list of PDB files of the subunits (SU.list):

```
1wcmA_1_342.pdb
1wcmA_347_661.pdb
1wcmA_665_873.pdb
1wcmA_879_1057.pdb
```

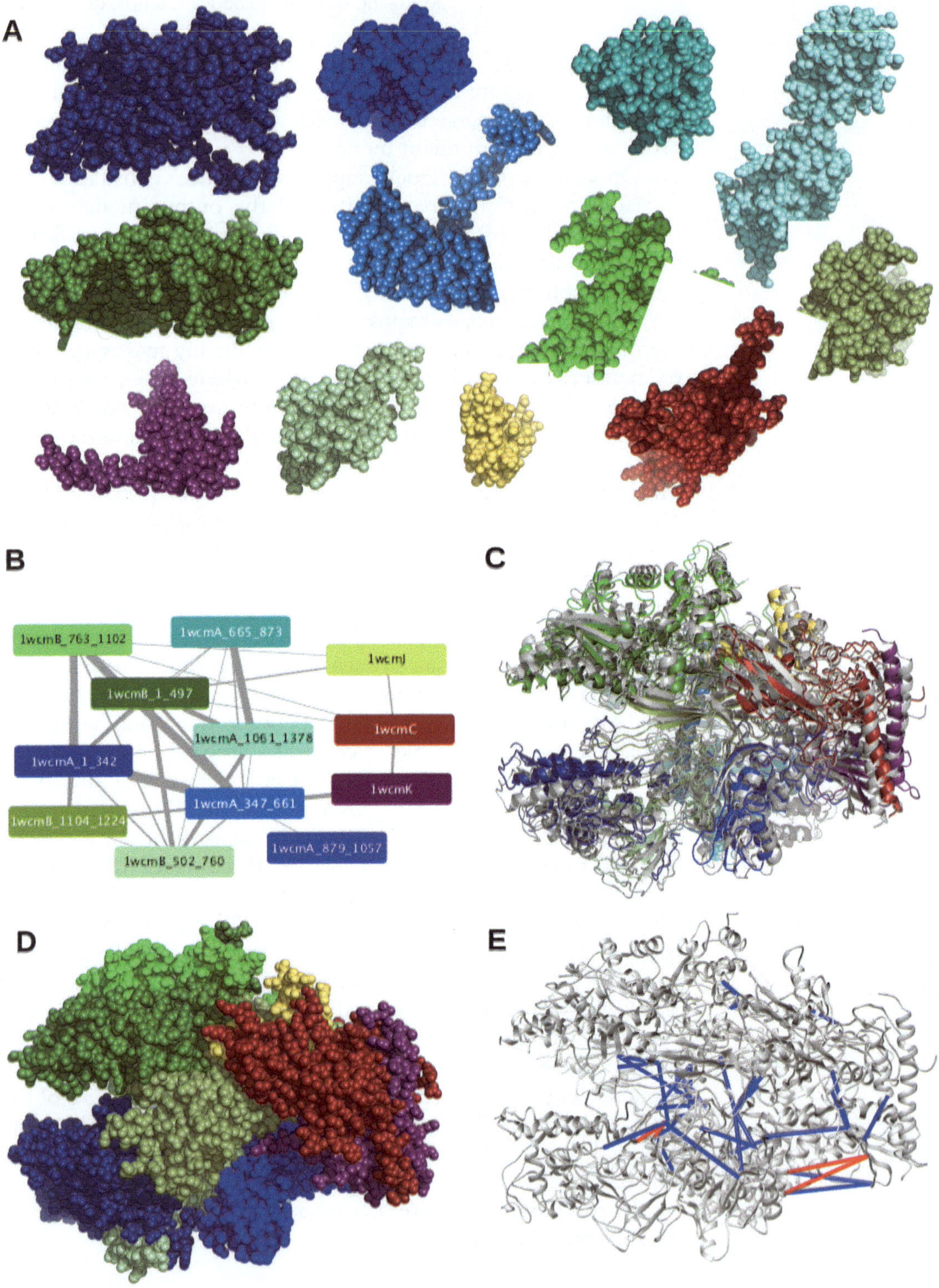

Fig. 1 Assembly of RNA polymerase II. (**a**) Input subunits: Rbp1 and Rpdb2 domains are in blue and green colors, respectively; Rpb3, Rpb10, and Rpb11 are in red, yellow, and purple, respectively. (**b**) Interaction network is based on cross-linking data; the width of the edge is based on the number of cross-links. (**c**) Assembly with best RMSD (4.3 Å, same subunit colors) vs. crystal structure (gray). (**d**) Assembly with best RMSD in the spacefill view. (**e**) Cross-links mapped on the assembly with best RMSD. The cross-links with Cα–Cα distance below 25 Å are depicted in blue; red indicates longer distances

```
1wcmA_879_1057.pdb
1wcmA_1061_1378.pdb
1wcmB_1_497.pdb
1wcmB_502_760.pdb
1wcmB_763_1102.pdb
1wcmB_1104_1224.pdb
1wcmC.pdb
1wcmJ.pdb
1wcmK.pdb
```

We also prepare cross-linking file (dist_restraints) in the following format:

```
ResId1 Chain1 ResId2 Chain2 min_dist max_dist
1057 B 199 C 0 30
1092 A 830 A 0 30
```

where ResId and Chain correspond to cross-linked residue number and chain identifiers. In addition, the user needs to specify minimal and maximal distance thresholds for the cross-link. Here we use 0 and 30 Å for standard cross-linker length (BS3 and DSS). In total, there were ~65 inter-subunit cross-links for the 12 input subunits in the two cross-linking datasets (Fig. 1b).

3.1.2 All Pairs Docking with PatchDock

In principle, input transformations can be generated using any protein-protein docking algorithm. Here we used PatchDock [22] to generate the pairs of docked configurations (http://bioinfo3d.cs.tau.ac.il/PatchDock/). PatchDock is an efficient rigid docking method that maximizes geometric shape complementarity. Protein flexibility is accounted for by a geometric shape complementarity scoring function, which allows only a small amount of steric clashes at the interface. A typical docking run will take several minutes on a single CPU, enabling rapid generation of all pairwise candidate docking configurations for the input subunits. PatchDock can use the cross-links already in the pairwise docking stage, producing more relevant transformations for CombDock. To run PatchDock on a pair of proteins, we generate a parameter file and run it as follows:

```
> buildParamsXlinks.pl subunit1PDBfile subunit2PDBfile
> patch_dock.Linux params.txt docking.res
```

where params.txt is the generated parameter file by buildParamsXlinks.pl and docking.res is the output file. In addition, PatchDock produces best.res file, where it outputs transformations for models with highest ratio of satisfied restraints.

It is possible to run PatchDock in parallel for all the relevant pairs on multiple cores or on a cluster. Once pairwise docking is done, we prepare transformation files in the following format by selecting the relevant columns in the best.res file:

```
# score rmsd transformation
1 2141 57.03 0.97134 -0.65503 -3.07278 220.67010 143.81566 57.82494
2 2050 58.70 1.15360 -0.56251 -3.11197 239.22656 142.45685 38.92773
3 2031 62.68 -1.03373 0.01782 -1.84279 196.96078 195.85928 24.61943
```

where the columns are the transformation number, geometric shape complementarity score, rmsd relative to starting orientation (used for validation only), and transformation represented by three rotational and three translational parameters (*see* **Note 1**). The transformation files should be named according to the subunit PDB filenames. For example, transformations between 1wcmA_1_342.pdb and 1wcmC.pdb will be in the file best_1wcmA_1_342_plus_1wcmC.

3.1.3 Running CombDock

To run CombDock on the subunit file SU.list with cross-links in the dist_restraints file and transformation files in the results folder, simply type:

```
> CombDock.Linux SU.list results/best_ 1000 1000 dist_restraints -r 0.5
```

where SU.list is the file with subunit names, results/best_ is the prefix of the transformation files, the first 1000 is the maximal number of transformations to read for each pair from the transformation file, the second 1000 is the number of subcomplexes to store at each assembly stage, dist_restraints is the file with distance restraints, and 0.5 is the required minimal ratio of satisfied restraints. The transformation numbers significantly affect the runtime of the program. Therefore, it is recommended to run CombDock with 100 input transformations first, followed by increasing it to 1000 transformations to get better sampling. Assembly of the 12 subunits of RNA polymerase II with the parameters above took ~4 days on the node with 42 cores. CombDock will automatically define distance constraints to enforce sequence connectivity between consecutive domains that are given as separate subunits (*see* **Note 2**).

3.1.4 CombDock Output

The results of the assembly are given in the combdock.res file, where each line corresponds to one possible solution. The clustered solutions are given in the clustered.res file. The second column is the size of the cluster. We recommend sorting the results by cluster size, as larger clusters indicate there were multiple assembly combinations leading to similar solutions. To produce PDB files corresponding to the assembled model type:

```
> sort -nrk2 clusters.res > clusters_sorted.res
> prepareComplex.pl SU.list clusters_sorted.res 1 10
```

This will produce PDB files for the first ten models from the clusters_sorted.res file.

In our assembly results for RNA polymerase II, the model closest to the crystal structure had an RMSD of 4.3 Å (Fig. 1c, d). It had all but three cross-links satisfied (Fig. 1e). Other top scoring models were also close to crystal structure with RMSD in the range of 13–16 Å.

3.2 Macromolecular Assembly with DockStar

The input to DockStar is the set of protein structural models participating in the multimolecular complex. Each of the individual structures has a set of candidate poses (rotations and translations in the 3D space) associated with it. If available, the input also includes information on maximal and minimal distances, between specified amino acids on different subunits and the interaction graph of the entire assembly. The goal of the algorithm is to predict the full (or partial) structure of the entire complex by selecting at most a single pose for each structural unit, such that the resulting complex has optimal fitting score between the participating subunits. First, for each pair of poses belonging to interacting subunits, a binding score is calculated based on the number of satisfied cross-link restraints and knowledge-based binding interface potential. Then, a globally optimal solution for the entire complex is calculated by formulating the global binding score optimization as an ILP task. In such a solution, each subunit is either assigned a pose or does not appear in the resulting complex ("missing" subunit). The algorithm is tuned to produce a ranked set of K user-requested highest scoring solutions for the entire complex.

In the case when the input pose generation is based on the "star" method (*see* Subheading 3.2.2), the ILP stage solves only star-shaped subcomplexes of the entire complex. In such a case, the interaction graph of the whole complex is partitioned into such (overlapping) subcomplexes with star-shaped spanning trees, and the top scoring solutions for these subcomplexes are merged to detect global solutions of the entire complex.

For details of the algorithm, the reader is referred to [19].

3.2.1 Inputs

The DockStar server http://bioinfo3d.cs.tau.ac.il/DockStar/ requires as input the PDB files of the individual subunits participating in the assembly as well as a set of candidate poses (3D Euclidean transformations) associated with each subunit. In addition, a restraints file indicating maximal and minimal distances between pairs of residues of participating subunits can be uploaded. The restraints file is optional. The user should indicate the number of ranked global solutions that he/she is interested to receive and also check a box, if he/she is interested in, the so-called partial results.

The latter option takes into account that the submitted candidate poses (transformations) file might not include a near-native pose or, even worse, that the submitted subunit model is incorrect. Thus, the algorithm in its "partial results" option returns an optimally scoring solution that might not include all the submitted subunits of the assembly. For details, *see* [19].

3.2.2 Input Pose Generation

DockStar accepts candidate input poses of the individual subunits regardless of the methods these poses have been generated, as long as they adhere to the technical input requirements. Each subunit will typically have tens or even hundreds of candidate poses. Each pose is represented by a seven-dimensional vector, where the first entry is the number of the pose in the list, entries 2–4 represent the rotation in radians, and entries 5–7 represent the translation in Angstroms (*see* **Note 1**). Nevertheless, in order to be mathematically sound, the candidate input poses should be consistent with each other, namely, generated with regard to the same 3D Cartesian reference frame. In [19] two key experimentally sound scenarios have been suggested to ensure that the candidate poses have been generated in the same reference frame. The first scenario assumes that a homology model of the full protein complex is available and the candidate poses are generated by candidate structural alignments of the different subunits to this homolog complex. These alignments can be executed, e.g., by DALI (*see* Chapter 3 in this book) or MultiProt [23]. A similar scenario applies when a cryo-EM map of the full complex is available and the candidate poses are generated by fitting the various subunits into the cryo-EM map of the complex.

The second scenario assumes knowledge of the complex interaction graph. In this case, one subunit, preferably the one with the maximal number of neighbors in the interaction graph, is chosen as an *anchor* and its neighbors are docked to it, resulting in candidate poses for a star-shaped complex, with the anchor subunit at the center of the star. In [19] this pairwise docking of the anchor to its neighbors is done by first applying PatchDock [22] (http://bioinfo3d.cs.tau.ac.il/PatchDock/) and then rescoring and refining the 1000 top-ranked PatchDock poses by FiberDock [24] (http://bioinfo3d.cs.tau.ac.il/FiberDock/) to choose a predefined number of top poses for each subunit vis-à-vis the anchor. In this scenario, the anchor subunit should be represented by a single pose with the all-zero transformation, namely, 1 0 0 0 0 0 0.

In the case the interaction graph of the complex is not star-shaped, it is divided into overlapping star-shaped subcomplexes, each of which is solved separately. Then top solutions of subcomplexes that share a subunit are merged with the shared subunit as a new anchor. All the poses in the merged subcomplex are recalculated vis-à-vis the reference frame of the new shared anchor. The merge step is handled by http://bioinfo3d.cs.tau.ac.il/DockStar/merge_solutions.

3.2.3 Running DockStar

The DockStar server allows upload of the individual subunits with their associated poses/transformations files one by one, or together as a zipped file.

If available, the cross-link-induced experimental restraints are included in a special restraints file, which includes a line for each restraint specifying the subunit, chain, and identity of each of the cross-linked residues, as well the minimum and maximum distance between their C-alpha centers. Obviously, similar distance restraints obtained by other experimental methods can be incorporated in this file, as well. The technical details of the exact submission format are outlined on the server page as well as in its "help" section http://bioinfo3d.cs.tau.ac.il/DockStar/help.

3.2.4 DockStar Output

The output is a ranked list of the top-requested solutions. Each line of the output includes the rank of the solution, the score it has received, the number of fulfilled distance restraints, and the file of the assembled complex in PDB format. The score is a combined score, which takes into account the number of satisfied restraints as well as a statistical pairwise atomic potential (*see* [19]).

3.2.5 The TRiC/CCT Example

We shall demonstrate how DockStar has been applied for the task of detecting the assembly order of the subunits of the TRiC/CCT chaperonin [25], as this is an efficient and somewhat less intuitive exploitation of the homology model scenario. The eukaryotic TRic/CCT chaperonin is composed of two octameric rings, where there are eight different subunits, each appearing once in one of the rings. Although the individual subunits in each ring are different, they exhibit about 30% of sequence identity and are structurally almost indistinguishable in lower resolution cryo-EM maps. To resolve this issue [26, 27], collected cross-link data and applied time-consuming combinatorial methods to detect the correct order of the subunits. By using DockStar, the correct order of the subunits can be detected in about 10–15 min of CPU time by the following sequence of steps:

1. Model the individual subunits according to a template subunit from a known homologous complex [28] which has 30–40% sequence homology with the individual target subunits.
2. Treat the place of each subunit in the homolog complex as a "placeholder," and align each of the eight subunits from **step 1** in that place. Each such alignment provides a "pose/transformation" as a DockStar input. Do it for both rings.
3. Use DockStar with the distance restraints imposed by the interunit cross-links on both rings to resolve the correct permutations of the subunits in both rings (*see* the details in [19] and its supplementary material therein).

4 Conclusions

The two methods presented here address the challenging task of multi-subunit complex assembly. With the advance of cryo-EM and cross-linking mass spectrometry, we expect the methods will be highly relevant in structure modeling applications. The two applications presented here, RNA polymerase II and TRiC/CCT chaperonin, showcase the applicability of the methods in different scenarios. Both methods are available from http://bioinfo3d.cs.tau.ac.il/.

5 Notes

1. The input transformations to CombDock and DockStar are represented by six parameters: three rotation parameters around *x*, *y*, and *z* axes (in this order) in radians and three translation parameters in Angstroms.
2. For correct mapping of distance restraints and constraints onto your input subunits, it is important to provide input PDB files with consistent numbering of residues and a unique chain identifier for each protein. If protein domains are divided into more than one input subunit, they should have the same chain identifier. In this case, both PatchDock and CombDock will automatically add a distance constraint to enforce chain connectivity.

Acknowledgments

The work of D.S. is supported by the Israel Science Foundation (1466/18), Binational Science Foundation (2016070), and the Ministry of Science and Technology (80802). The work of H.J.W. was supported by the I-Core program of the Budgeting and Planning Committee and the Israel Science Foundation (Center No. 1775/12) and by Len Blavatnik and the Blavatnik Family Foundation.

References

1. Robinson CV, Sali A, Baumeister W (2007) The molecular sociology of the cell. Nature 450(7172):973–982. https://doi.org/10.1038/nature06523. nature06523 [pii]
2. Ideker T, Krogan NJ (2012) Differential network biology. Mol Syst Biol 8:565. https://doi.org/10.1038/msb.2011.99
3. Fraser JS, Gross JD, Krogan NJ (2013) From systems to structure: bridging networks and mechanism. Mol Cell 49(2):222–231. https://doi.org/10.1016/j.molcel.2013.01.003
4. Ritchie DW (2008) Recent progress and future directions in protein-protein docking. Curr Protein Pept Sci 9(1):1–15

5. Ryan CJ, Cimermancic P, Szpiech ZA, Sali A, Hernandez RD, Krogan NJ (2013) High-resolution network biology: connecting sequence with function. Nat Rev Genet 14 (12):865–879. https://doi.org/10.1038/nrg3574
6. Lensink MF, Velankar S, Kryshtafovych A, Huang SY, Schneidman-Duhovny D, Sali A, Segura J, Fernandez-Fuentes N, Viswanath S, Elber R, Grudinin S, Popov P, Neveu E, Lee H, Baek M, Park S, Heo L, Rie Lee G, Seok C, Qin S, Zhou HX, Ritchie DW, Maigret B, Devignes MD, Ghoorah A, Torchala M, Chaleil RA, Bates PA, Ben-Zeev E, Eisenstein M, Negi SS, Weng Z, Vreven T, Pierce BG, Borrman TM, Yu J, Ochsenbein F, Guerois R, Vangone A, Rodrigues JP, van Zundert G, Nellen M, Xue L, Karaca E, Melquiond AS, Visscher K, Kastritis PL, Bonvin AM, Xu X, Qiu L, Yan C, Li J, Ma Z, Cheng J, Zou X, Shen Y, Peterson LX, Kim HR, Roy A, Han X, Esquivel-Rodriguez J, Kihara D, Yu X, Bruce NJ, Fuller JC, Wade RC, Anishchenko I, Kundrotas PJ, Vakser IA, Imai K, Yamada K, Oda T, Nakamura T, Tomii K, Pallara C, Romero-Durana M, Jimenez-Garcia B, Moal IH, Fernandez-Recio J, Joung JY, Kim JY, Joo K, Lee J, Kozakov D, Vajda S, Mottarella S, Hall DR, Beglov D, Mamonov A, Xia B, Bohnuud T, Del Carpio CA, Ichiishi E, Marze N, Kuroda D, Roy Burman SS, Gray JJ, Chermak E, Cavallo L, Oliva R, Tovchigrechko A, Wodak SJ (2016) Prediction of homoprotein and heteroprotein complexes by protein docking and template-based modeling: a CASP-CAPRI experiment. Proteins 84 (Suppl 1):323–348. https://doi.org/10.1002/prot.25007
7. Andre I, Bradley P, Wang C, Baker D (2007) Prediction of the structure of symmetrical protein assemblies. Proc Natl Acad Sci U S A 104 (45):17656–17661. https://doi.org/10.1073/pnas.0702626104. 0702626104 [pii]
8. Berchanski A, Eisenstein M (2003) Construction of molecular assemblies via docking: modeling of tetramers with D2 symmetry. Proteins 53(4):817–829. https://doi.org/10.1002/prot.10480
9. Pierce B, Tong W, Weng Z (2005) M-ZDOCK: a grid-based approach for Cn symmetric multimer docking. Bioinformatics 21(8):1472–1478. https://doi.org/10.1093/bioinformatics/bti229
10. Comeau SR, Camacho CJ (2005) Predicting oligomeric assemblies: N-mers a primer. J Struct Biol 150(3):233–244. https://doi.org/10.1016/j.jsb.2005.03.006
11. Schneidman-Duhovny D, Inbar Y, Nussinov R, Wolfson HJ (2005) Geometry-based flexible and symmetric protein docking. Proteins 60 (2):224–231. https://doi.org/10.1002/prot.20562
12. Karaca E, Melquiond AS, de Vries SJ, Kastritis PL, Bonvin AM (2010) Building macromolecular assemblies by information-driven docking: introducing the HADDOCK multi-body docking server. Mol Cell Proteomics 9:1784. https://doi.org/10.1074/mcp.M000051-MCP201. M000051-MCP201 [pii]
13. van Zundert GCP, Rodrigues J, Trellet M, Schmitz C, Kastritis PL, Karaca E, Melquiond ASJ, van Dijk M, de Vries SJ, Bonvin A (2016) The HADDOCK2.2 web server: user-friendly integrative modeling of biomolecular complexes. J Mol Biol 428(4):720–725. https://doi.org/10.1016/j.jmb.2015.09.014
14. van Zundert GCP, Melquiond ASJ, Bonvin A (2015) Integrative modeling of biomolecular complexes: HADDOCKing with cryo-electron microscopy data. Structure 23(5):949–960. https://doi.org/10.1016/j.str.2015.03.014
15. Esquivel-Rodríguez J, Yang YD, Kihara D (2012) Multi-LZerD: multiple protein docking for asymmetric complexes. Proteins 80 (7):1818–1833
16. Kuzu G, Keskin O, Nussinov R, Gursoy A (2014) Modeling protein assemblies in the proteome. Mol Cell Proteomics 13:887–896. https://doi.org/10.1074/mcp.M113.031294
17. Russel D, Lasker K, Webb B, Velazquez-Muriel J, Tjioe E, Schneidman-Duhovny D, Peterson B, Sali A (2012) Putting the pieces together: integrative modeling platform software for structure determination of macromolecular assemblies. PLoS Biol 10(1):e1001244. https://doi.org/10.1371/journal.pbio.1001244
18. Inbar Y, Benyamini H, Nussinov R, Wolfson HJ (2005) Prediction of multimolecular assemblies by multiple docking. J Mol Biol 349(2):435–447. https://doi.org/10.1016/j.jmb.2005.03.039. S0022-2836(05)00317-7 [pii]
19. Amir N, Cohen D, Wolfson HJ (2015) DockStar: a novel ILP-based integrative method for structural modeling of multimolecular protein complexes. Bioinformatics 31(17):2801–2807
20. Trnka MJ, Baker PR, Robinson PJ, Burlingame AL, Chalkley RJ (2014) Matching cross-linked peptide spectra: only as good as the worse identification. Mol Cell Proteomics 13 (2):420–434. https://doi.org/10.1074/mcp.M113.034009

21. Chen ZA, Jawhari A, Fischer L, Buchen C, Tahir S, Kamenski T, Rasmussen M, Lariviere L, Bukowski-Wills JC, Nilges M, Cramer P, Rappsilber J (2010) Architecture of the RNA polymerase II-TFIIF complex revealed by cross-linking and mass spectrometry. EMBO J 29(4):717–726. https://doi.org/10.1038/emboj.2009.401
22. Schneidman-Duhovny D, Inbar Y, Nussinov R, Wolfson HJ (2005) PatchDock and SymmDock: servers for rigid and symmetric docking. Nucleic Acids Res 33(Web Server issue): W363–W367. https://doi.org/10.1093/nar/gki481. 33/suppl_2/W363 [pii]
23. Shatsky M, Nussinov R, Wolfson HJ (2004) A method for simultaneous alignment of multiple protein structures. Proteins 56:143–156
24. Mashiach E, Nussinov R, Wolfson HJ (2010) FiberDock: a web server for flexible induced-fit backbone refinement in molecular docking. Nucleic Acids Res 38(Web Server): W457–W461. https://doi.org/10.1093/nar/gkq373
25. Martín-Benito J, Grantham J, Boskovic J, Brackley KI, Carrascosa JL, Willison KR, Valpuesta JM (2007) The inter-ring arrangement of the cytosolic chaperonin CCT. EMBO Rep 8 (3):252–257
26. Kalisman N, Adams CM, Levitt M (2012) Subunit order of eukaryotic TRiC/CCT chaperonin by cross-linking, mass spectrometry, and combinatorial homology modeling. Proc Natl Acad Sci U S A 109(8):2884–2889
27. Leitner A, Joachimiak LA, Bracher A, Mönkemeyer L, Walzthoeni T, Chen B, Pechmann S, Holmes S, Cong Y, Ma B, Ludtke S (2012) The molecular architecture of the eukaryotic chaperonin TRiC/CCT. Structure 20(5):814–825
28. Shomura Y, Yoshida T, Iizuka R, Maruyama T, Yohda M, Miki K (2004) Crystal structures of the group II chaperonin from Thermococcus strain KS-1: steric hindrance by the substituted amino acid, and inter-subunit rearrangement between two crystal forms. J Mol Biol 335 (5):1265–1278

Chapter 13

Biological Assembly Comparison with VAST+

Thomas Madej, Aron Marchler-Bauer, Christopher Lanczycki, Dachuan Zhang, and Stephen H. Bryant

Abstract

The VAST+ algorithm is an efficient, simple, and elegant solution to the problem of comparing the atomic structures of biological assemblies. Given two protein assemblies, it takes as input all the pairwise structural alignments of the component proteins. It then clusters the rotation matrices from the pairwise superpositions, with the clusters corresponding to subsets of the two assemblies that may be aligned and well superposed. It uses the Vector Alignment Search Tool (VAST) protein–protein comparison method for the input structural alignments, but other methods could be used, as well. From a chosen cluster, an "original" alignment for the assembly may be defined by simply combining the relevant input alignments. However, it is often useful to reduce/trim the original alignment, using a Monte Carlo refinement algorithm, which allows biologically relevant conformational differences to be more readily detected and observed. The method is easily extended to include RNA or DNA molecules. VAST+ results may be accessed via the URL https://www.ncbi.nlm.nih.gov/Structure, then entering a PDB accession or terms in the search box, and using the link [VAST+] in the upper right corner of the Structure Summary page.

Key words Protein complex, Molecular assembly, Structure comparison, Structure alignment, Secondary structure element

1 Introduction

Quite some years ago, in the earlier days of structural bioinformatics, among the hot topics were the prediction of protein three-dimensional (3D) structure, exploration of the protein fold universe, and understanding the molecular evolution of proteins. For all these studies, the recognition of distant homologues or analogous folds via protein-to-protein structural similarity is necessary. Efficient and useful protein-to-protein structure comparison methods were developed, such as DALI, SSAP, Vector Alignment Search Tool (VAST), CE, MATRAS, TM-align, and others [1–6]. Of course, it was also clear that the comparison of biological assemblies of proteins (and including other molecule types) to one another is important, although the range of complexity of available biological assemblies was limited in the early days of the Protein Data Bank

Zoltán Gáspári (ed.), *Structural Bioinformatics: Methods and Protocols*, Methods in Molecular Biology, vol. 2112,
https://doi.org/10.1007/978-1-0716-0270-6_13, © The Author(s) 2020

(PDB). Comparison of assemblies is necessary for the study of molecular interactions and interfaces in atomic-level detail. Good computing performance is highly desirable, as the public structure database PDB right now contains nearly 150,000 structures and continues to grow.

Although it may seem that the comparison of biological molecular assemblies is not much more complicated than the comparison of two molecules, there is an extra degree of complexity introduced because assemblies can be very large, including several thousand or more residues/nucleotides, much larger than size of typical protein–protein comparisons. The fact that the molecules cannot be ordered in any canonical way is another complication. The MM-align algorithm (MultiMer-align [7]) solves these problems by using dynamic programming to achieve good performance and handles the comparison of assemblies by essentially comparing many pairs of large individual "artificial" molecules. The ordering problem is taken care of by considering every possible order of the protein chains, which results in many large individual and "artificial" proteins. The SCPC (Structural Comparison of Protein Complexes [8]) method detects similarities between substructures using secondary structure elements (SSEs). The individual protein chains are compared, using the SSE decompositions. This gives a collection of similar pairs, which are then further agglomerated into larger similar substructures, using a scoring function that cross-checks SSE positions across the constituent pairs. There is also 3D Complex, which represents assemblies by graphs and uses a graph matching algorithm to assess similarity between assemblies [9], producing a hierarchical classification of complexes.

In this chapter, we describe VAST+, a biological assembly comparison algorithm [10] that uses our original VAST [3] protein–protein comparison method. As will be apparent, there is no strict dependence on the VAST algorithm per se; any other pairwise-molecule structure comparison method could be used to provide the needed input to the VAST+ program. Besides the identification of similarities between biological assemblies, we have also made an effort to detect and emphasize meaningful *dissimilarities*, as described in the methods section about "refined alignment." By considering dissimilarities between assemblies, one can more easily visualize state transitions and important conformational changes. In general, there will be many dissimilarities between assemblies that differ in many details, and automatically detecting and annotating those that may be biologically significant is a major challenge.

2 Methods

2.1 Clustering by Rotation Matrix Distance

Here is the intuition behind the method. In Fig. 1, there are two "assemblies," ABC (blue) and a similar copy ABC (red). If we do a superposition of the A's only, then we translate the center of mass (centroid) of the red-A to the centroid of the blue-A and rotate, R1. The amount of rotation is given by the angle between two lines: one line is determined by the centroid of the blue-A and the centroid of the blue assembly, c1; the other line goes through the centroid of the red-A and the centroid of the red assembly, c2.

To superpose the two assemblies, i.e., red-ABC and blue-ABC simultaneously, we translate c2 to c1 and then rotate (R2); again, the degree of rotation is determined by the angle between the translated lines. A translation does not change the angle between lines, and therefore we see that R1 must be equal to R2.

Of course, this is harder to visualize in 3D, but the same argument applies. The three superpositions of red-A and blue-A, red-B and blue-B, and red-C and blue-C represent the comparisons between single protein chains. The superposition of red-ABC to blue-ABC corresponds to the comparison of two molecular assemblies.

There is only one problem that arises, which is shown in Fig. 2. In this situation, we see that the rotation matrix for the A's, after translation, is the identity matrix and likewise for the B's. But the "interface" between A and B in the two assemblies is different! To handle this inconvenience, we simply introduce another numerical restriction, the "orientation check." Namely, in order to cluster rot(A, A′) and rot(B, B′) together, not only do we require that the Euclidean distance between rot(A, A′) and rot(B, B′) be small enough, but also the vector from the centroid of A to the centroid of B needs to be in roughly the same direction as the corresponding

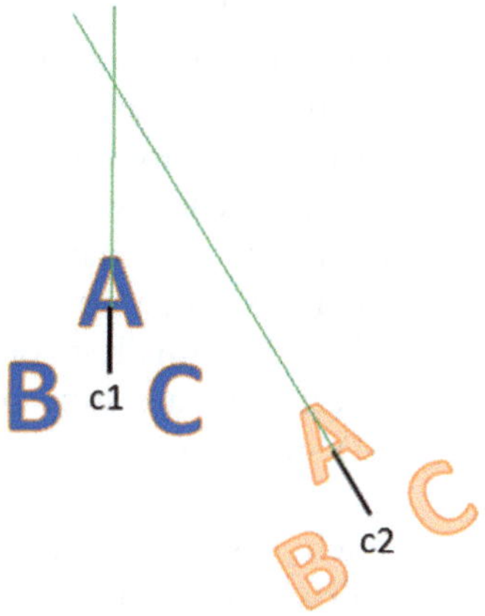

Fig. 1 The angle between the lines gives the rotation required to superpose both the red-A and blue-A and also the entire "assemblies" red-ABC and blue-ABC

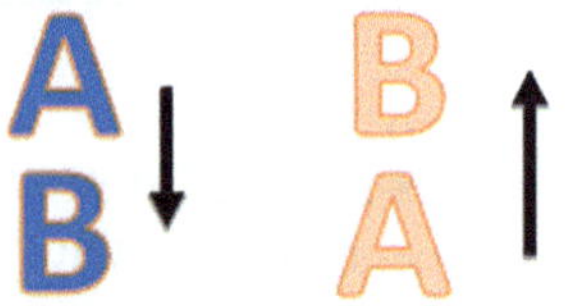

Fig. 2 The rotations to superpose red-A with blue-A and red-B with blue-B are both the identity rotation, but red-AB and blue-AB cannot be superposed together because of a difference in orientation

vector between A′ and B′, after the translation and rotation. In the example in Fig. 2, these vectors point in opposite directions.

To determine a reasonable clustering threshold, we took random pairs of arbitrarily chosen rotation matrices obtained from structure comparisons and calculated the Euclidean distance between them. The mean distance was about 2.4, and only about 1.7% of the pairs had a distance less than 1.0. We chose 1.0 as a threshold; this is the simple Euclidean distance in nine dimensions and there are no units.

In outline, the algorithm is thus as follows. Given two protein assemblies to compare, first compute all pairwise structural alignments between the component protein chains. For each pair of similar proteins, there is a rotation matrix for the superposition. The rotation matrices for the pairs are also going to include, with minor deviations, the rotations that superpose entire similar subassemblies, by the preceding argument. Cluster the rotation matrices using complete-linkage clustering, with Euclidean distance between the matrices (i.e., distance <1.0), and using the "orientation check." The resulting clusters correspond to mappings between protein chains in one assembly and protein chains in the other assembly. An overall alignment for the assemblies can be obtained by simply combining the pairwise structural alignments that we started with for the component chains. In the next section, we will refer to this as an "original alignment" of the assemblies. We can get a superposition of the assemblies via the original alignment. Each of the clusters gives a different alignment; from among these various possibilities, choose one according to whatever desirable criteria. It makes the most sense to consider the clusters with the most protein chains aligned (i.e., the largest clusters), and then further select among these, e.g., by largest total number of residues aligned or smallest root mean square deviation (RMSD; all RMSDs referred to are superposition RMSDs, i.e., obtained after optimal 3D superposition).

Consider the case of comparing two hemoglobin tetramers. Each has four chains, ABCD and A′B′C′D′. The chains are all pairwise comparable, so there are rotation matrices rot(A, A′), rot (A, B′), rot(A, C′), etc., which are 16 in all. After the clustering, we get four clusters of size 4, and these correspond exactly to all the

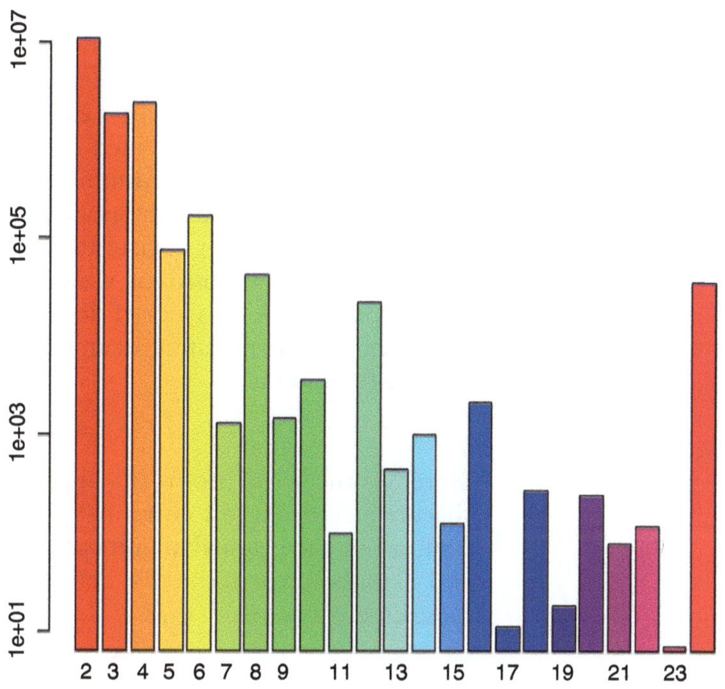

Fig. 3 Counts of similar assemblies in the VAST+ database. The *x*-axis contains the number of molecules in the complexes from dimers to 24-mers. The *y*-axis is on a logarithmic scale, because the dimer-dimer, trimer-trimer, and tetramer-tetramer counts are so dominant

possible alignments of the tetramers. Some of these will map the alpha chains to alpha and beta to beta, and some will map alpha to beta and beta to alpha. The ones mapping alpha to alpha and beta to beta will have slightly larger alignments with a better RMSD, and one will be chosen as the representative assembly alignment, correctly.

If two biological assemblies include nucleotide chains, with structural alignments computed between them, then there is no problem in applying the rotation matrix clustering to include the alignments between the nucleotide chains. Thus, the method is easily extended to more complicated molecular assemblies such as ribosomes.

Figure 3 displays the number of similar assemblies in the VAST+ database, in terms of dimer-dimer, trimer-trimer, etc., on up through 24-mer to 24-mer similarities.

2.2 Refined Alignment

As you can guess, the assembly superposition obtained by using the pairwise alignments/superpositions of the component molecules, i.e., the "original" alignment, amounts to an average superposition. In many, or even most, cases, this is satisfactory, e.g., for closely related structures, the RMSD of the assembly superposition may be

1.0 Å or less, which is better than the resolution of almost all structures in the database. However, in some cases, the average superposition obscures our understanding of the assembly similarity. This will be apparent in the examples given below.

The purpose of the "refined alignment" algorithm is to process the original alignment by trimming it, in order to more easily view any biologically relevant differences between the assemblies. The trimmed pieces, which deviate from a "common core" of the two assemblies, are the positions where the conformations differ to a greater degree. Here is a brief description of the refined alignment algorithm. We formulate it as an optimization problem, where the scoring function is

$$f(S) = N - E$$

Here S is the set of atoms (C-alphas, really paired/aligned atoms) that are included in the current alignment. The current alignment is a subset of the original alignment. Then N is simply the size of S, and E is the error term. We define the error term E as a sum of indexed error terms $E(i, j)$:

$$E = \sum(S)\ E(i,j)$$

The "$\Sigma(S)$" means "sum over all positions i, j in S," and $E(\mathrm{i}, j) = \Delta(i, j) - T + 1$ if $\Delta(i, j) > T$, where T is the tolerance. The deltas are given by

$$\Delta(i,j) = |d(i,j) - d(i',j')|$$

where $d(i, j)$ = (Euclidean) distance between atoms i and j in structure 1, $d(i', j')$ = distance between i' and j' in structure 2, and i is aligned with i', j with j'. The deltas are measuring the difference in distance between equivalent pairs of positions in the two structures. It works well to choose the tolerance T to be the RMSD of the original alignment/superposition. If we have a current alignment S and add a residue position to it, then N will increase by 1, but the error term E will also increase, depending on how many deltas associated with the position are bigger than the original RMSD and also by how much. We can see that there is going to be a tendency to shrink the original alignment by removing "erroneous" residues.

The scoring function $f(S) = N - E$ is like those appearing in the classical, hard optimization problems. When trying to maximize it, there is a tension between choosing atoms to add to S to increase N and choosing to increase the error term, E. If we set the error term, $E(i, j)$ = infinity whenever $\Delta(i, j) > T$, then the problem amounts to finding a maximum independent set (MIS) on a type of graph embedded in three dimensions. The MIS on general graphs is a classical NP-complete problem, and there is no known efficient algorithm to solve it. It is not at all obvious if the MIS for general graphs can be transformed to the problem involving our special class of graphs, so it is unclear whether our problem is NP-complete or not. Nonetheless, a reasonable approach to solving our problem is to use a heuristic. We use a Monte Carlo-type algorithm, a Gibbs sampler. To further simplify the problem, we distinguish the SSEs belonging to the structures. Conveniently, VAST computes the protein-to-protein alignments using SSEs. Corresponding to each aligned SSE, there is a contiguous segment associated with it, in each of the two structures, which may extend past the endpoints of the SSE and into adjacent loop regions. A "move" in the MC algorithm consists of replacing a segment by a different one. When we replace a segment in one structure, we also replace the corresponding segment in the other structure, so that the replacement is consistent with the original alignment. Notice that we can delete a segment simply by replacing it with the empty segment. The error term for replacing a segment is simply the sum of the error terms for the positions that are added, minus the sum of the error terms for the positions removed. Then the Gibbs sampler proceeds in the standard way: choose a segment at random, calculate the scores for all the possible replacements of that segment, assign each possible replacement the Boltzmann weight, and then choose the replacement segment probabilistically according to the weights [11].

3 Examples

An excellent example is provided by comparison between the R (relaxed) and T (tense) states of aspartate transcarbamoylase, e.g., PDB structures 4kh1 and 1rae. 4kh1 is the R-state structure of an ATCase from *E. coli* [12], whereas 1rae is a T-state CTP-ligated ATCase also from *E. coli* [13]. The ATCase assembly consists of two catalytic trimers and three regulatory dimers, for a total of 12 protein chains [14]. Comparing the individual protein chains pairwise, between the 4kh1 and 1rae structures, we see that the pairs with similar folds align very well and superpose with excellent RMSDs of under 1.5 Å.

By clustering the rotation matrices and simply combining all the alignments of the 12 component protein chains, we get an original alignment of the entire assembly involving 2613 residues with a 5.4 Å RMSD. Structure 3D graphical viewers such as iCn3D https://github.com/ncbi/icn3d may have an "alternate structure" feature to flip between the structures in the superposed state. When viewed in iCn3D at a good angle, by flipping between the structures, the relative motion of the more rigid halves of the assemblies is readily seen.

However, the refined alignment makes the differences between the T- and R-states even more apparent. For example, in hand,

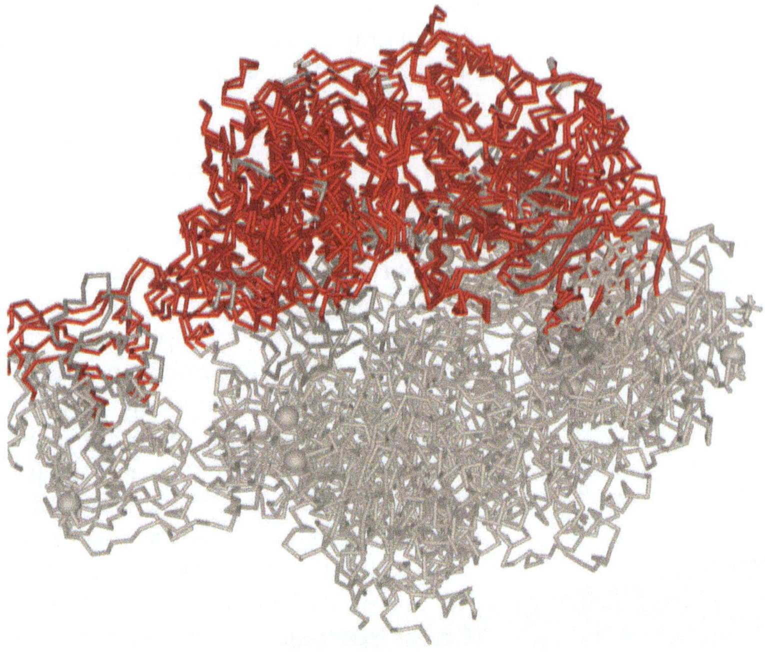

Fig. 4 Refined alignment/superposition of aspartate transcarbamoylase T- and R-state structures (PDB accessions 1rae and 4kh1). The red part displays the six-chain subcomponents that are superposed, which includes one catalytic trimer and one chain from each of the three regulatory dimers

4kh1 and 1rae, the refined alignment has 1113 aligned residues at a RMSD of 2.3 Å. The most striking difference between the T-state and the R-state structures is an expansion of about 11 Å along the threefold axis [14]. The refined alignment redefines the superposition relative to one of the catalytic trimers and includes only six chains, namely, a catalytic trimer and one chain from each of the three regulatory dimers. This helps to make the large-scale difference between the T- and R-states more obvious (*see* Fig. 4). When displayed using the iCn3D graphical viewer, alternating the structures with the "a" key now shows the lower part moving vertically relative to the fixed upper part. *See* Fig. 5 for an overview of how to access the VAST+ webserver using this particular example.

The refinement algorithm can also detect much finer-grained conformational differences. The hemagglutinin influenza virus example involves PDB structures 3sdy and 1mqm. The 3sdy structure is a broadly neutralizing antibody bound to the influenza A H3 hemagglutinin [15]. The 1mqm structure is the hemagglutinin for a potential avian progenitor of the 1968 Hong Kong pandemic influenza virus [16]. All of the protein-to-protein superpositions between the hemagglutinin molecules are excellent, complete chain alignments at under 1.0 Å RMSD. When we combine all these to get our original alignment, we have 1470 residues aligned with an

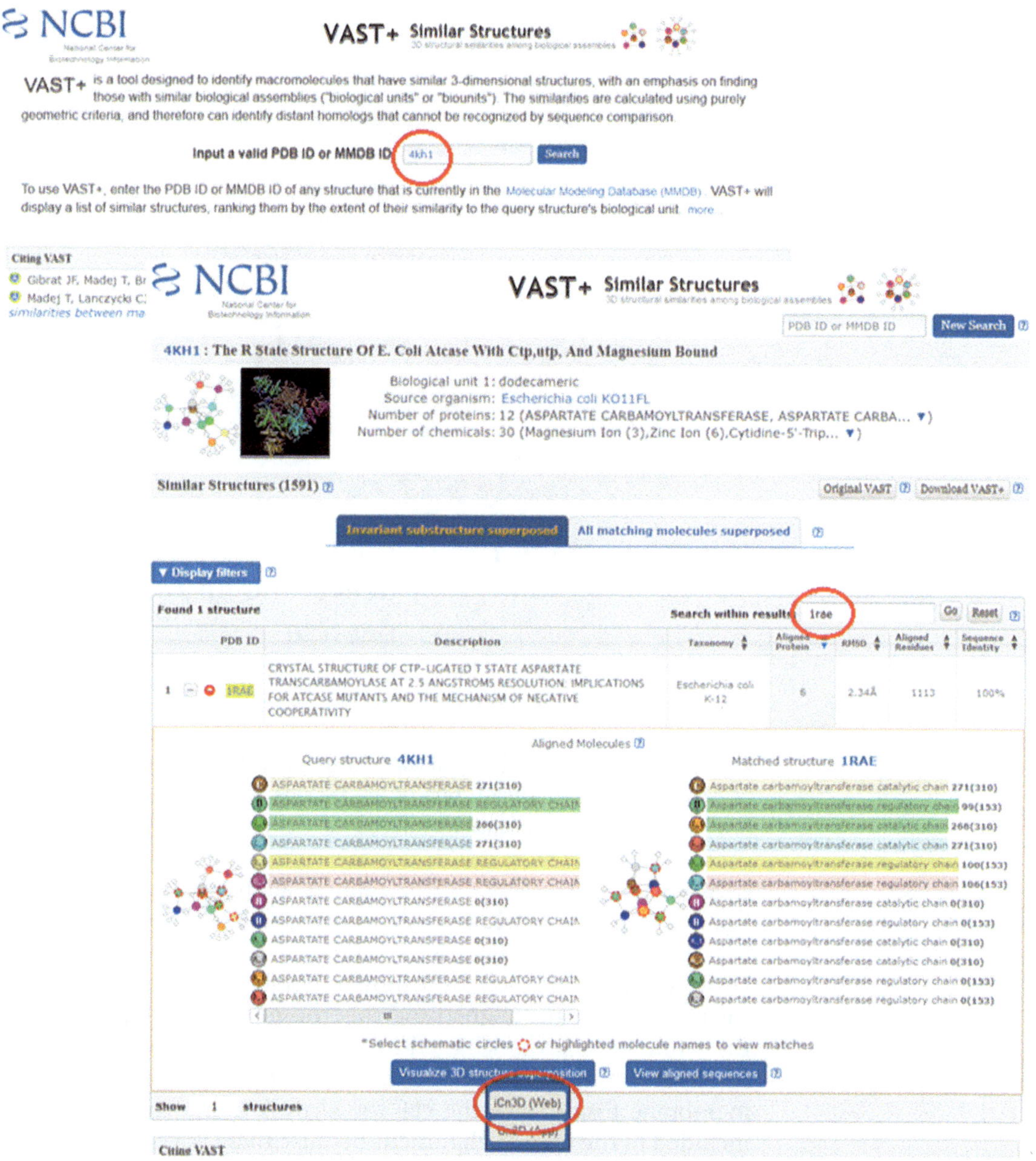

Fig. 5 From the "VAST+ Similar Structures" webpage, one can enter the PDB accession for a structure of interest (topmost red circle). This goes to the main VAST+ page, which will present a list of similar assemblies in the bottom panel. There is a search box which can be used to filter the results using simple search terms, e.g., keywords appearing in the titles or taxonomy. In this example, we are specifically interested in the PDB accession 1rae and so enter it in the search box (middle red circle). When we expand the entry for that result, we see the list of protein chains from each structure (4kh1 and 1rae), and the ones that correspond in the precomputed superposition are highlighted in the same color. An interactive graphical view can be obtained by launching the iCn3D viewer (bottom red circle)

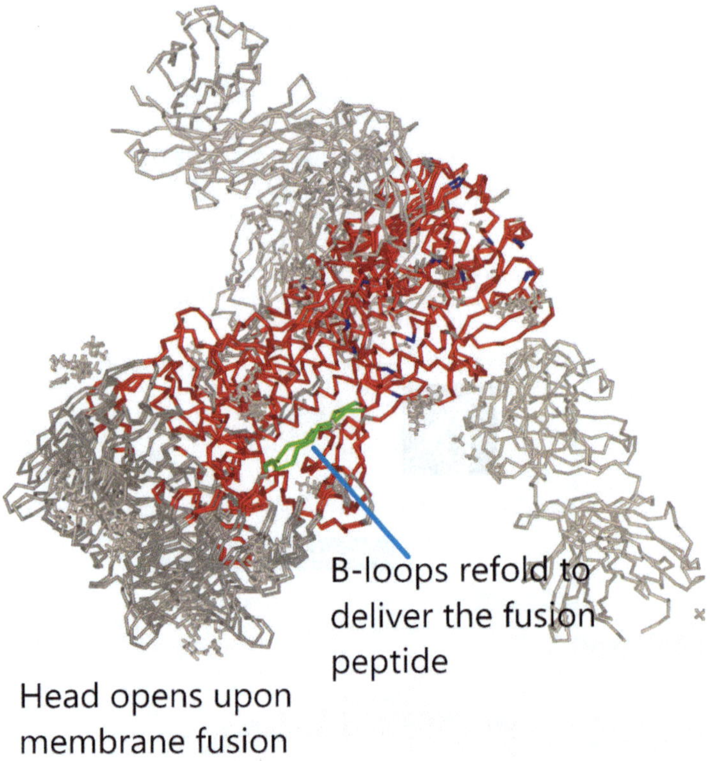

Fig. 6 Refined alignment of influenza virus hemagglutinin assemblies; PDB accessions 3sdy and 1mqm. One of the B-loops is highlighted in green. The head and B-loop annotations were added manually

excellent 1.5 Å RMSD. The refined alignment is trimmed to 865 residues at 0.6 Å RMSD. As in the aspartate transcarbamoylase example, we see that regions involving larger-scale movements become unaligned, i.e., are trimmed (*see* Fig. 6). Again in Fig. 6, the red parts are the aligned and superposed refined alignment, and the white parts are unaligned in the refinement. The head of the hemagglutinin assembly, which is that part which opens upon membrane fusion with the cell the virus is trying to infect, is not included in the refined alignment because there is a conformational difference. Note that the white chain traces at the top and right-hand side, and in the background, are the antibodies in the 3sdy structure, which are not present in 1mqm and hence not included in the original alignment. However, besides the change at the head of the hemagglutinins, finer-grained details are also detected, such as the three B-loops. The B-loops change conformation in order to deliver the fusion peptide [15]. This is a good example, where the protein-to-protein alignments are excellent and align complete chains, but by comparing the assemblies and using the refinement method, we can detect subtle but important dissimilarities that are biologically relevant.

4 Summary

The VAST+ algorithm is an efficient, simple, and elegant solution to the problem of comparing the atomic structures of biological assemblies. It uses the VAST protein–protein comparison method for the underlying structural alignments, but other methods could be used, as well. The original VAST was designed to detect similarity between protein folds and was not very concerned with subtle differences in the structures. By using VAST+ and comparing entire biological assemblies, we can automatically detect important and subtle biologically relevant differences in the structures. Moreover, as examples like the influenza hemagglutinin show, it is only by comparing entire assemblies that we can detect and clarify relationships between the structures.

Acknowledgments

This research was supported by the Intramural Research Program of the NIH, National Library of Medicine at National Institutes of Health/DHHS. Funding for open access charge: Intramural Research Program of the National Library of Medicine, National Institutes of Health.

References

1. Holm L, Sander C (1993) Protein structure comparison by alignment of distance matrices. J Mol Biol 233(1):123–138
2. Orengo CA, Taylor WR (1996) SSAP: sequential structure alignment program for protein structure comparison. Methods Enzymol 266:617–635
3. Gibrat JF, Madej T, Bryant SH (1996) Surprising similarities in structure comparison. Curr Opin Struct Biol 6(3):377–385
4. Shindyalov IN, Bourne PE (1998) Protein structure alignment by incremental combinatorial extension (CE) of the optimal path. Protein Eng 11(9):739–747
5. Kawabata T, Nishikawa K (2000) Protein structure comparison using the Markov transition model of evolution. Proteins 41 (1):108–122
6. Zhang Y, Skolnick J (2005) TM-align: a protein structure alignment algorithm based on the TM-score. Nucleic Acids Res 33 (7):2302–2309
7. Mukherjee S, Zhang Y (2009) MM-align: a quick algorithm for aligning multiple-chain protein complex structures using iterative dynamic programming. Nucleic Acids Res 37 (11):e83
8. Koike R, Ota M (2012) SCPC: a method to structurally compare protein complexes. Bioinformatics 28(3):324–330
9. Levy ED, Pereira-Leal JB, Chothia C, Teichmann SA (2006) 3D complex: a structural classification of protein complexes. PLoS Comput Biol 2(11):e155
10. Madej T, Lanczycki CJ, Zhang D, Thiessen PA, Geer RC, Marchler-Bauer A, Bryant SH (2014) MMDB and VAST+: tracking structural similarities between macromolecular complexes. Nucleic Acids Res 42(Database issue):D297–D303
11. Tanner MA (1998) Tools for statistical inference: methods for the exploration of posterior distributions and likelihood functions, Springer series in statistics. Springer-Verlag, New York
12. Cockrell GM, Zheng Y, Guo W, Peterson AW, Truong JK, Kantrowitz ER (2013) New paradigm for allosteric regulation of Escherichia coli aspartate transcarbamoylase. Biochemistry 52(45):8036–8047

13. Kosman RP, Gouaux JE, Lipscomb WN (1993) Crystal structure of CTP-ligated T state aspartate transcarbamoylase at 2.5 A resolution: implications for ATCase mutants and the mechanism of negative cooperativity. Proteins 15(2):147–176
14. Lipscomb WN, Kantrowitz ER (2012) Structure and mechanisms of Escherichia coli aspartate transcarbamoylase. Acc Chem Res 45 (3):444–453
15. Ekiert DC, Friesen RH, Bhabha G, Kwaks T, Jongeneelen M, Yu W, Ophorst C, Cox CF, Korse HJ, Brandenburg B, Vogels R, Brakenhoff JP, Kompier R, Koldijk MH, Cornelissen LA, Poon LL, Peiris M, Koudstaal W, Wilson IA, Goudsmit J (2011) A highly conserved neutralizing epitope on group 2 influenza A viruses. Science 333(6044):843–850
16. Ha Y, Stevens DJ, Skehel JJ, Wiley DC (2003) X-ray structure of the hemagglutinin of a potential H3 avian progenitor of the 1968 Hong Kong pandemic influenza virus. Virology 309(2):209–218

Open Access This chapter is licensed under the terms of the Creative Commons Attribution 4.0 International License (http://creativecommons.org/licenses/by/4.0/), which permits use, sharing, adaptation, distribution and reproduction in any medium or format, as long as you give appropriate credit to the original author(s) and the source, provide a link to the Creative Commons license and indicate if changes were made.

The images or other third party material in this chapter are included in the chapter's Creative Commons license, unless indicated otherwise in a credit line to the material. If material is not included in the chapter's Creative Commons license and your intended use is not permitted by statutory regulation or exceeds the permitted use, you will need to obtain permission directly from the copyright holder.

Chapter 14

BioMagResBank (BMRB) as a Resource for Structural Biology

Pedro R. Romero, Naohiro Kobayashi, Jonathan R. Wedell, Kumaran Baskaran, Takeshi Iwata, Masashi Yokochi, Dimitri Maziuk, Hongyang Yao, Toshimichi Fujiwara, Genji Kurusu, Eldon L. Ulrich, Jeffrey C. Hoch, and John L. Markley

Abstract

The Biological Magnetic Resonance Data Bank (BioMagResBank or BMRB), founded in 1988, serves as the archive for data generated by nuclear magnetic resonance (NMR) spectroscopy of biological systems. NMR spectroscopy is unique among biophysical approaches in its ability to provide a broad range of atomic and higher-level information relevant to the structural, dynamic, and chemical properties of biological macromolecules, as well as report on metabolite and natural product concentrations in complex mixtures and their chemical structures. BMRB became a core member of the Worldwide Protein Data Bank (wwPDB) in 2007, and the BMRB archive is now a core archive of the wwPDB. Currently, about 10% of the structures deposited into the PDB archive are based on NMR spectroscopy. BMRB stores experimental and derived data from biomolecular NMR studies. Newer BMRB biopolymer depositions are divided about evenly between those associated with structure determinations (atomic coordinates and supporting information archived in the PDB) and those reporting experimental information on molecular dynamics, conformational transitions, ligand binding, assigned chemical shifts, or other results from NMR spectroscopy. BMRB also provides resources for NMR studies of metabolites and other small molecules that are often macromolecular ligands and/or nonstandard residues. This chapter is directed to the structural biology community rather than the metabolomics and natural products community. Our goal is to describe various BMRB services offered to structural biology researchers and how they can be accessed and utilized. These services can be classified into four main groups: (1) data deposition, (2) data retrieval, (3) data analysis, and (4) services for NMR spectroscopists and software developers. The chapter also describes the NMR-STAR data format used by BMRB and the tools provided to facilitate its use. For programmers, BMRB offers an application programming interface (API) and libraries in the Python and R languages that enable users to develop their own BMRB-based tools for data analysis, visualization, and manipulation of NMR-STAR formatted files. BMRB also provides users with direct access tools through the NMRbox platform.

Key words BioMagResBank, BMRB, NMR data, NMR-STAR, Visualization, Deposition, Retrieval, Search, Data formats, Analysis, Structures, PDB, NMR spectra, Time-domain data, Python, R

Zoltán Gáspári (ed.), *Structural Bioinformatics: Methods and Protocols*, Methods in Molecular Biology, vol. 2112, https://doi.org/10.1007/978-1-0716-0270-6_14, © Springer Science+Business Media, LLC, part of Springer Nature 2020

1 Introduction

The Biological Magnetic Resonance Data Bank (BioMagResBank or BMRB) [1] has served for the past 30 years as the primary archive for spectral and derived data generated by nuclear magnetic resonance (NMR) spectroscopy of biological systems. The BMRB archive is unique among biophysical data banks in that the archive contains primary time-domain data obtained by NMR spectrometers, processed spectra, spectral peak characteristics, assigned spectral peak chemical shifts, and derived data such as relaxation parameters, pK_a values, and atomic coordinates for certain smaller molecules not covered by the Protein Data Bank Archive [2] (URL: http://wwpdb.org). BMRB has developed technology for annotating and processing the assigned chemical shift data archived at BMRB and the chemical shift and constraint data underlying NMR-based structures archived at the PDB. The field of biomolecular NMR is evolving continuously, and newly developed NMR techniques [3, 4] have the potential to enable NMR studies of various biological molecular systems, including larger proteins, nucleic acids, molecular machines, and membrane-bound biopolymers.

BMRB also offers support for studies of metabolomics and natural products through a library of a variety of 1D and two-dimensional (2D) NMR spectra of pure compounds (including metabolites, natural products, drugs, and compounds used for screening in drug discovery) and through its adoption of the ALATIS compound and atom identifiers [5], which are universal and based solely on the three-dimensional (3D) structure of the compound and the InChI convention. In addition, for a growing number of small molecules, BMRB is providing spin matrices in the GISSMO convention [6], which enables accurate simulation of spectra at any field strength. The combination of unique ALATIS naming and parameterized spectra offers the users of BMRB data a distinctive benefit in terms of robustness and reproducibility.

The importance of publicly accessible and persistent data archives such as BMRB for sustainable and reproducible research is embodied in the FAIR principles [7] espoused by the wwPDB [8], which are that data should be findable, accessible, interoperable, and reusable.

BMRB consists of the main archive at the University of Wisconsin-Madison and branches at UConn Health, Osaka, Japan (PDBj-BMRB), and Florence, Italy. BMRB is a core member of the Worldwide Protein Data Bank (wwPDB) along with the RCSB PDB, PDBe, and PDBj [2]. Currently, the core archives of the wwPDB consist of the PDB archive and the BMRB archive [8, 9]. In 2019, the Electron Microscopy Data Bank (EMDB) located at EMBL-EBI will become a core wwPDB member, and

EMDB will become the third core archive. The core archives each share a common data format driven by a data dictionary, which enables coordinated searching of the combined resources.

BMRB mirror sites are supported at Osaka University, Japan, and at CERM in Florence, Italy, with the Osaka facility also being a data deposition and processing site. BMRB collaborates closely with the other groups in the wwPDB (RCSB PDB, PDBe, and PDBj). Over the years, BMRB has benefitted from interactions with many groups in the NMR community, including the CCPN group (now at the University of Leicester), the Northeast Structural Genomics (NESG) group, the Center for Eukaryotic Structural Genomics (CESG), the NMR metabolomic groups, and the National Magnetic Resonance Facility at Madison (NMRFAM). A wealth of experimental data from NMR studies linked to high-quality 3D structures of proteins was deposited to BMRB by centers funded by the NIH Protein Structure Initiative (PSI) [10]. Although the pace of structural NMR studies deposited declined after the end of the PSI in 2015, structural NMR data continue to comprise approximately half of BMRB depositions.

In addition, BMRB is a member of the Center for NMR Data Processing and Analysis, provider of the NMRbox [11] platform. NMRbox is a cloud-based computing platform providing the bio-NMR community access to existing NMR software tools and computational resources. Project goals include improving NMR data reproducibility, facilitating depositions to BMRB and other public databases, and developing new data analysis tools. As such, BMRB currently provides machine-to-machine (M2M) access to some services (e.g., CS-Rosetta structure determination) and is developing a computer-assisted deposition service that will gather data, metadata, and workflow information to facilitate and enrich depositions into BMRB for enhanced reproducibility. This chapter covers the usage of extant NMRbox-integrated BMRB tools where appropriate.

2 Resources

As an open access digital data resource that applies the FAIR data principles [7], BMRB houses and distributes the experimental and derived data from NMR experiments carried out on biologically relevant molecular systems. The archive consists of six main data depositories: (1) quantitative NMR spectral parameters for proteins, peptides, nucleic acids, carbohydrates, and ligands or cofactors (e.g., assigned chemical shifts, coupling constants, and peak lists) and derived data (e.g., relaxation parameters, residual dipolar couplings, hydrogen exchange rates, and pK_a values); (2) time-domain spectral data from NMR experiments used to assign spectral resonances and determine the structures of biological macromolecules; (3) an archive of atomic coordinates for small molecules not accepted by the wwPDB; (4) a database for NMR constraints processed from original author depositions available from the PDB; (5) an archive of CS-Rosetta structures derived from BMRB chemical shift entries [12, 13]; and (6) a growing database of ^{1}H and ^{13}C 1D and 2D NMR spectra (including time-domain data) and assigned chemical shifts for over 1000 biological small molecules. Validation reports for BMRB chemical shift entries and MolProbity [14] validation reports for all PDB entries are available on the BMRB website. BMRB provides a variety of software services for querying the archive: for enabling interactive data visualizations of the archival data, user supplied data, and combinations of both; for carrying out file format conversions; for validating data; and for high-throughput calculation of structures using CS-Rosetta and HTCondor [15] in collaboration with the Center for High-Throughput Computing (CHTC) and the Open Science Grid (OSG). The data in the BMRB archives are linked to the literature citations related to the entries and to a number of public databases through BLAST sequence homology searches that are updated weekly. BMRB acquires data through depositor submissions by means of three deposition systems: (1) OneDep [16], through the wwPDB OneDep website https://deposit.wwpdb.org/; (2) ADIT-NMR, through either the Madison or Osaka BMRB branch (for other NMR data beyond those accepted by OneDep, as well as nonstructural NMR data); and (3) SMSDep, through the Osaka BMRB branch (for NMR-derived structures of molecules that do not fit the guidelines of the PDB archive). Additional data acquisition methods are (1) transfer of metabolite spectral data collected at the National Magnetic Resonance Facility at Madison (NMRFAM) to the Madison BMRB and (2) in-house generation of validation and structural data using third-party software like AVS, PANAV, and CS-Rosetta. A new version of ADIT-NMR, called BMRBdep, extends the capabilities of the original deposition system for biomacromolecules and supports the deposition of NMR data from small molecules of biological importance (e.g., metabolites, natural products, drugs). The Madison and Osaka branches of BMRB carry out processing and annotation of entries deposited at their sites. NMR data associated with structures deposited through the OneDep system are transferred to either of two BMRB annotation units, and BMRB annotators at these sites curate as the data as needed and convey any changes back to the OneDep team.

BMRB uses the NMR-STAR [17] data format to represent experiments, spectral and derived data, and supporting metadata. NMR-STAR was constructed along the object-relational data model using a subset of the Self-Defining Text Archival and Retrieval (STAR) specification [18]. The growth of the biological NMR field and the development of new experimental technologies have mandated the revision and enlargement of the NMR-STAR ontology [17]. BMRB provides users with tools to facilitate editing and handling of NMR-STAR files, whose use is explained below. The NMR-STAR ontology enhances the reusability (a FAIR goal) of NMR data by providing ample information on the experimental data being archived.

In terms of findability and accessibility, public uninterrupted access to BMRB services is provided through the BMRB website at http://www.bmrb.wisc.edu and its mirror websites in Japan (https://bmrb.pdbj.org) and Italy (http://bmrb.cerm.unifi.it). Due to ongoing development, the websites' appearances may differ somewhat from the screenshots shown in this chapter. Nevertheless, the functionality of the tools and services described here will be maintained, and the website documentation will cover any future updates. The BMRB unit in Osaka (PDBj-BMRB) also maintains a website with access to other useful tools and extra documentation including a Japanese version at https://bmrbdep.pdbj.org (DOI: 10.1002/pro.3273). PDBj-BMRB has been improving the interoperability of BMRB archives with a particular focus on semantic web standards, represented by XML, RDF formats, of which data archives are accessible from each BMRB site (DOI: 10.1186/s13326–016–0057-1). For enhanced interoperability, further access to BMRB data is provided through BMRB's application programming interface (API), described in the "Services for Programmers" section. Help is available through the "bmrbhelp" mailing list, by sending a message to bmrbhelp@bmrb.wisc.edu.

3 Methods

This section describes the methodology for accessing and using the tools and services available for each of four areas, data deposition, data retrieval, data analysis, and programmer services. Most of the included figures correspond to screenshots from the BMRB website at the time of writing this chapter. The website's documentation pages will be kept up to date with any new developments, both in terms of new resources and updates to existing ones.

3.1 Data Deposition

BMRB accepts deposition containing many kinds of experimental and derived NMR data, including time-domain and processed data files. BMRB's main deposition system (ADIT-NMR at the time of writing but soon to be replaced by BMRBdep which is described below) has been designed with the aim of providing researchers with a cloud-based data repository for an in-progress NMR project, allowing users to enter metadata and results as they are generated.

Since its inception, BMRB has worked closely with the PDB for the deposition and archiving of structural NMR data of biological macromolecules. Consequently, BMRB is considered a structural database, and many journals require depositions to both PDB and BMRB as a prerequisite for publication of an NMR-based structure. This close relationship has resulted in the integration of BMRB and PDB depositions for structural NMR data within the OneDep software system [16], which handles deposition of coordinates and associated data into the PDB archive, as well as chemical shifts, restraints, and associated data files into the BMRB archive.

3.1.1 PDB OneDep

The wwPDB partners, including BMRB, joined forces in creating OneDep, which replaced disparate depositions previously used by the wwPDB partners. The OneDep system unifies the deposition and annotation systems across all wwPDB deposition centers and focuses on improving data quality and completeness in the PDB archive while supporting growth in the number of depositions and increases in size and complexity of the structures deposited. The OneDep system at https://deposit.wwpdb.org serves as a single access point for biomolecular NMR data: if the user indicates that coordinates will be deposited, data will be collected through the OneDep site in the depositor's geographical zone (RCSB PDB, PDBe, or PDBj); if no coordinates are associated, the user is transferred to the BMRB deposition system in the depositor's geographical zone (BMRB or PDBj-BMRB). Alternatively, data not involving coordinates can be deposited by directly accessing either the BMRB or PDBj-BMRB site.

Depositions made through OneDep will generate both a PDB and a BMRB entry from the depositor's data, but currently NMR data beyond those accepted by the PDB archive (chemical shifts and restraints are required; NOE peak lists are recommended) cannot be entered in this way and require a separate deposition at one of the BMRB sites. In response to feedback from the NMR community, the wwPDB has pledged in principle to integrate BMRB's new deposition system (BMRBdep) with OneDep to allow NMR researchers to directly deposit more complete NMR data.

The wwPDB website provides ample documentation and tutorials on the use of OneDep. We provide here a summary of the instructions found at https://www.wwpdb.org/deposition/tutorial that are related to NMR depositions. Instructions for depositing coordinates are found at the same URL. As noted above, OneDep sends the depositor directly to the BMRB deposition system if no coordinates are being deposited. Even if the user is depositing coordinates, it is strongly recommended to access the BMRB deposition system after depositing the structure to PDB and to use the provided BMRB ID to continue the deposition of NMR data (e.g., time-domain data, spectra, relaxation parameters, pK_a values, and other derived data). Once OneDep is fully integrated with BMRBdep, this step will not be necessary. As explained in Subheading 3.1.2, a practical and efficient way to deposit NMR structures would be to open a BMRB deposition early in the experimental work and gradually upload the data as the experiment proceeds, in effect using the BMRB depository as a lab notebook. This guarantees a complete BMRB deposition when depositing the final set of data.

At the start of deposition, OneDep will ask the depositor to provide information about the experimental methods employed to determine the structure. If "Solution NMR" or "Solid-state NMR" has been selected as the experimental method, you will be asked whether you are depositing coordinates. If "No" is selected, you will see that BMRB is the only requested accession code, and a "Deposit NMR data at BMRB" button will appear at the bottom of the screen. Once you click the button, you will then be redirected to the BMRB deposition start page. Please note that if you make a BMRB-only deposition, later deposition of associated coordinates will require that you complete a new PDB-only deposition. If "Yes" is selected in response to the coordinates question, then you will be asked to upload three mandatory files (coordinates in PDBx/mmCIF format [19], restraints in NEF [20] or NMR-STAR format, and chemical shifts in NMR-STAR format) and will be encouraged to upload another (peak lists in NMR-STAR format).

During file uploads, the following checks are performed: (1) the coordinates file is checked to ensure that each model has the same chemistry (identical atoms); (2) the chemical shift values are checked for outliers; and (3) the atom nomenclature in the chemical shift and coordinates files is compared for consistency.

If any warnings or errors are generated by these checks, they will be reported as follows:

Warning messages: Warnings encountered upon file upload will be presented in the "File format validation for model coordinates and data files" window. For NMR entries, warning messages provide information about chemical shift values outside of acceptable ranges. Warning messages are provided for depositors to review and either negate or correct as appropriate.

Error messages: Errors encountered upon file upload will be presented in two places: (1) on the diagnostic screen (headed by a graphic of red gears) that appears after the "Populate" or "Repopulate" button on the "File upload" page has been pressed and (2) on the "Upload Summary" page of the deposition interface. For NMR entries, error messages highlight atom nomenclature issues that must be corrected. If an error is present, new coordinates and/or chemical shift files must be uploaded before the deposition can be completed.

Entering NMR Data into the Deposition Interface

For NMR depositions, it is best to enter information starting from the top of the left-hand navigation panel and working downward sequentially, page by page, as some values on later pages (lower on the navigation panel) are dependent on information entered on earlier pages (higher on the navigation panel).

In particular, the chemical shift connection page cannot be completed before mandatory data items in the NMR experimental section and NMR software section have been completed. The chemical shift connection page links data used to assign the chemical shifts with information contained in other NMR sections, i.e., chemical shift filenames, chemical shift references, NMR samples, sample conditions, NMR experiments, and software. In addition, the spectral peak list section becomes mandatory when NMR peak list files are uploaded.

3.1.2 ADIT-NMR and BMRBdep

The ADIT-NMR deposition system was originally developed in collaboration with the RCSB-PDB, built upon PDB's ADIT system for deposition of X-ray structures. The newly developed deposition system, BMRBdep, reproduces the ADIT-NMR functionality with a more responsive, easier-to-use interface, adding the capacity to handle deposition of small-molecule NMR data. Both deposition systems are driven by the NMR-STAR data dictionary, which automatically supports any changes in the NMR-STAR format.

BMRB encourages researchers to start a deposition early and to use ADIT-NMR/BMRBdep as a lab notebook that is filled in as the work proceeds. Incomplete sessions are not deleted, so a deposition session will remain accessible for a long time (up to 2 years since the last update). Please note that older deposition sessions may be required to be upgraded if the dictionary has changed significantly. The completed NMR-STAR files should then be ready for deposition.

Users from Asian countries (excluding Oceania) are encouraged to deposit to the regional BMRB mirror site PDBj-BMRB: http://deposit.bmrb.pdbj.org. Please note that depositions started at PDBj-BMRB cannot be continued on the UW-Madison server, and vice versa.

While ADIT-NMR is still the main deposition server at the time of writing, it will soon be replaced by the BMRBdep server. As a result, the procedures described below are for the new BMRBdep system, which was released recently. You can start a BMRBdep deposition at https://bmrbdep.bmrb.wisc.edu. PDBj-BMRB also provides the BMRBdep deposition service from the same URL, https://deposit.bmrb.pdbj.org, with appropriate changeover time.

The deposition procedure follows the following steps:

Step 1: Preparation for data deposition

Step 2: Creation of a BMRBdep session

Step 3: Upload of data files

Step 4: Entering relevant data

Step 5: Previewing and depositing the entry

Step 6: Receiving a report from BMRB/PDB

Step 7: Hold and release of the entry

Step 1: Preparation for Data Deposition

Before proceeding with your deposition, it will be useful to have on hand the following information:

- Chemical description of the molecules in the system studied.
- Residue sequences for polymers.
- Sequence database reference for the biological molecule(s).
- Atom and bond lists for ligands and nonstandard residues.
- List of contents for at least a representative sample.
- List of experimental conditions (temperature, pH, etc.).
- A list of names to use for each sample and each set of experimental conditions.
- ASCII file(s) containing chemical shift assignments or coupling constants, preferably in NMR-STAR format, but ASCII files containing tables with tab- or comma-delimited fields will be accepted.

Note that since a user can return to their deposition at any time, it is not necessary to have all or even any of the data mentioned above to begin a deposition, although they will be needed to complete it. The BMRB website provides tools for generating NMR-STAR files. To create NMR-STAR chemical shift assignment files, the user can access the template generators (available at http://www.bmrb.wisc.edu/software/tablegen/. *See* Fig. 1) or the STARch file converter, which takes different file formats and converts them into NMR-STAR (*see* Subheading 3.3.1).

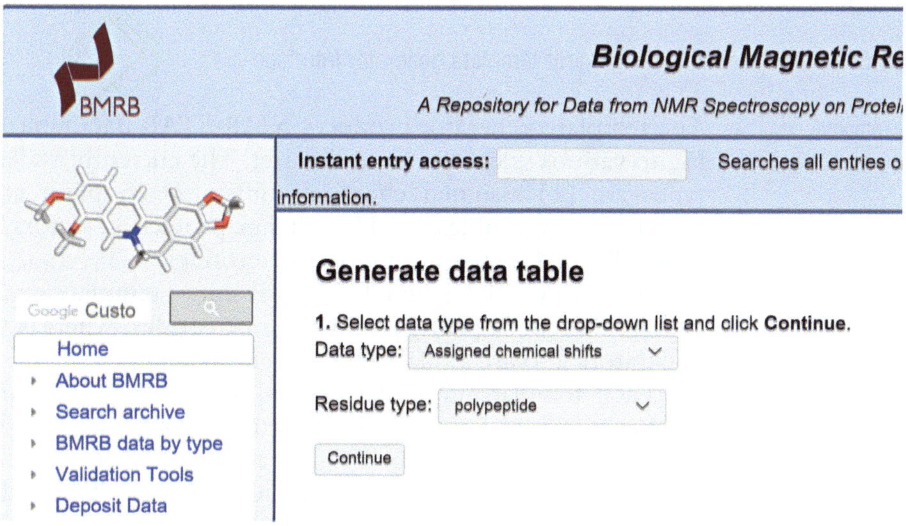

Fig. 1 NMR-STAR template generator initial interface

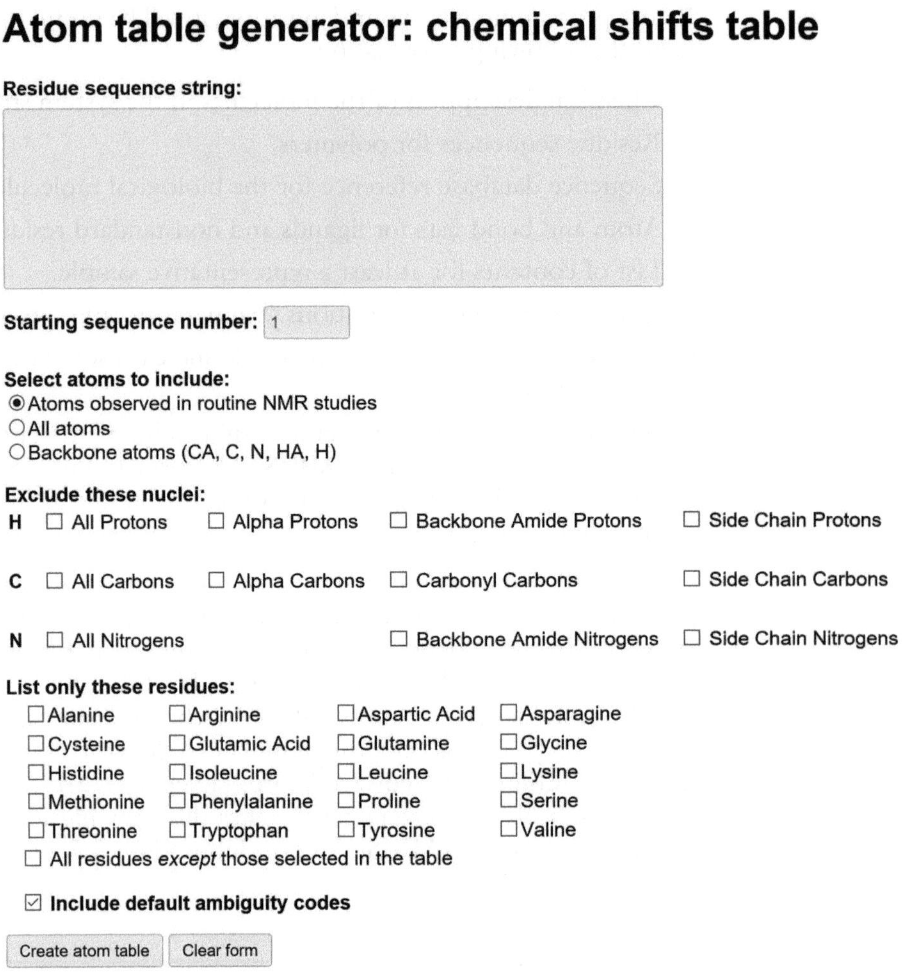

Fig. 2 NMR-STAR chemical shift assignment template generator interface

The BMRB Template Generator

The template generator produces NMR-STAR data tables according to various selections from the user. The currently available data types are (1) assigned chemical shifts, (2) coupling constants, (3) H-exchange tables, (4) H-exchange protection factors, (5) heteronuclear NOE values, and (6) heteronuclear T_1, T_{1rho}, and T_2 values. The type 1 (assigned chemical shifts) template generator is available for both proteins and polynucleotides, whereas template generators 2–6 are available only for proteins.

Each template generator has an input screen for entering a residue sequence string (in single letter format), and the user can make other selections, such as the atoms to include per residue. Figure 2 shows the screen template for chemical shift assignments.

Step 2: Creation of an BMRBdep Session

To begin a BMRBdep session, a user must simply enter their e-mail address and a reference name for their deposition and specify the deposition type (new, from an existing entry, or from an uploaded file). By optionally filling out their ORCID ID, some of the fields in the deposition will be automatically populated using the data available in the ORCID database.

If the user wants to start the deposition from an existing NMR-STAR file or released BMRB entry, they can do so by selecting the appropriate "deposition type" and selecting the NMR-STAR file or BMRB ID to use to start the deposition. The NMR-STAR file can be edited before the upload. The best option is to use the JavaScript NMR-STAR viewer tool (*see* Subheading 3.3.4). Using this tool will ensure that your modifications do not violate the NMR-STAR format. Alternatively, it is possible to edit the file in a plain text editor like Notepad. In the future, an enhancement to the BMRBdep system will add a feature to existing depositions that will allow the user to start a new deposition pre-filled with the data contained within the existing deposition.

After clicking the new deposition button, the deposition session will be created and saved in the user's browser, and an e-mail will be sent with a link used to verify the user's e-mail. This verification link must be clicked prior to entering information about the deposition. To end a session, the user may click the "End session" button. A depositor can leave the site or close the browser at any time. As long as the original e-mail has been saved, the user can get back to the deposition session in the future by clicking the link in the original deposition e-mail. In the case of a lost e-mail, the user should contact BMRB help at bmrbhelp@bmrb.wisc.edu.

Step 3: Upload of Deposition Data Files

Once the session is initiated, the user is taken to the "data files" page (Fig. 3), where the user uploads one or more data files associated with the entry. After file upload, the user must select one or more data types contained within the file. The action of

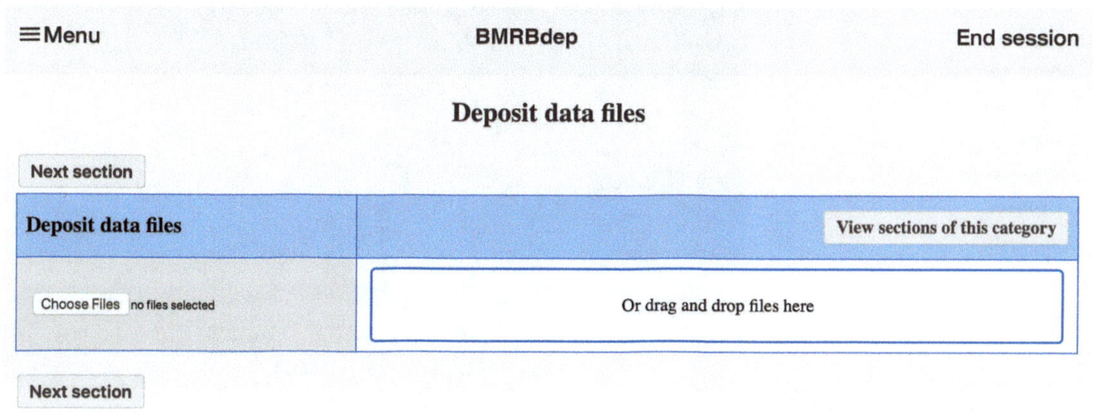

Fig. 3 BMRBdep interface for data files deposition

selecting the appropriate data types will enable additional data input fields in the deposition system related to the uploaded data types. The user can return to this page at any time to upload additional data files, and the interface will automatically and instantly update to reflect the changes.

As explained above, BMRB uses the NMR-STAR format for all stored NMR data. File conversion instructions are described in **Step 1**. It is not necessary to have the data in NMR-STAR format before deposition, but it is very helpful.

Step 4: Entering Relevant Data

After uploading data files and proceeding, the layout of BMRBdep will be as shown in Fig. 4. To progress through the deposition, the user simply clicks the "Next section" or "Previous section" buttons above and below the data entry fields. Alternatively, they can navigate through the deposition and see which sections still need to be completed by opening the left panel menu. To do this, the user clicks the three-horizontal-line icon (called a hamburger menu) on the top left of the page. The navigation panel is illustrated in Fig. 4.

The navigation panel indicates sections that still need to be completed with a circled exclamation mark symbol and indicates completed sessions with a check mark symbol.

To view help information for any data field, clicking on the data field name opens a help box. (This is indicated to the user by the mouse cursor turning into a question mark when hovering over this section of the interface.) Within any given category, a mandatory data field will have a red asterisk (*) next to it. In addition, the data input field will be highlighted with a light pink color if its value must be entered or if an invalid value has been entered.

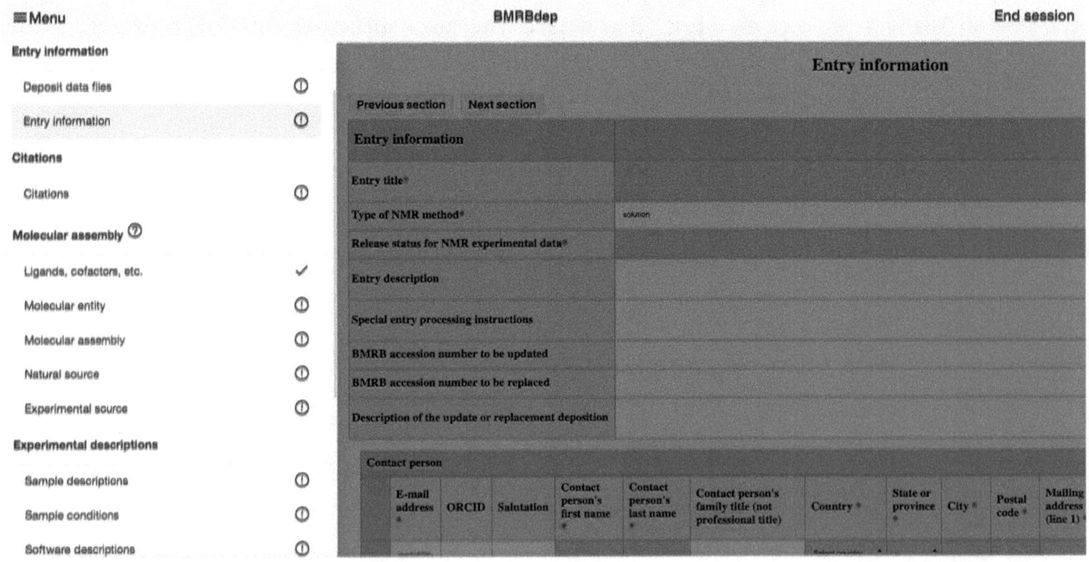

Fig. 4 The BMRBdep navigation menu

Changes are automatically saved as they are made. The save operation has completed when the progress bar at the top of the screen disappears. If the user closes their browser before the data have uploaded to the server, unsaved data are preserved locally in the browser cache. Opening the page again (as long as the "End session" button has not been clicked) will allow the system to save all the changes to the server.

Step 5: Previewing and Depositing the Entry

After data entry is completed, the user can progress to the deposition step by clicking the "Deposit entry" button in the left navigation menu. After selecting that button, the user will be shown any mandatory fields that remain to be filled out and will be given an option to review the full deposition before submission. Once the deposition is complete and passes initial validation, the user will be given the option to submit the entry to the BMRB.

Step 6: Receiving a Report from BMRB/PDB

After a deposition has been submitted through BMRBdep, the authors will receive through e-mail a short notice of receipt with the BMRB accession number. BMRB's annotators will check and validate the entry and then, usually within a few days, will send the depositor the full annotation reports from BMRB with comments for clarifications and/or updates, including the appropriate AVS and PANAV reports for the assignments of protein systems. A link to the processed entry in NMR-STAR format is included in the letter, and any corrections/updates can be sent to BMRB annotators.

Step 7: Hold and Release of the Entry

When the deposition is completed, the deposited data will be put on "Hold" status, until the publication of the associated paper or the date specified by the user (up to 1 year following deposition). The on-hold status of any entry can be confirmed by accessing "Entries on Hold" from the "Search Archive" section of the BMRB website's navigation menu (http://www.bmrb.wisc.edu/data_library/held.shtml). BMRB and PDB accept revisions of entries at any time prior to release.

Upon publication of the associated paper or, in the absence of a publication, 1 year after deposition, the entry data will be released to the public on the BMRB and, if applicable, PDB archives. Users are encouraged to notify BMRB/PDB when the paper associated with a deposition is published so that its release is timely.

3.1.3 PDBj-BMRB SMSDep

In recognition of the fact that scientists have no place to archive information about NMR structures of biomolecules that fall outside the guidelines of the PDB (e.g., small cyclic peptides), the BMRB will consider accepting coordinate sets representing 3D structural models provided that the following criteria are met:

- The molecule falls outside the guidelines of the PDB (i.e., the molecule is a peptide with 23 or fewer residues, a polynucleotide with three or fewer residues, a polysaccharide with three or fewer sugar residues, or a natural product).
- The molecule is of biological interest.
- The structural model(s) are based on experimental NMR data.
- The coordinates are accompanied by a representation of the covalent structure of the molecule (atom connectivity), the assigned NMR chemical shifts for the molecule, and the structural restraints used in generating the structural model.

For depositions meeting these criteria, BMRB encourages authors to submit to PDBj-BMRB SMSDep (https://smsdep.pdbj.org), in addition to the primary (time-domain) data, peak lists, NOEs, and other relevant information.

3.2 Data Retrieval

Users have different options for accessing and downloading data from BMRB. Search options include a powerful "instant search" tool that searches the databases according to keywords entered by the user, an "advanced search" search page that lets the user control both the search criteria and the output requested (particularly useful for downloading subsets of the databases), and a "search grid" that provides single-click access to common searches by data types (chemical shifts, relaxation values, restraints, etc.). Data can also be downloaded from a web-accessible FTP server or kept up to date with the BMRB archive over time by connecting to the BMRB RSYNC server.

3.2.1 Instant Search

The BMRB "instant search" bar is present in the header of every BMRB web page. This search bar will search BMRB entries as you type and shows an automatically updating list of matching BMRB entries. The user can either press "enter" to go to a dedicated search results page or click on one of the suggested results to go directly to the entry summary page for the matched BMRB entry. Alternatively, entering a BMRB entry ID and pressing enter will take you directly to that entry.

The search is performed against multiple relevant fields in a BMRB entry: the citation authors, the entry title, the chemical formula and InChI key of any ligands, database IDs associated with the entry (PDB, PubChem, etc.), and a variety of other commonly searched fields. By hovering the mouse cursor over the suggested results, additional information about the entry will be displayed, including a description of exactly which field was matched.

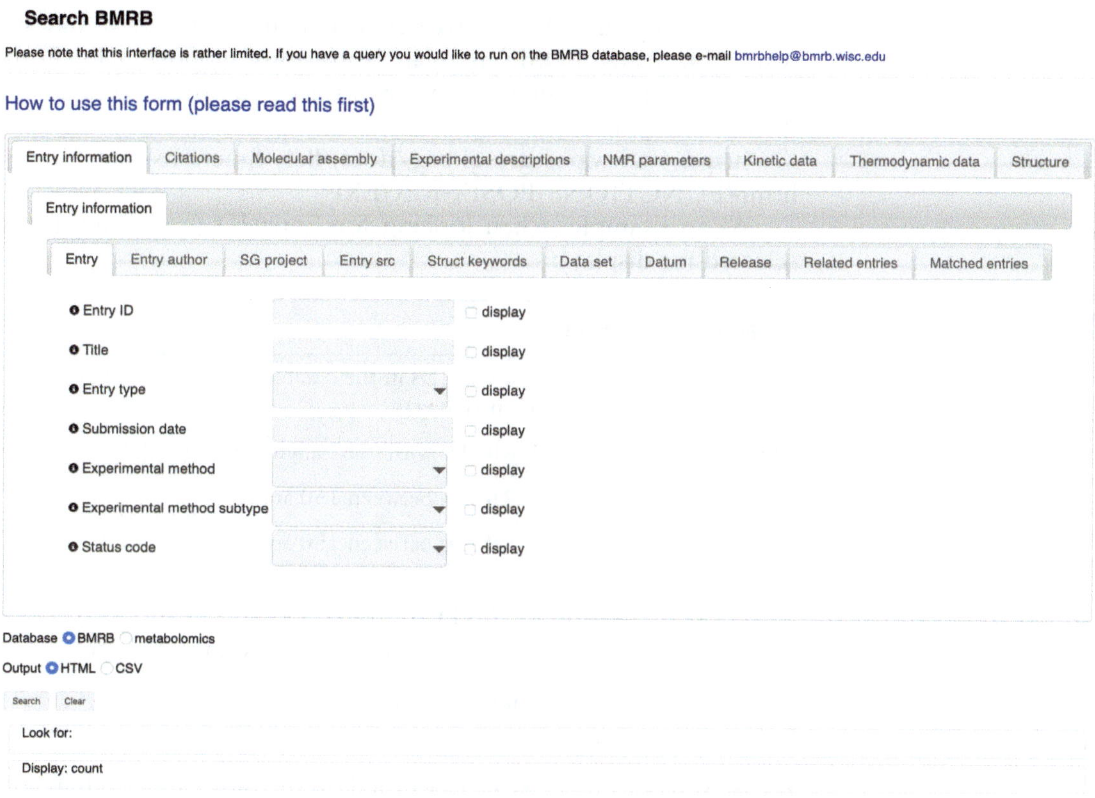

Fig. 5 The advanced search interface

3.2.2 Advanced Search

BMRB provides an "advanced search" interface that is well suited for generating tabular subsets of the databases, useful for data science projects. The interface provides access to most fields in either the BMRB macromolecule or small-molecule (metabolomics) database (represented by their corresponding NMR-STAR tag).

NMR-STAR 3.2 tags map to database tables and columns as `_Table.Column`, e.g., the value of `_Entry.ID` is stored in the "ID" column of the "Entry" table. In order to search for an _Entry.ID, select the top-level Entry information tab (section), then the middle-level Entry information tab (group), and then the Entry tab at the third level (Fig. 5).

For each searchable tag, there is a "search term" text box and a "display" checkbox (Fig. 5). Select "display" to include the tag in the search result table. This arrangement provides the flexibility of using a tag as part of the search criteria or displaying the tag as a column in the resulting table or both: tags to display do not have to be the same ones searched on, and they need not be in the same tables either. If no tags are selected for display, the result will be the count of matching database rows.

Currently, all fields are treated as case-insensitive text and are searchable using POSIX regular expressions supported by PostgreSQL database engine, as explained in the PostgreSQL documentation:

https://www.postgresql.org/docs/8.4/functions-matching.html#FUNCTIONS-POSIX-REGEXP

As an example, when looking for Entry.ID (and Entry.ID is selected for display):

Search term	Result
.*	List of all BMRB IDs in the database (entries that have anything in ID tag)
15.*	List of BMRB IDs with "15" anywhere in the ID
^15.$	List of BMRB IDs between 150 and 159
^15[0–3]$	List of BMRB IDs between 150 and 153

All search terms are AND'ed. together.

Please note that the limitations of this interface include:

- No support for numeric comparisons or range searches.
- No OR searches.
- An SQL JOIN clause is performed on the tables to search in and the tables to display. This may affect the results as JOIN, which may exclude some rows.
- Join to bullet above some regular expressions supported by PostgreSQL do not work, for example, "advanced" REs starting with ***:(?options).

If you have a query you would like to run on the BMRB database that is not supported by this interface, contact bmrbhelp@bmrb.wisc.edu.

Search results are returned either as a webpage (in a separate window/tab) or a comma-delimited file. Note that BMRB's metabolomics (i.e., small molecule) archive is maintained as a separate database; you can search in either macromolecule or metabolomics database (but not both at the same time).

3.2.3 BMRB Query Grid

BMRB's query grid interface provides a mechanism for retrieving sets of entries that fit the criteria of predefined queries from the BMRB archive. It is accessible from the "Search Archive" sub-menu in the website navigation panel (Fig. 6). In the main query grid page, the query criteria are described by the headings for the rows and columns in each grid (*see* Fig. 7). For example, from the main grid, if one clicks on the link in the box in the grid that is at the intersection of the column labeled "DNA" and the row labeled "^{1}H Chemical Shifts," they will be taken to a page containing a listing of

Fig. 6 Search archive sub-menu in the navigation panel

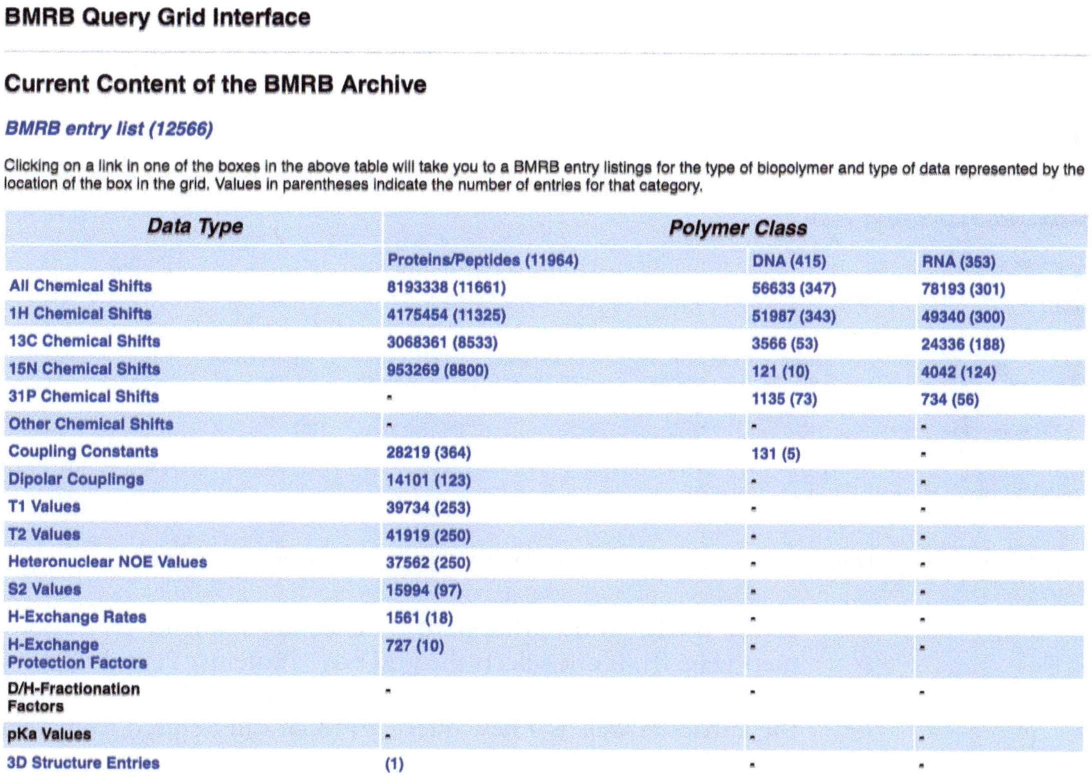

BMRB Query Grid Interface

Current Content of the BMRB Archive

BMRB entry list (12566)

Clicking on a link in one of the boxes in the above table will take you to a BMRB entry listings for the type of biopolymer and type of data represented by the location of the box in the grid. Values in parentheses indicate the number of entries for that category.

Data Type	*Polymer Class*		
	Proteins/Peptides (11964)	DNA (415)	RNA (353)
All Chemical Shifts	8193338 (11661)	56633 (347)	78193 (301)
1H Chemical Shifts	4175454 (11325)	51987 (343)	49340 (300)
13C Chemical Shifts	3068361 (8533)	3566 (53)	24336 (188)
15N Chemical Shifts	953269 (8800)	121 (10)	4042 (124)
31P Chemical Shifts	-	1135 (73)	734 (56)
Other Chemical Shifts	-	-	-
Coupling Constants	28219 (364)	131 (5)	-
Dipolar Couplings	14101 (123)	-	-
T1 Values	39734 (253)	-	-
T2 Values	41919 (250)	-	-
Heteronuclear NOE Values	37562 (250)	-	-
S2 Values	15994 (97)	-	-
H-Exchange Rates	1561 (18)	-	-
H-Exchange Protection Factors	727 (10)	-	-
D/H-Fractionation Factors	-	-	-
pKa Values	-	-	-
3D Structure Entries	(1)	-	-

Fig. 7 Main query grid page

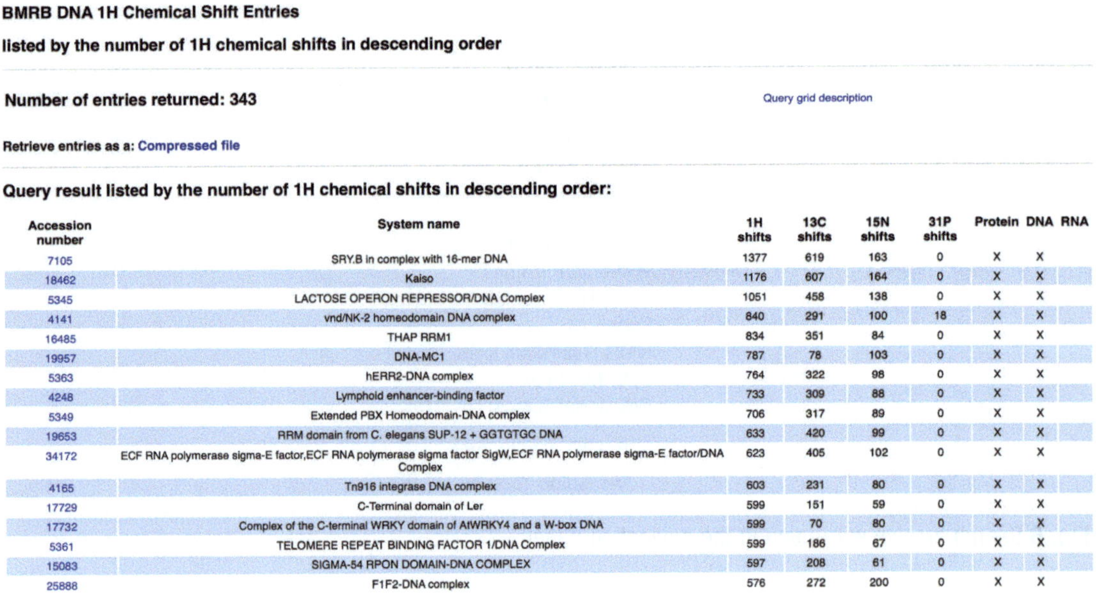

BMRB DNA 1H Chemical Shift Entries

listed by the number of 1H chemical shifts in descending order

Number of entries returned: 343 Query grid description

Retrieve entries as a: Compressed file

Query result listed by the number of 1H chemical shifts in descending order:

Accession number	System name	1H shifts	13C shifts	15N shifts	31P shifts	Protein	DNA	RNA
7105	SRY.B in complex with 16-mer DNA	1377	619	163	0	X	X	
18462	Kaiso	1176	607	164	0	X	X	
5345	LACTOSE OPERON REPRESSOR/DNA Complex	1051	458	138	0	X	X	
4141	vnd/NK-2 homeodomain DNA complex	840	291	100	18	X	X	
16485	THAP RRM1	834	351	84	0	X	X	
19957	DNA-MC1	787	78	103	0	X	X	
5363	hERR2-DNA complex	764	322	98	0	X	X	
4248	Lymphoid enhancer-binding factor	733	309	88	0	X	X	
5349	Extended PBX Homeodomain-DNA complex	706	317	89	0	X	X	
19653	RRM domain from C. elegans SUP-12 + GGTGTGC DNA	633	420	99	0	X	X	
34172	ECF RNA polymerase sigma-E factor,ECF RNA polymerase sigma factor SigW,ECF RNA polymerase sigma-E factor/DNA Complex	623	405	102	0	X	X	
4165	Tn916 integrase DNA complex	603	231	80	0	X	X	
17729	C-Terminal domain of Ler	599	151	59	0	X	X	
17732	Complex of the C-terminal WRKY domain of AtWRKY4 and a W-box DNA	599	70	80	0	X	X	
5361	TELOMERE REPEAT BINDING FACTOR 1/DNA Complex	599	186	67	0	X	X	
15083	SIGMA-54 RPON DOMAIN-DNA COMPLEX	597	208	61	0	X	X	
25888	F1F2-DNA complex	576	272	200	0	X	X	

Fig. 8 Listings obtain by selecting "DNA" and "^{1}H Chemical Shifts" on the main query grid page

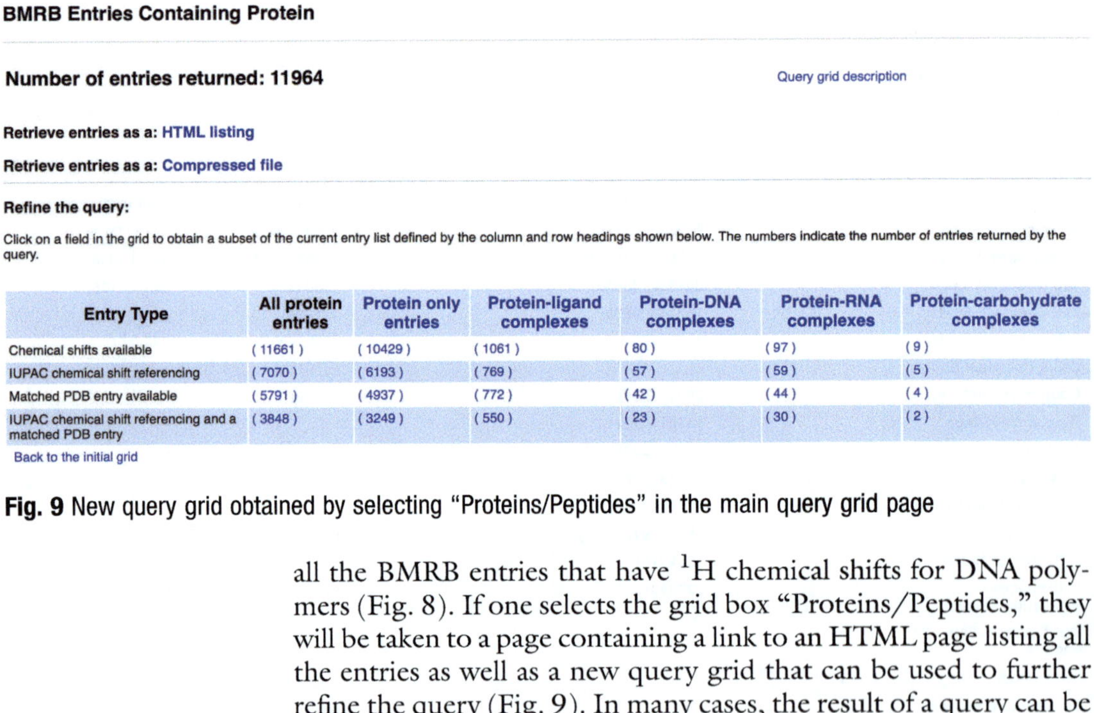

BMRB Entries Containing Protein

Number of entries returned: 11964 Query grid description

Retrieve entries as a: HTML listing

Retrieve entries as a: Compressed file

Refine the query:

Click on a field in the grid to obtain a subset of the current entry list defined by the column and row headings shown below. The numbers indicate the number of entries returned by the query.

Entry Type	All protein entries	Protein only entries	Protein-ligand complexes	Protein-DNA complexes	Protein-RNA complexes	Protein-carbohydrate complexes
Chemical shifts available	(11661)	(10429)	(1061)	(80)	(97)	(9)
IUPAC chemical shift referencing	(7070)	(6193)	(769)	(57)	(59)	(5)
Matched PDB entry available	(5791)	(4937)	(772)	(42)	(44)	(4)
IUPAC chemical shift referencing and a matched PDB entry	(3848)	(3249)	(550)	(23)	(30)	(2)

Back to the initial grid

Fig. 9 New query grid obtained by selecting "Proteins/Peptides" in the main query grid page

all the BMRB entries that have ^{1}H chemical shifts for DNA polymers (Fig. 8). If one selects the grid box "Proteins/Peptides," they will be taken to a page containing a link to an HTML page listing all the entries as well as a new query grid that can be used to further refine the query (Fig. 9). In many cases, the result of a query can be displayed with the entries sorted by the number of ^{1}H, ^{13}C, and ^{15}N chemical shifts or BMRB accession number or PDB code. By selecting the "Compressed file" link, it is possible to download a single compressed file that contains all the entries returned by the current query.

Other query grids by data type are available under the "BMRB Data by Type" item in the navigation panel. The current queryable data types are:

- Macromolecular types.
- NMR spectral parameters.
- Restraints with atomic coordinates and chemical shifts.
- Kinetics.
- Thermodynamics.
- Small-molecule structures.
- Time-domain sets.
- Solid-state NMR.
- Unfolded proteins.
- Binding data.
- Entries relating to human diseases.
- CS-Rosetta structures for BMRB entries.
- Human genes.

By selecting an item in the sub-menu, the user gets directed to a query grid interface for the corresponding data type.

Cautionary Notes

- The results obtained for any query will depend on the quality of the annotation for each entry. If ligands are not reported for a protein entry but do exist, the entry will still be included in the "Protein only" query result. In general, entries with accession numbers below 4000 may appear in inappropriate query results for these reasons.
- In some cases, it is important to think about how a query may have been constructed. If one selects "DNA" from the main query grid, all of the entries where DNA was reported in the molecular system studied will be returned whether or not there is any quantitative data for the DNA. However, by selecting the link in the grid box corresponding to "DNA" and "All Chemical Shifts," the entries containing DNA and where chemical shifts for the DNA were reported will be returned.
- Currently a few BMRB entries (~25) in an older format are not included in the current queries.

The NMR Restraints Grid

A special query grid is the "NMR Restraints Grid" (Fig. 10), also accessible from the "Search Archive" sub-menu, under "NMR Restraints from PDB MR Files" (Fig. 6). The NMR Restraints Grid provides access to the restraints data from the PDB NMR structures contained in the BMRB archive.

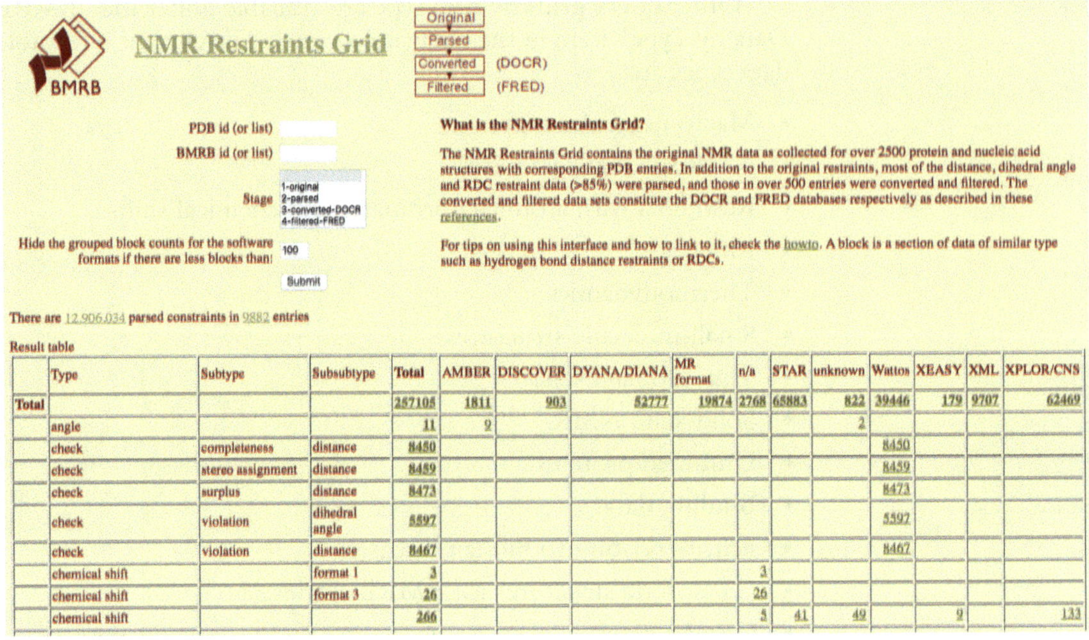

Fig. 10 NMR Restraints Grid interface

In addition to the original restraints, most of the distance, dihedral angle, and RDC restraint data (>85%) were parsed, and those in over 500 entries were converted and filtered. The converted and filtered data sets constitute the DOCR and FRED databases, respectively [21].

To obtain the restraints for a specific PDB ID, follow these steps:

1. Start by selecting the NMR Restraints Grid so that the default options are used.
2. Specify a PDB entry to find data by entering its four-character code (e.g., 1A24).
3. Select 0 for "Hide the grouped block counts for the software formats if there are less blocks than."
4. Submit by clicking on the Submit button.
5. A summary table with all the types of restraints archived for the entry will appear. The user can select different sets of restraints by clicking in the corresponding numbers or select all by clicking in the number under the row and column labeled "Total."
6. In the resulting table, select one of the blocks by clicking on the number in the cell where the column is labeled mrblock_id. For example, select the block ID for where the stage column reads 3-converted-DOCR and the columns program, type, subtype, and format read XPLOR/CNS, distance, NOE, and ambi (Block ID 468407).

7. After selecting the mrblock_id, the NOE distance restraints of entry 1A24 from the DOCR database will be shown and can be saved in a ZIP file.

Other ways of downloading data are possible. Just check the "How to" page for the NMR Restraints Grid at http://restraintsgrid.bmrb.wisc.edu/NRG/wattos/MRGridServlet/html/howto.html.

3.2.4 Data Download

In general, data can be downloaded from BMRB through the website, either from the search tools described above or by clicking on appropriate download links from entry pages. BMRB users can also access data from our databases through our FTP server at http://www.bmrb.wisc.edu/ftp/pub/bmrb/.

The following directories are available at the FTP site:

- Derived data.
- Entry directories.
- Entry lists.
- Internal data.
- Metabolomics.
- NMR-STAR dictionary.
- NMR PDB integrated data.
- PDB MolProbity.
- Relational tables.
- Secondary metabolomics.
- Sequence libraries.
- Software.
- Statistics
- Time domain.
- Validation reports.

Data in these directories comes in different formats, depending on the type of information accessed (e.g., entry data can be downloaded in NMR-STAR, RDF, and/or XML format). Entry pages in the website link directly to the FTP server for associated files, like time-domain data. The relational tables are used to recreate a local copy of the database using PostgreSQL and following the instructions at http://www.bmrb.wisc.edu/search/rdb31.shtml.

This can also be done with the small-molecule (metabolomics) database.

For users or labs that would like to keep a local updated copy of the BMRB databases, BMRB provides access to a public *rsync* server (for LINUX/UNIX-based systems). Contact BMRB through bmrbhelp@bmrb.wisc.edu for help on setting up a local database and keep it updated through *rsync*.

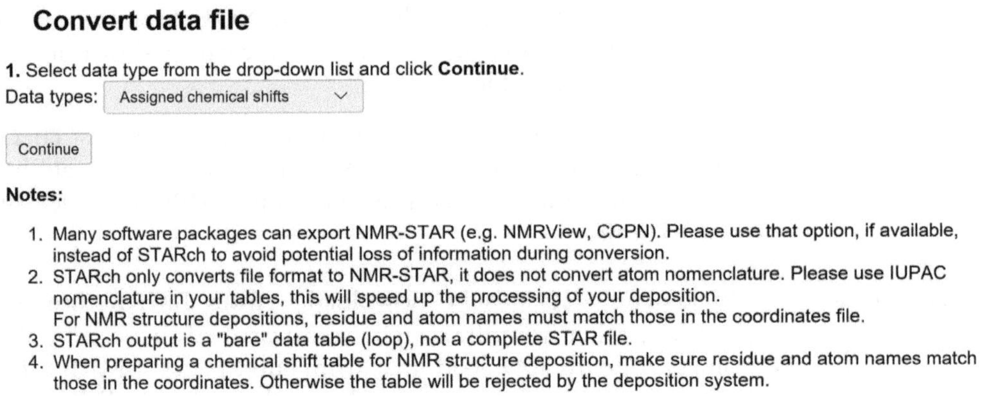

Fig. 11 The STARch file data converted interface

3.3 Data Analysis

3.3.1 Data Conversion

The STARch File Converter

The STARch file converter (http://www.bmrb.wisc.edu/software/starch) can take chemical shift tables in several formats (*see* Fig. 11), including ASCII tables in comma-separated or tab-delimited formats, and convert them into NMR-STAR version 3.2 format. For other data types (e.g., dipolar couplings, RDCs, relaxation values, etc.), STARch can convert NMR-STAR version 2.1 or ASCII tables (comma-separated or tab-delimited) into NMR-STAR 3.2.

Information and recommendations for STARch users are as follows:

- Many software packages can export NMR-STAR (e.g., NMRView, CCPN). For these packages, the export option should be used, rather than STARch, to avoid potential loss of information during conversion.
- STARch only converts a given file format to NMR-STAR; it does not convert atom nomenclature. The user is advised to use IUPAC nomenclature in all tables, which will speed up the processing of the deposition.
- For NMR structure depositions, residue and atom names must match those in the coordinates file.
- STARch output is a "bare" data table (loop), not a complete STAR file.
- When preparing a chemical shift table for NMR structure deposition, make sure residue and atom names match those in the coordinates. Otherwise, the table will be rejected by the deposition system.

3.3.2 Validation

During data curation, BMRB's annotators validate the data files deposited. Under the "Validation Tools" item in the navigation panel, users will find links to many publicly available validation tools, as well as the possibility to download data validation software.

Depositors of protein depositions are strongly encouraged to use validation software packages (in particular the wwPDB validation server and the PSVS (AVS) server) to check their NMR experimental data and structure files before uploading them.

3.3.3 CS-Rosetta High-Throughput Structure Estimation

CS-Rosetta is a software package that attempts to make de novo protein structure predictions from known chemical shifts and optionally RDC and NOE values [12, 13]. Using a library of protein fragments from proteins with experimentally determined structures, it uses Monte Carlo methods to make its structural predictions. Due to the computational effort involved in the Monte Carlo structure determination step, large computational resources can dramatically decrease the amount of time required to run CS-Rosetta. BMRB has utilized its computational resources, as well as those of the CHTC and OSG, to develop a web service that allows for the deposition of chemical shifts (and optionally RDC and NOE values) and returns the results of a CS-Rosetta run in an interactive fashion via a web page. Users who have submitted to the server are e-mailed status during the CS-Rosetta run. When it completes, they are e-mailed a link where they can inspect the generated structures.

When viewing the CS-Rosetta results, an interactive graph showing the RMS distance to the lowest energy structure and the relative energy of all structures is displayed (Fig. 12). Clicking on a point in the graph will pull up that specific structure in an interactive JSmol viewer [22].

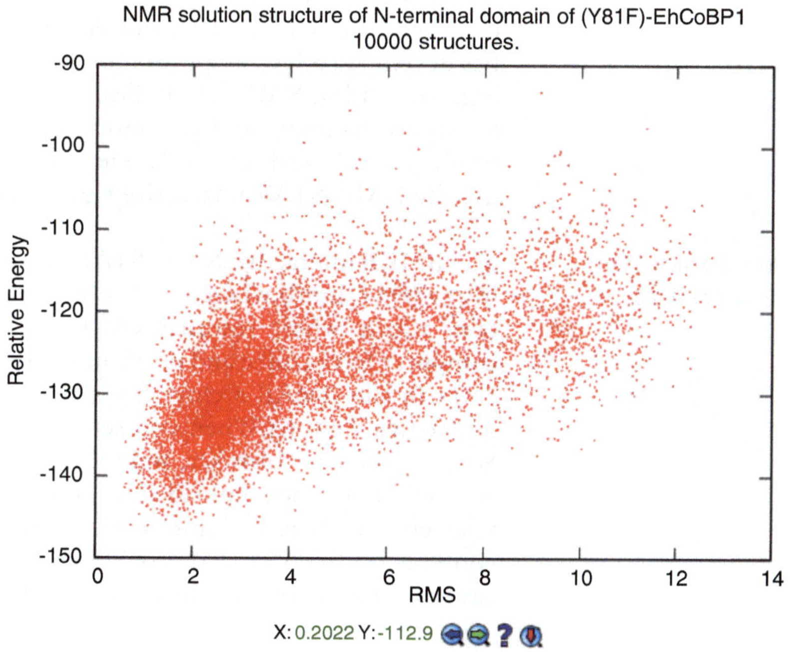

Fig. 12 Calculated CS-Rosetta structures for BMRB entry 19193

In addition to a standard web interface, the CS-Rosetta server also provides an API that can be used to automatically make a submission to the server (http://www.bmrb.wisc.edu/tools/automated_csrosetta.shtml). BMRB has also developed a tool embedded in the NMRbox that allows a user to select input files via a GUI and then submits them to the CS-Rosetta server via this API. Moreover, the NMRFAM PINE [23] server has the option to automatically submit the chemical shifts to the CS-Rosetta server using the API.

The BMRB CS-Rosetta server is available at the following URL: https://csrosetta.bmrb.wisc.edu.

3.3.4 Working with NMR-STAR Files

As described above, NMR-STAR [17] is the archival and exchange format used by BMRB. NMR-STAR is available as input and/or output by several software packages that deal with the harvesting and processing of biomolecular data (CCPN [24], NMRView, TALOS [25], NMRFAM-SPARKY [26], PINE [23], ARECA [27], PONDEROSA [28], Integrative NMR [29], CSI [30], NMRFx [31], RCI [32], ABACUS [33], and PDBstat [34]) and with chemical shift prediction (SHIFTX2 [35] and SHIFTS [36]). NMR-STAR also is used as a data exchange format by the NMRbox project [11].

The NMR-STAR v3.2 ontology [17] provides an extensive controlled vocabulary for the description of NMR spectroscopic studies of biological systems. The ontology includes the description of experiments, the data generated, and the derived results such as molecular structures, dynamics, and functional properties. New NMR techniques and experiments are being developed continuously, and the NMR-STAR ontology grows consequently. BMRB has developed a JavaScript tool for interactively visualizing, validating, and editing NMR-STAR files, as well as a Python programming library for handling and generating NMR-STAR files programmatically (*see* Subheading 3.4.2). Here we describe how to access and use the NMR-STAR interactive viewer tool.

Using the NMR-STAR Interactive Viewer

The BMRB interactive NMR-STAR visualizer is located at the URL http://www.bmrb.wisc.edu/dictionary/starviewer. The page contains a form where you may select an NMR-STAR file from your computer to visualize and perform minor edits. (Alternatively, each entry in the BMRB archive has a link to load its contents in the interactive viewer.) When you select a file, it is processed in JavaScript inside your browser, and the contents of the NMR-STAR file you select are not uploaded to BMRB servers. When a file is selected, the page will automatically render the NMR-STAR file contents on the page and provide several features that will make it easier to inspect or edit the file (Fig. 13):

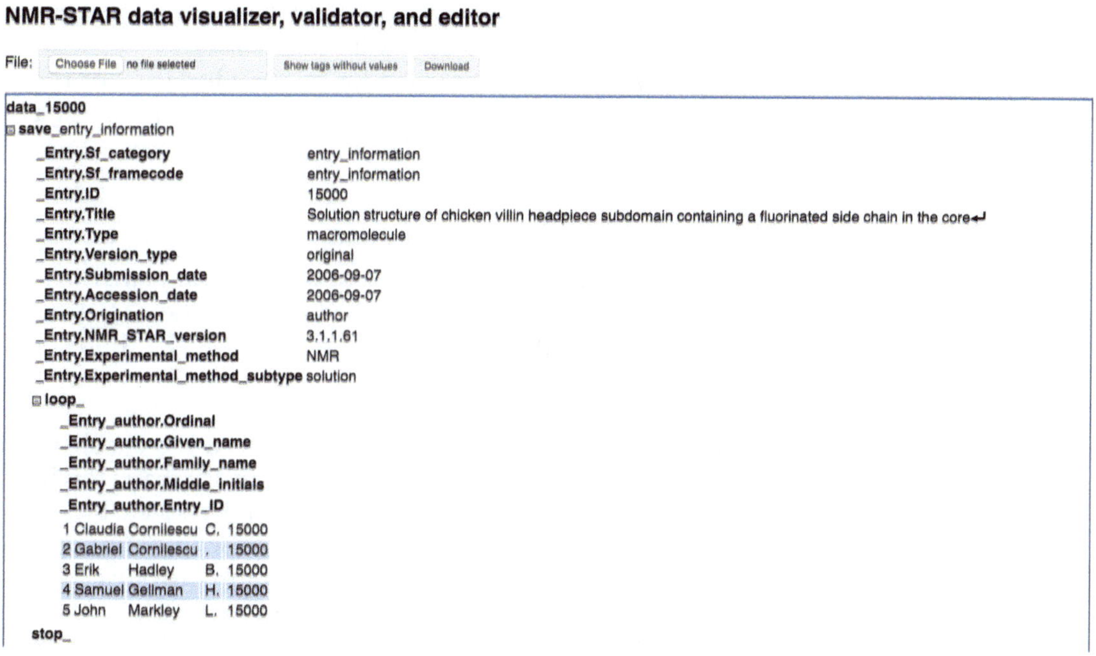

Fig. 13 The NMR-STAR viewer

- Any tags that do not contain values in the uploaded file are hidden by default in order to make the file more readable. You can toggle the display of empty tags by clicking the "Show tags without values" button at the top of the page.
- All tags can be hovered over with a mouse in order to view a description of what data the tag references.
- Data values are styled as plain text rather than an input field, but by clicking on any tag value, you can edit a value. Furthermore, if the tag is one that contains enumerations (suggested values) in the NMR-STAR dictionary, they will be auto-suggested as options when the first characters entered match an existing enumeration.
- NMR-STAR data blocks (called saveframes and loops) can be collapsed to make it easier to view different portions of the entry. In addition, data in loops are lightly color coded in order to facilitate seeing which column a given data value belongs to.
- Data type validation is performed instantaneously as changes are made. Tags with an invalid value (e.g., an invalid date) highlight in red to warn of the error.
- In the NMR-STAR format, it is possible for a tag to reference data elsewhere in the file. Where these links are present, a hyperlink will display that can be clicked to jump directly to the referenced data.

When finished editing the file, click the "Download" button at the top of the page to save a copy with your changes.

Note that this tool is intended for minor edits and not major changes, as creating or deleting data rows is not supported. For more significant modification of NMR-STAR files, *see* the PyNMR-STAR tool (Subheading 3.4.2).

3.4 Programming Tools

The BMRB team has developed several tools for software developers accessing the BMRB databases or manipulating NMR-STAR files. Libraries and other tools described in this section are accessible through the BMRB GitHub page at https://github.com/uwbmrb.

3.4.1 BMRB API

BMRB has developed and deployed a RESTful API to enable rapid access to the most up-to-date version of the database for both metabolomics and macromolecule entries. This API enables programmatic access to the BMRB database from scripts and applications. Many internal BMRB tools use this API to provide access to the most up-to-date version of the BMRB database; PyNMR-STAR, PyBMRB, the Instant Search, the NMR-STAR interactive viewer, and other BMRB tools all use the API to retrieve data.

The API supports a variety of different query types to facilitate research, and new ones are being added in response to community feedback. Examples of supported queries are listed below:

- Given a list of chemical shift values, return a list of BMRB entries that contain one or more of the queried shifts, ordered by number of shifts matched and the closeness of the match.
- Return all chemical shifts in the BMRB, optionally filtered by residue, atom type, and chemical shift value. Optionally, also return the pH and temperature at which the chemical shift was observed.
- Search a protein sequence in FASTA format against the BMRB archive and return a list of matching entries.
- Return all entries that contain a given value for a given tag (e.g., entries that contain the "solid-state" tag).
- Return information about the raw experimental data available for a given entry.
- Return information about a given BMRB entry in either NMR-STAR or JSON format.

Many other query types are also available. For the most up-to-date list of query types, as well as instructions and extensive documentation, please view the GitHub page for the project: https://github.com/uwbmrb/BMRB-API.

3.4.2 PyNMR-STAR: NMR-STAR Handling Python Programming Library

BMBR entries are internally represented as NMR-STAR files, a custom file type for NMR study records that was modeled on the STAR format [18]. This format enables the capture of nearly all NMR-related data and is self-describing and text based, which means it is editable in any standard text editor. Despite that, editing the file while maintaining syntactical validity requires some degree of knowledge of the format. To ease the process of reading, creating, and modifying NMR-STAR files, BMRB has developed and released a Python module (at time of publishing, compatible with Python versions 2.6–3.7) called PyNMR-STAR. This software library significantly eases working with NMR-STAR files—it ensures that all files generated are syntactically valid; it provides tools to validate records against the BMRB NMR-STAR schema; and it provides a wide variety of additional features to ease working with NMR-STAR files.

There is extensive documentation and an introduction to working with the NMR-STAR format at the GitHub page for the project: https://github.com/uwbmrb/PyNMRSTAR.

3.4.3 PyBMRB and RBMRB: Data Retrieval and Visualization Libraries for Python and R

The information-rich data content in BMRB is a useful resource for understanding the properties of atoms in amino acids and nucleic acids in different sequences, conformational states, or sample conditions. Database-wide chemical shift statistics of a particular atom help us to understand the range of its properties. The numerous tags in the NMR-STAR dictionary precisely capture valuable metadata, including sample conditions, chemical shift referencing, and experiment type. The information-rich data content in NMR-STAR files is both human- and machine-readable format.

To help users visualize the data in different ways, BMRB has developed tools with similar functionalities in two languages popular among the bioinformatics community: Python and R. Under normal circumstances, one needs to download and parse the data before visualizing the data either as histogram or NMR spectrum. The BMRB-API supports direct access to the BMRB archive by tools in both Python (PyBMRB) and R (RBMRB) that enable database-level and entry-level data visualizations without the need for downloading and parsing steps. Both PyBMRB and RBMRB use the same visualization tool "plotly" in the backend, which generates interactive and portable visualizations. Interactive graphics visualizations open in any web browser and can be exported as a static image with a single click.

1. Database-Level Visualizations

   Histograms are quite useful for understanding the effects of secondary structure on the chemical shifts of atoms in amino acids and nucleic acids. The chemical shift distributions of atoms in BMRB are sometimes bimodal or trimodal, which indicates that a particular atom is sensitive to secondary

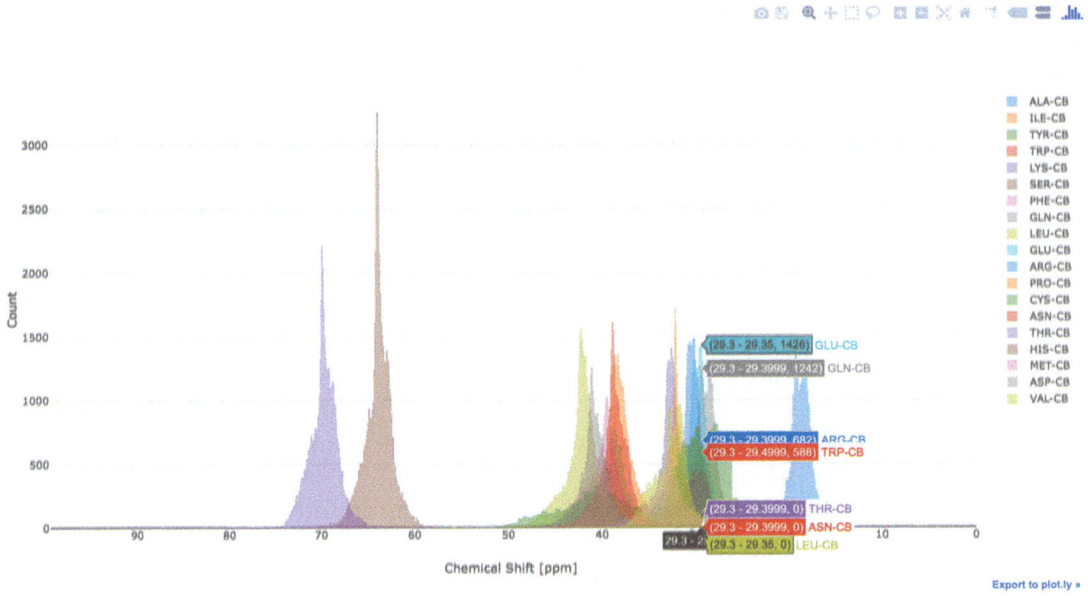

Fig. 14 Chemical shift histogram for CB atoms in BMRB, generated using the PyBMRB library as explained in the text

structural elements like helix, beta sheet, and coil. On the other hand, chemical shift outliers that are far away from expected values may indicate the presence of paramagnetic/metallic element or potential binding sites.

PyBMRB and RBMRB can generate chemical shift histogram of any amino acid or any atom or atom type using a single command. Figure 14 shows the chemical shift histogram of CB atoms generated by the command `hist(atom = "CB")` using PyBMRB v1.2.6. Similar histogram can be generated using RBMRB v2.1.2 with the command `chemical_shift_hist (atm = "CB")`. Chemical shift correlations between any two atoms in an amino acid can be visualized by a 2D histogram. Figure 15 shows the chemical shift correlation between CB and N atoms in cysteine. Additional parameters are available to filter the data for chemical shift outliers using standard deviation-based statistical filtering and to convert the histograms into normalized density plots. PyBMRB v1.2.6 has an additional option that can be used to generate a conditional histogram for a particular atom in a residue filtered against a set of preassigned chemical shifts for other atoms in the same residue.

2. Entry-Level Visualizations

^{1}H–^{15}N correlations from HSQC or other NMR experiments are commonly used in studies of folding, aggregation, ligand binding, or protein-protein interactions. PyBMRB and RBMRB have the functionality of simulating ^{1}H–^{15}N correlations from assigned chemical shifts in BMRB entries. The

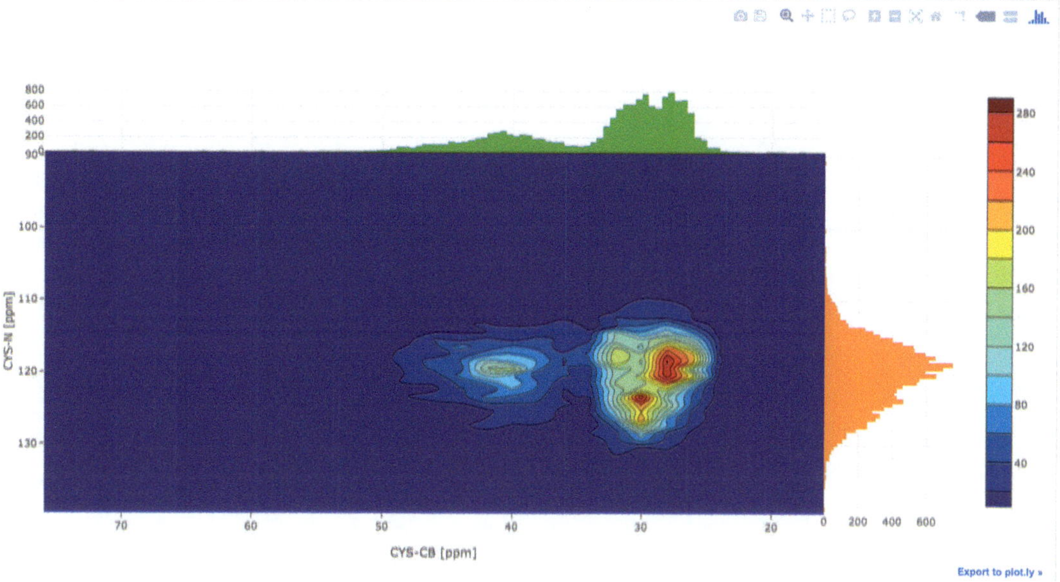

Fig. 15 Interactive diagram showing the chemical shift correlation between CB and N atoms in cysteine across BMRB, generated using the PyBMRB library

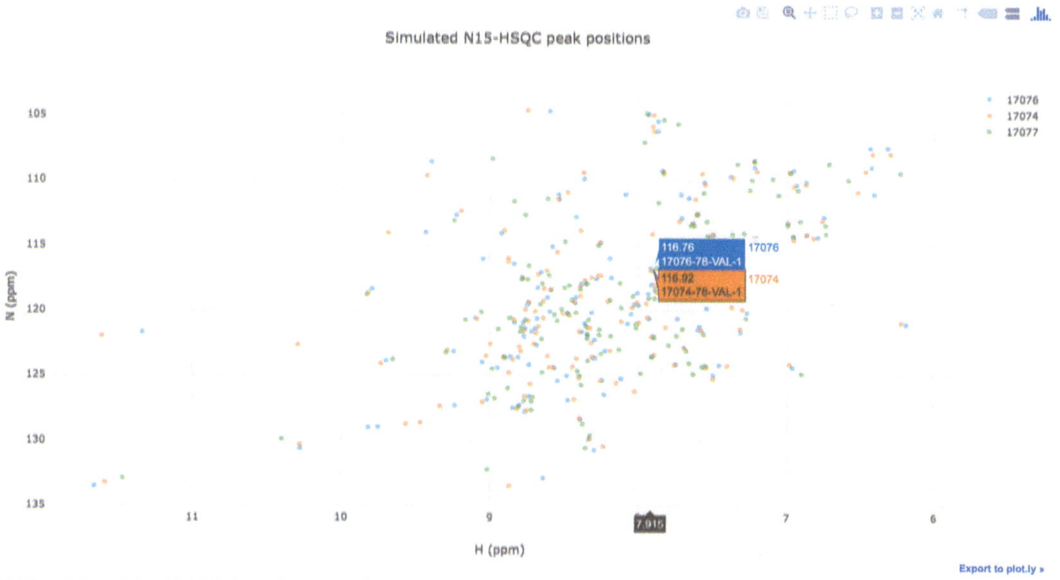

Fig. 16 PyBMRB-generated simulated HSQC overlapping spectra comparing three BMRB entries

software can overlay ^{1}H–^{15}N correlations from multiple entries, and chemical shift changes can be easily tracked by connecting the peaks from resides in same position in the sequence. Figure 16 compares ^{1}H–^{15}N correlations for the enzyme arsenate reductase from three different BMRB entries; the visualization was generated by PyBMRBv1.2.6 with the command:

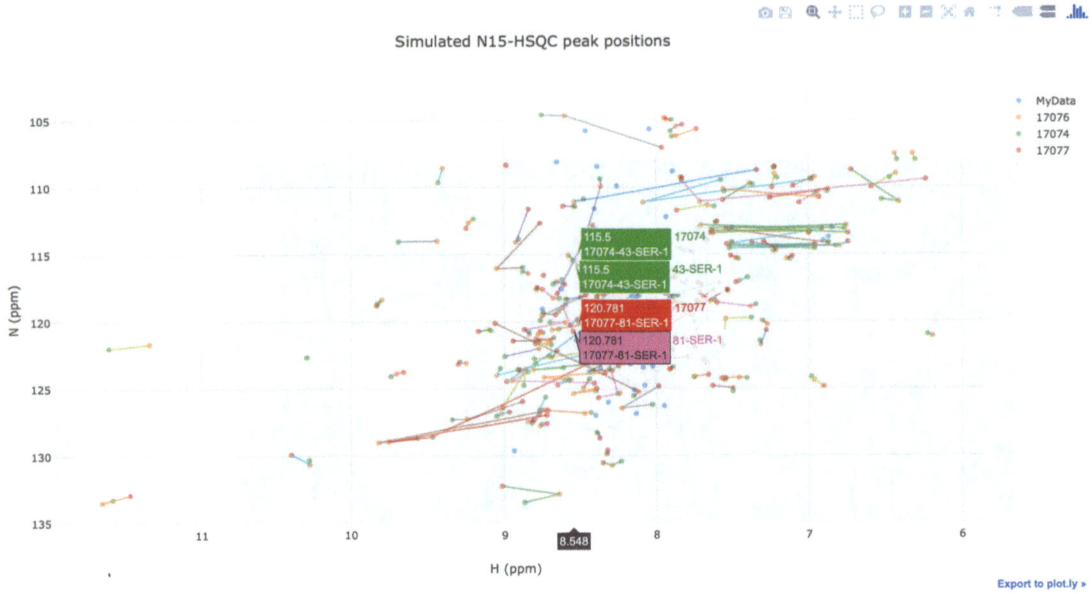

Fig. 17 This simulated HSQC overlapping spectrum comparing a user-uploaded spectrum with three similar entries in BMRB was generated using PyBMRB

```
n15hsqc(bmrbid = [17074,17076,17077]
```

The software can be used in a local computer to compare data in an NMR-STAR file with data from a BMRB entry and to track the chemical shift changes using sequence numbers. Figure 17 shows such a comparison generated by PyBMRBv1.2.6 with the command:

```
'n15hsqc(bmrbid = [17074,17076,17077], filename = 'test.str'),
```

which compares local data (filename) with data from three BMRB entries (17074, 17076, and 17077).

3. Installation and Availability

The source code and installation instructions for both packages are available from the BMRB GitHub repository (https://github.com/uwbmrb/RBMRB, https://github.com/uwbmrb/PyBMRB). These packages are also made available in their corresponding software package repositories, which enables users to easily install the packages using a single command. RBMRB v2.1.2 is available in CRAN repository (https://CRAN.R-project.org/package=RBMRB), and PyBMRB v1.2.6 is available in Python Package Index repository (https://pypi.org/project/pybmrb/). Documentation and example files are available in respective packages.

Jupyter Notebooks

Owing to their ease of use and reproducibility, Jupyter Notebooks (https://jupyter.org) are being widely adopted within the sciences; they provide a single environment that links computation and documentation in an interactive fashion. Both PyBMRB and RBMRB can be used in a notebook environment. The BMRB home page provides a link to a sample PyBMRB Jupyter Notebook, which was created by converting the PyBMRB GitHub repository into an interactive notebook by using a third-party server (https://mybinder.org), which does not require any installation and simply opens in a web browser. BMRB users can use this notebook to play with visualization tools without installing them.

References

1. Ulrich EL, Akutsu H, Doreleijers JF et al (2008) BioMagResBank. Nucleic Acids Res 36:D402–D408. https://doi.org/10.1093/nar/gkm957
2. Berman HM, Westbrook J, Feng Z et al (2000) The Protein Data Bank. Nucleic Acids Res 28:235–242. https://doi.org/10.1093/nar/28.1.235
3. Ladizhansky V (2017) Applications of solid-state NMR to membrane proteins. Biochim Biophys Acta Proteins Proteom 1865:1577–1586. https://doi.org/10.1016/j.bbapap.2017.07.004
4. Sugiki T, Kobayashi N, Fujiwara T (2017) Modern technologies of solution nuclear magnetic resonance spectroscopy for three-dimensional structure determination of proteins open avenues for life scientists. Comput Struct Biotechnol J 15:328–339. https://doi.org/10.1016/j.csbj.2017.04.001
5. Dashti H, Westler WM, Markley JL et al (2017) Unique identifiers for small molecules enable rigorous labeling of their atoms. Sci Data 4:170073. https://doi.org/10.1038/sdata.2017.73
6. Dashti H, Wedell JR, Westler WM et al (2018) Applications of parametrized NMR spin systems of small molecules. Anal Chem 90:10646–10649. https://doi.org/10.1021/acs.analchem.8b02660
7. Wilkinson MD, Dumontier M, Aalbersberg IJ et al (2016) The FAIR Guiding Principles for scientific data management and stewardship. Sci Data 3:160018. https://doi.org/10.1038/sdata.2016.18
8. Wwpdb (2019) Protein data Bank: the single global archive for 3D macromolecular structure data. Nucleic Acids Res 47:D520–D528. https://doi.org/10.1093/nar/gky949
9. Berman H, Henrick K, Nakamura H (2003) Announcing the worldwide Protein Data Bank. Nat Struct Biol 10:980. https://doi.org/10.1038/nsb1203-980
10. Gifford LK, Carter LG, Gabanyi MJ et al (2012) The protein structure initiative structural biology knowledgebase technology portal: a structural biology web resource. J Struct Funct Genom 13:57–62. https://doi.org/10.1007/s10969-012-9133-7
11. Maciejewski MW, Schuyler AD, Gryk MR et al (2017) NMRbox: a resource for biomolecular NMR computation. Biophys J 112:1529–1534. https://doi.org/10.1016/j.bpj.2017.03.011
12. Shen Y, Lange O, Delaglio F et al (2008) Consistent blind protein structure generation from NMR chemical shift data. Proc Natl Acad Sci U S A 105:4685–4690. https://doi.org/10.1073/pnas.0800256105
13. Shen Y, Vernon R, Baker D et al (2009) De novo protein structure generation from incomplete chemical shift assignments. J Biomol NMR 43:63–78. https://doi.org/10.1007/s10858-008-9288-5
14. Chen VB, Arendall WB, Headd JJ et al (2010) MolProbity: all-atom structure validation for macromolecular crystallography. Acta Crystallogr D 66:12–21. https://doi.org/10.1107/S0907444909042073
15. Thain D, Tannenbaum T, Livny M (2005) Distributed computing in practice: the Condor experience. Concurrency Comput Pract Ex 17:323–356. https://doi.org/10.1002/cpe.938
16. Young JY, Westbrook JD, Feng Z et al (2017) OneDep: unified wwPDB system for deposition, biocuration, and validation of macromolecular structures in the PDB archive. Structure 25:536–545. https://doi.org/10.1016/j.str.2017.01.004
17. Ulrich EL, Baskaran K, Dashti H et al (2018) NMR-STAR: comprehensive ontology for

representing, archiving and exchanging data from nuclear magnetic resonance spectroscopic experiments. J Biomol NMR 73:5–9. https://doi.org/10.1007/s10858-018-0220-3

18. Hall SR, Spadaccini N (1994) The Star File – detailed specifications. J Chem Inf Comput Sci 34:505–508. https://doi.org/10.1021/ci00019a005
19. Westbrook JD, Fitzgerald PM (2003) The PDB format, mmCIF, and other data formats. Methods Biochem Anal 44:161–179
20. Gutmanas A, Adams PD, Bardiaux B et al (2015) NMR Exchange Format: a unified and open standard for representation of NMR restraint data. Nat Struct Mol Biol 22:433–434. https://doi.org/10.1038/nsmb.3041
21. Doreleijers JF, Nederveen AJ, Vranken W et al (2005) BioMagResBank databases DOCR and FRED containing converted and filtered sets of experimental NMR restraints and coordinates from over 500 protein PDB structures. J Biomol NMR 32:1–12. https://doi.org/10.1007/s10858-005-2195-0
22. Hanson RM, Prilusky J, Renjian Z et al (2013) JSmol and the next-generation web-based representation of 3D molecular structure as applied to proteopedia. Isr J Chem 53:207–216. https://doi.org/10.1002/ijch.201300024
23. Bahrami A, Assadi AH, Markley JL et al (2009) Probabilistic interaction network of evidence algorithm and its application to complete labeling of peak lists from protein NMR spectroscopy. PLoS Comput Biol 5:e1000307. https://doi.org/10.1371/journal.pcbi.1000307
24. Vranken WF, Boucher W, Stevens TJ et al (2005) The CCPN data model for NMR spectroscopy: development of a software pipeline. Proteins 59:687–696. https://doi.org/10.1002/prot.20449
25. Cornilescu G, Delaglio F, Bax A (1999) Protein backbone angle restraints from searching a database for chemical shift and sequence homology. J Biomol NMR 13:289–302
26. Lee W, Tonelli M, Markley JL (2015) NMRFAM-SPARKY: enhanced software for biomolecular NMR spectroscopy. Bioinformatics 31:1325–1327. https://doi.org/10.1093/bioinformatics/btu830
27. Dashti H, Tonelli M, Lee W et al (2016) Probabilistic validation of protein NMR chemical shift assignments. J Biomol NMR 64:17–25. https://doi.org/10.1007/s10858-015-0007-8
28. Lee W, Kim JH, Westler WM et al (2011) PONDEROSA, an automated 3D-NOESY peak picking program, enables automated protein structure determination. Bioinformatics 27:1727–1728. https://doi.org/10.1093/bioinformatics/btr200
29. Lee W, Cornilescu G, Dashti H et al (2016) Integrative NMR for biomolecular research. J Biomol NMR 64:307–332. https://doi.org/10.1007/s10858-016-0029-x
30. Hafsa NE, Arndt D, Wishart DS (2015) CSI 3.0: a web server for identifying secondary and super-secondary structure in proteins using NMR chemical shifts. Nucleic Acids Res 43:W370–W377. https://doi.org/10.1093/nar/gkv494
31. Norris M, Fetler B, Marchant J et al (2016) NMRFx Processor: a cross-platform NMR data processing program. J Biomol NMR 65:205–216. https://doi.org/10.1007/s10858-016-0049-6
32. Berjanskii MV, Wishart DS (2007) The RCI server: rapid and accurate calculation of protein flexibility using chemical shifts. Nucleic Acids Res 35:W531–W537. https://doi.org/10.1093/nar/gkm328
33. Grishaev A, Steren CA, Wu B et al (2005) ABACUS, a direct method for protein NMR structure computation via assembly of fragments. Proteins 61:36–43. https://doi.org/10.1002/prot.20457
34. Tejero R, Snyder D, Mao B et al (2013) PDBStat: a universal restraint converter and restraint analysis software package for protein NMR. J Biomol NMR 56:337–351. https://doi.org/10.1007/s10858-013-9753-7
35. Han B, Liu Y, Ginzinger SW et al (2011) SHIFTX2: significantly improved protein chemical shift prediction. J Biomol NMR 50:43–57. https://doi.org/10.1007/s10858-011-9478-4
36. Xu XP, Case DA (2002) Probing multiple effects on N-15, C-13 alpha, C-13 beta, and C-13′ chemical shifts in peptides using density functional theory. Biopolymers 65:408–423. https://doi.org/10.1002/bip.10276

Chapter 15

Integrating Molecular Simulation and Experimental Data: A Bayesian/Maximum Entropy Reweighting Approach

Sandro Bottaro, Tone Bengtsen, and Kresten Lindorff-Larsen

Abstract

We describe a Bayesian/Maximum entropy (BME) procedure and software to construct a conformational ensemble of a biomolecular system by integrating molecular simulations and experimental data. First, an initial conformational ensemble is constructed using, for example, Molecular Dynamics or Monte Carlo simulations. Due to potential inaccuracies in the model and finite sampling effects, properties predicted from simulations may not agree with experimental data. In BME we use the experimental data to refine the simulation so that the new conformational ensemble has the following properties: (1) the calculated averages are close to the experimental values taking uncertainty into account and (2) it maximizes the relative Shannon entropy with respect to the original simulation ensemble. The output of this procedure is a set of optimized weights that can be used to calculate other properties and distributions of these. Here, we provide a practical guide on how to obtain and use such weights, how to choose adjustable parameters and discuss shortcomings of the method.

Key words Conformational ensemble, MD simulations, Integrative structural biology

1 Introduction

Experimental determination of biomolecular structure and dynamics is an important and difficult problem in molecular biology. A large variety of techniques to tackle this problem exist, including X-ray/neutron diffraction and scattering experiments, nuclear magnetic resonance (NMR) spectroscopy, cryo-electron microscopy, and a plethora of other techniques. These experiments often result in noisy and incomplete data, making it non-trivial to solve the inverse problem of reconstructing structural and dynamical molecular properties from experiments alone [1].

Computer simulations based, e.g., on physics-derived or knowledge-based models can in principle provide a detailed thermodynamic description for arbitrary molecular systems. Performing such simulations is, however, often not sufficient, due

Zoltán Gáspári (ed.), *Structural Bioinformatics: Methods and Protocols*, Methods in Molecular Biology, vol. 2112,
https://doi.org/10.1007/978-1-0716-0270-6_15, © Springer Science+Business Media, LLC, part of Springer Nature 2020

to inaccuracies of the molecular models (force fields) and the high computational cost associated with extensive simulations.

For these reasons, there exist a number of integrative approaches in which simulations and experiments are combined [2–6]. In some of these approaches, a physical model (i.e., the force field) is complemented by a set of experimental restraints favoring molecular conformations that ***individually*** match experimental data. When studying flexible molecular systems that populate multiple conformations, however, this approach leads to wrong results, because it drives the simulation towards "intermediates" that are not representative of any of the relevant states (Fig. 1) [6–11]. Such problems may be particularly relevant when the

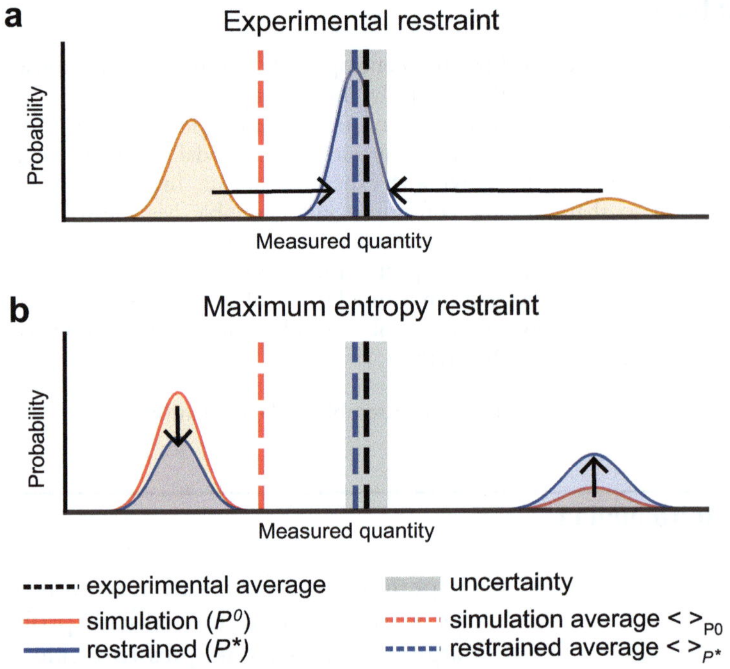

Fig. 1 Schematic example showing two different strategies to restrain simulations using experimental data. When performing a molecular simulation, samples from a *prior* distribution, P^0, are generated, using, for example, the Boltzmann distribution from a Molecular Dynamics (MD) simulation. When a calculated average $\langle \cdot \rangle_{P^0}$ does not match the experimental measurement, it is possible to use the experimental data to modify the prior distribution, resulting in new, optimized probability distribution P^* (also called the *posterior*). The new average $\langle \cdot \rangle_{P^*}$ matches the experimental data with some level of uncertainty. Different strategies to derive the posterior distribution are possible. (**a**) A common choice is to require all individual molecular conformations to match the experimental data within uncertainty. (**b**) In the ME formalism, P^* is instead the minimal modification to P^0 that brings the calculated averages to match the experimental data, resulting in the optimal combination between simulations and experiments

systems are structurally very heterogeneous, and when the experimental measurements have nonlinear dependencies of the conformational properties.

Maximum-entropy (ME) [12] approaches treat experimental data as time/ensemble averages, and make it possible to combine the physico-chemical information derived from the simulation with experimental knowledge. In its basic implementation, however, the ME formalism does not take uncertainty and noise into account. Recently, it has, however, been shown how to generalize ME to take into account the uncertainty associated with the experimental data [10, 13, 14].

There are two principally different ways of combining the experimental data and molecular force field to generate the ME ensemble [11]. One set of methods uses the experimental data directly as restraints *during* the simulations, thus generating samples directly from the target probability distribution. An advantage of this approach is that one can focus sampling efforts only on the most relevant regions of conformational landscape, but comes at the cost of both additional complexity in simulation software and a requirement that the experimental data can be calculated rapidly from molecular conformations and with analytical gradients. The second approach uses standard simulation methods to generate a conformational ensemble, which is then reweighted *afterwards* using the experimental data to generate a weighted ensemble representing the target probability distribution P^*. The advantages of this approach include its simplicity, the fact that it can easily be combined with numerous methods for enhanced sampling, and that one can use rather complex models for calculating experimental observables [15]. It is this second method that is the focus of this paper.

Thus, we here describe a procedure to apply the ME approach to existing simulations sampled from some prior probability distribution. The procedure is in essence identical to the Bayesian inference of ensembles (BioEn) [10, 16, 17], also originally called ensemble refinement of SAXS (EROS) [18, 19], that consists in finding the set of weights that maximize a functional that ranks configuration space distributions. Here, we explicitly make use of the ME formalism: this makes it possible to simplify considerably the minimization problem [13, 20]. For ease of reference, we refer to our approach as Bayesian/MaxEnt (BME) reweighting. Note that equivalent or similar reweighting schemes have been used to construct conformational ensembles in biomolecular contexts [21–27].

We begin by briefly describing the underlying theoretical problem, and then exemplify the procedure on a two-dimensional toy model. We then proceed with providing a step-by-step guide for two examples showing how to combine (1) NMR data with MD simulations of a single-stranded RNA tetranucleotide, and

(2) SAXS data with both atomistic and coarse-grained simulations of a dynamic, two-domain protein. In these examples we also show how to use the optimized weights to calculate other structural properties, thereby providing a more accurate description of the system of interest. Throughout the examples we use our Python library called BME, which is freely available at https://github.com/KULL-Centre/BME under the GNU GPLv3 license, and where the reader may also find detailed step-by-step guides and examples.

2 Theoretical Background

We consider the case in which one samples molecular conformations, $\mathbf{x}$, from a prior distribution $P^0(\mathbf{x})$. Sampling can be performed using, e.g., MD simulation with an atomistic force field or Monte Carlo simulations with a coarse-grained model. In practice, our model P^0 is only an approximation of the "true" (but generally unknown) probability distribution P^{TRUE}. Depending on the system, P^{TRUE} may be characterized by a single dominant state (as in a structured protein) or by several distinct states with different populations, as in single-stranded RNA or intrinsically disordered proteins. Because of model inaccuracies, P^0 and P^{TRUE} may differ: In these cases averages calculated from simulations $\langle F^{calc}\rangle$ may not agree with corresponding experimental measurements $F^{\exp}$.

In the BME approach, one seeks a new probability distribution P^* with the following properties:

- It maximizes the relative Shannon entropy:

$$S_{\mathrm{REL}}(P||P^0) = -\int d\mathbf{x}P(\mathbf{x})\log\left[\frac{P(\mathbf{x})}{P^0(\mathbf{x})}\right] \tag{1}$$

- It matches m experimental restraints $F_i^{\exp}$ within a tolerance, ϵ_i, determined via some error model (see further below):

$$\langle F_i^{calc} + \epsilon_i\rangle = F_i^{\exp} \qquad i = 1 \ldots m \tag{2}$$

- It is normalized:

$$\int d\mathbf{x}P(\mathbf{x}) = 1 \tag{3}$$

Note that there are in principle no restrictions on the number (m) and type of experimental restraints: one can, e.g., combine hundreds of ^3J scalar couplings with NOE data and SAXS measurements at the same time. In practical cases discussed here m is in the order of 10^1–10^3. The relative entropy $S_{\mathrm{REL}}(P||P^0)$ is the negative

Kullback-Leibler divergence [7, 28, 29]: The probability distribution that maximizes the relative entropy can then be considered as the smallest modification to P^0, where the notion of distance in probability distribution space is given by the Kullback-Leibler divergence. A direct link to Bayesian statistics is provided by the observation that the Maximum Entropy distribution is the most probable probability distribution compatible with the data [30].

Here, we consider the discrete case where a finite number of configurations $\mathbf{x}_1 \ldots \mathbf{x}_n$ have been sampled from the prior distribution P^0. The integrals in Eq. 1–3 can then be written as summations over the n configurations with corresponding weights $w_1^0 \ldots w_n^0$, so that $\langle F_i^{calc} \rangle = \sum_{j=1}^{n} w_j F_i(\mathbf{x}_j)$.

For methods such as standard MD or MC simulations that generate samples directly from the Boltzmann distribution defined by the force field, the initial weights are uniform ($w_j^0 = 1/n; j = 1 \ldots n$). When using biasing techniques such as umbrella sampling [31] or metadynamics [32], they are non-uniform, and have to be estimated using standard techniques prior to using BME (*see* **Note 1**).

It can be shown [7, 12, 13, 29] that the weights $\{w_1^* \ldots w_n^*\}$ that satisfy Eqs. 1–3 are given by

$$w_j^* = \frac{1}{Z(\boldsymbol{\lambda}^*)} w_j^0 \exp\left[-\sum_i^m \lambda_i^* F_i(\mathbf{x}_j)\right] \tag{4}$$

where the normalization Z is defined as

$$Z(\boldsymbol{\lambda}^*) = \sum_{j=1}^{n} w_j^0 \exp\left[-\sum_i^m \lambda_i^* F_i(\mathbf{x}_j)\right] \tag{5}$$

and $\boldsymbol{\lambda}^* = \lambda_1^* \ldots \lambda_m^*$ is a set of Lagrange multipliers (one per experimental restraint).

When assuming that uncertainties are modeled by independent Gaussian distributions, i.e., $P(\epsilon_i) \propto \exp\left(-\frac{\epsilon_i^2}{2\theta\sigma_i^2}\right)$, the Lagrange multipliers are determined by minimizing the following function [13, 29]:

$$\Gamma(\boldsymbol{\lambda}) = \log\left(Z(\boldsymbol{\lambda})\right) + \sum_i^m \lambda_i F_i^{\exp} + \frac{\theta}{2} \sum_i^m \lambda_i^2 \sigma_i^2 \tag{6}$$

Here, σ_i is the uncertainty on the restraint $F_i^{\exp}$ and includes experimental errors and inaccuracies introduced by the calculation of the experimental quantity from a structure (i.e. the forward model).

Because this combined uncertainty is not always known accurately, a global scaling parameter, θ, is introduced [18]. When θ is large, all σ are multiplied by a large factor, and in the limit $\theta \to \infty$ this corresponds to no confidence in the experimental data and reverts to the prior distribution. Conversely, a perfect match with experimental data is achieved when $\theta = 0$. Note that when $\sigma = 0$ Eq. 6 reduces to the maximum-entropy solution with no error treatment.

It has been shown that the optimal weights w_j^* obtained in this way correspond to the weights that minimize the function [10, 18]:

$$\mathcal{L}(w_1 \ldots w_n) = \frac{m}{2}\chi^2(w_1 \ldots w_n) - \theta S_{\mathrm{REL}}(w_1 \ldots w_n) \tag{7}$$

In this equation, the reduced χ^2 quantifies the agreement with the experiments:

$$\chi^2(w_1 \ldots w_n) = \frac{1}{m}\sum_i^m \frac{\left(\sum_j^n w_j F_i(\mathbf{x}_j) - F_i^{\mathrm{EXP}}\right)^2}{\sigma_i^2} \tag{8}$$

and the relative entropy term $S_{\mathrm{REL}} = -\sum_j^n w_j \log\left(\frac{w_j}{w_j^0}\right)$ measures the deviation from the initial weights w^0.

Few items are worth highlighting. First, the function $\mathcal{L}$ in Eq. 7 can be interpreted as a "pseudo free-energy", where χ^2 plays the role of enthalpy, S_{REL} is the entropy, and the parameter θ is the temperature. At high temperature (large θ) the entropy dominates, while in the limit $\theta \to 0$ all that matters is to minimize the deviation between experiments and simulation. We note also that θ is introduced as a global scaling parameter for all the data, and thus scales the uncertainty of all data uniformly.

An important practical point to note is that while Eq. 7 is more easily interpretable, it may in practice be difficult to minimize if the number of weights, determined by the number of conformations n, is large. One approach previously employed has thus been to cluster the conformations prior to reweighting, thus reducing the number of weights that need to be determined, but at the same time also losing details present in the original ensemble. Since the Bayesian and Maximum Entropy with error formulations are mathematically equivalent, it is thus in some, but not all [17], cases more convenient to minimize $\Gamma(\boldsymbol{\lambda})$ in Eq. 6 rather than Eq. 7, because the number of experimental measurements, m, is typically much smaller than the number of frames, n.

3 Toy Model

We illustrate the application and outcome of the above-described reweighting procedure on a two-dimensional toy model (Fig. 2a). We construct a model with three states (S1, S2, S3) defined by $P^{\mathrm{TRUE}}(x, y)$, that here represents the "true" probability

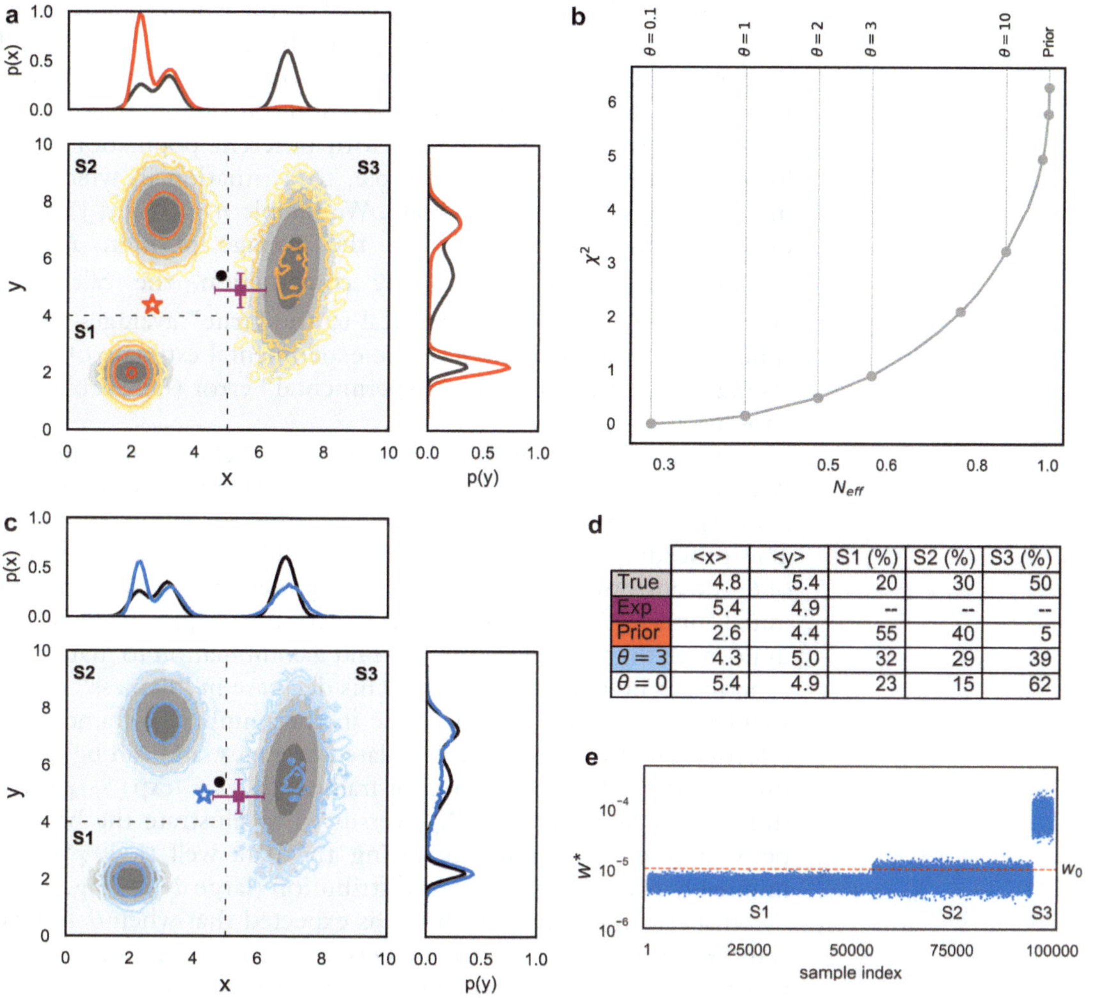

	<x>	<y>	S1 (%)	S2 (%)	S3 (%)
True	4.8	5.4	20	30	50
Exp	5.4	4.9	--	--	--
Prior	2.6	4.4	55	40	5
$\theta = 3$	4.3	5.0	32	29	39
$\theta = 0$	5.4	4.9	23	15	62

Fig. 2 BME reweighting illustrated using a 2D model. (**a**) $P^{TRUE}(x, y)$ (gray scale) and prior distribution $P^0(x, y)$ (orange/red). Both distributions are characterized by three states (S1, S2, S3), but with different populations. The boundaries between the states are shown as dashed lines and are used to calculate the probabilities for the three macrostates, but are not needed in the reweighting analysis. The marginal distributions along x and y are shown as black or red lines. The average x, y position calculated from P^0 (red star) is not compatible with the one calculated from P^{TRUE} (black dot), nor with a hypothetical experimental measurement of $\langle x \rangle$ and $\langle y \rangle$ with associated uncertainty (dark purple square). (**b**) The effective fraction of frames left after reweighting is shown versus χ^2 for different values of the parameter θ. In this case $\chi \approx 1$ is obtained using $\theta = 3$. (**c**) The histogram calculated using the optimized weights w^* with $\theta = 3$ is shown in blue and overlaid on P^{TRUE} (gray scale). The new average (blue star) is, per construction, in better agreement with the experimentally measured average position, and also closer to the "true" average. (**d**) Table reporting the average x, y position and the population of the three states S1, S2, S3 for the "true" model, calculated from the prior P^0, and after reweighting using $\theta = 3$ and $\theta = 0$. (**e**) Values of the optimized weights w^*. The state corresponding to the weights is indicated in the labels, and for visualization purposes the samples were sorted so that samples from the same state are shown together

distribution (shown in gray/black). We then assume that it is possible to measure, with some error, the average of the x and y coordinates, shown as a dark purple square with error bars in Fig. 2a. We then construct a prior distribution $P^0(x, y)$ that has the same three states as P^{TRUE}, but with different populations (red lines). P^0 corresponds, for example, to a situation in which the molecular force field is inaccurate. We sample $n = 10^5$ (x, y) coordinates from P^0, and calculate the average position $F_1^{calc} = \sum_{j=1}^{n} w_j^0 x_j$, $F_2^{calc} = \sum_{j=1}^{n} w_j^0 y_j$. By construction, the calculated averages (red star) are not identical to the "true" averages (black sphere), and the differences to the experimental estimate of these average are greater than the "experimental" error (here arbitrarily chosen).

Given the initial weights $w^0 = 1/n$ of each sample, and the "experimentally" measured values and uncertainties, we minimize $\Gamma(\lambda_1, \lambda_2)$ (Eq. 6) and find the optimal weights w^* defined via Eq. 4. This procedure is repeated for different values of θ. At high values of θ, the entropy term dominates and the weights are close to their initial (uniform) values. As θ is decreased, the weights become less uniform as they are reweighted to find a combination to match the experiment better (decrease χ^2). This decrease in "flatness" of the weights corresponds to a decrease in the number of frames that effectively contribute to the calculated averages, and can be quantified by the "effective fraction of frames" $N_{eff} = \exp(S_{\text{REL}})$. We thus find it useful to plot N_{eff} versus χ^2 to illustrate the balance between the requirement of fitting the data well (low χ^2) and minimally perturbing the prior distribution (large N_{eff}) (Fig. 2b).

Inspection of this plot shows as expected that when $\theta \rightarrow 0$ we achieve a very good agreement between simulation and "experiments" ($\chi^2 \rightarrow 0$). At the same time, we introduce a large perturbation to the prior probability distribution P^0, so that the relative entropy is a large, negative number and the effective fraction of frames becomes small. In the limit of large θ, instead, χ^2 approaches the initial value obtained when sampling from P^0, the new weights w^* are close to w^0 and thus the number of effective frames N_{eff} approaches 1. A practical solution to the trade-off between the two limits can be found by scanning different values of the parameter, starting from a large number, until a further decrease in θ does not result a significant decrease in the associated χ^2. Such a procedure, often termed finding the "elbow" of the curve (and similar to L-curve selection in other regularization techniques), provides a range of viable values for θ: in the toy model, for example, one could pick $\theta = 3$, that leads to a $\chi^2 \approx 1$ for both x and y coordinates. After fixing $\theta = 3$ we can observe the modification to the probability distribution introduced by reweighting (Fig. 2c). First, the calculated average (blue star) is, per construction, closer to the experimental average (within a level set by the uncertainty). Additionally, the reweighted (blue) and "true" distribution (gray) are more similar to one another than the prior and true distributions are.

In this toy model we have three distinct states, whose population can be calculated by summing the weights of the samples belonging to each region. In Fig. 2d we report these population for the "true" model, for the prior distribution P^0 and after reweighting with $\theta = 3$ and $\theta = 0$. We note that this kind of clustering into states is not a part of the actual reweighting procedure, but may be useful in the subsequent analyses. We can see that the population of state S1 decreases from 55% in the unreweighted ensemble to ≈25% in the reweighted one. The population of S3, instead, is increased from 5% to 40%, substantially closer to the "true" population of 50%. Note that it is in principle possible to obtain a better agreement with experiments by letting $\theta \to 0$ (Fig. 2d). In a realistic scenario this would not be advisable, since experimental quantities are not known with infinite precision, and setting $\theta = 0$ effectively corresponds to ignoring the uncertainties in the experiments and forward models.

Finally, it is instructive to plot the individual weights w^* for each sample (Fig. 2e). In agreement with the population shift described above, we can see that all samples belonging to state S1 are down-weighted with respect to the initial weights w^0, while the opposite effect happens to samples belonging to S3.

As is clear from the example above, the BME reweighting procedure enables the reconstruction of an ensemble that is aimed to be closer to the "true" ensemble by combining the imperfect prior model with the experimental data. Before proceeding to discuss applications in molecular simulations and structural biology, we note, however, that there might occur situations in which the reweighting approach described here would provide incomplete or wrong result. More precisely, we highlight four possible sources of problems:

- *Insufficient sampling.* If sampling is not exhaustive, relevant states are not explored, and thus it is not possible to estimate with any certainty their weights after reweighting. This observation is equivalent to the well-known problem of large uncertainties in estimating free energies between states with little overlap. In such cases a small θ (corresponding to a small N_{eff}) could be required to achieve a reasonable agreement between simulations and experiments. As a consequence, most of the optimized weights are vanishingly small, and a small set of weights dominates the ensemble. In this situation, longer simulations or the use of enhanced sampling techniques is necessary, and could, e.g., be guided by the structures whose weights are increased at intermediate values of θ. In practice, the prior that enters the reweighting approach is not the full distribution, P^0, but rather our estimate of this from the finite samples representing the starting ensemble. Best practices and metrics for the quality control of the reweighted ensembles have recently been discussed [33].

- *Inaccurate force field.* The reweighting approach relies on the accuracy of the prior distribution, in particular when the data is sparse and noisy. One can imagine, for example, the case of a uniform prior distribution over x, y (within some range) in the toy model. A small modification to this distribution would lead to a very good agreement with the experimental data, but would not be close to the "true" probability distribution. Indeed, the BME formalism is not guaranteed to give the best model possible. Instead, it provides the least biased model that takes into account all the knowledge we have of the system, encoded both in the (potentially inaccurate) force field and the (noisy and sparse) experimental data, and no more than this information. If more information were available or assumed (such as assuming that the ensemble is narrow), this should preferably be encoded and input into the model.
- *Inconsistent or wrong experimental data.* When data are inconsistent, the reweighting approach might fail in obtaining improved agreement with all experimental data. For example, in our previous work we have identified spectral overlaps by noticing that a subset of NOE distances could not be reweighted [20]. In certain cases, the presence of inconsistent data points could be detected via cross-validation. In general, however, there is no guarantee that erroneous data cannot be fitted with an erroneous ensemble, and indeed ensemble fitting can be prone to such "overfitting" to erroneous data. When very small values of θ are needed to fit the data, this can be a sign of either a poor prior, underestimation of the uncertainty in the data, or actual errors in the experiments.
- *Non-informative experimental data.* There might be situations in which the available experimental data cannot substantially correct the inaccuracies of the model. In the toy model, this would correspond, for example, to knowing the average y position but not x. In such a situation, the "true" population of state S3 could not be determined very accurately because the average x position carries information on the relative populations of S1+S2 with respect to S3. In practice for high-dimensional systems such as biomolecular ensembles, the situation is more complex. Indeed, most experimental measurements are sensitive to some, but not other aspects of the distribution of

conformations, and generally there are many more degrees of freedom than experimental observations. Indeed, it is this underdeterminism that necessitates the use of the prior model, but the user should be aware that not all data provide equal amounts of information.

In realistic cases, these problems can occur simultaneously, and can sometimes be difficult to disentangle. We also stress that BME is inherently an ensemble refinement procedure [18], and the successful use of this approach depends on the amount/quality of experimental data, on sampling, and on force field accuracy. Further discussions of problematic situations in ME approaches are also found in Refs. 6, 10, 29.

4 BME Software

4.1 Requirements, Download and Installation

The BME package requires Python$\geq$2.7 or Python$\geq$3.6 with NumPy [34] and SciPy [35] libraries, and it is freely available at https://github.com/KULL-Centre/BME. The package can either be downloaded as a zip file or cloned using git (www.git-scm.com). Using the software requires a basic understanding of the Python language. BME also does not provide functionality to execute or directly extract conformational properties (distances, angles, etc.) from molecular simulations; instead, the user is expected to have such properties calculated before using BME.

4.2 Combining NMR Data with MD Simulation of RNA

Following the above introduction of the BME method with the two-dimensional toy model, we now proceed to describe how to obtain the conformational ensemble of an RNA tetranucleotide using atomistic MD simulations in combination with experimental data from nuclear magnetic resonance (NMR) spectroscopy. The code to perform the analysis shown below and to produce all the figures in this example can be found in the notebook folder on the github repository (https://github.com/KULL-Centre/BME/tree/master/notebook), both as a Jupyter Notebook and in HTML format. The procedure can be summarized in four steps: (1) Data collection and preparation, (2) Minimizing Γ and parameter selection, (3) Cross validation, and (4) Interpretation of the weights and of the reweighted ensembles.

Step 1: Data Collection and Preparation: The first step is to collect and format experimental and simulation data necessary for BME reweighting. The experimental data file contains a list of averages and associated uncertainties. In our example of an RNA tetranucleotide with sequence `CCCC`, 26 ^{3}J scalar couplings have been measured, and are stored in a file with the following format:

```
# DATA=JCOUPLINGS PRIOR=GAUSS
C1-H1H2   1.0   1.5
C1-H2H3   3.6   1.5
C1-H3H4   8.7   1.5
⋮
C4-2H5P   1.1   1.5
```

The first line is a header that specifies the type of input data and the type of prior on the error. Currently, the BME software specifically supports the following data types: scalar couplings (`JCOUPLINGS`), nuclear Overhauser effect (`NOE`), chemical shifts (`CS`), small-angle X-ray scattering (`SAXS`), and generic distance restraints (`DIST`). *See* also **Note 2**. Only a Gaussian (`GAUSS`) error prior is implemented in the current version of the BME software. The first column is a user-defined label and in this case indicates the atoms involved in the measured ^{3}J scalar coupling. The second column is the experimental value and the third the associated uncertainty. Inequality restraints (e.g., upper/lower) NOE boundaries can be specified (*see* **Note 3**). Degenerate and ambiguous NOEs cannot currently be handled. In our example, experimental data are taken from a previous study [36], but one can also retrieve the data for a system of interest from public repositories such as the BioMagResBank (BMRB: www.bmrb.wisc.edu) or from .mr files in the protein data bank (PDB). The user is expected to process such data to be in the input format for BME.

The other information required for reweighting are the values of the experimental observables calculated from simulation. We here consider an extensive MD simulation of the `CCCC` RNA tetranucleotide taken from our previous study [20], and for each of the $n = 20,000$ frames we calculate the $m = 26$ scalar couplings using Karplus equations. The data is stored in a file with n rows and $m + 1$ columns:

1	$F_{1,1}^{\mathrm{CALC}}$	...	$F_{1,26}^{\mathrm{CALC}}$
2	$F_{2,1}^{\mathrm{CALC}}$	...	$F_{2,26}^{\mathrm{CALC}}$
⋮			
20000	$F_{20000,1}^{\mathrm{CALC}}$	...	$F_{20000,26}^{\mathrm{CALC}}$

The first column is user-defined and can, for example, be the frame number. The other columns report the calculated values of the experimental quantities, so that $F_{1,1}^{\mathrm{CALC}}$ is the scalar couplings `C1-H1H2` calculated for frame 1, $F_{1,2}^{\mathrm{CALC}}$ is `C1-H2H3` for the same frame, and so on. The number of frames n can be on the order of tens or hundreds of thousands: there are in principle no restrictions on n since the complexity of the problem is mostly determined by the number of experimental restraints m. Note also that the calculation of the experimentally probed quantities is performed only once, and any type of forward model can be used [37], as long as the calculated values can be written as a weighted average over the input configurations (*see* **Note 2**).

In Fig. 3a we show in gray the $m = 26$ experimental measurements, sorted by magnitude for visualization purposes. The averages calculated using $n = 20,000$ frames from the simulation are shown in red. Simulations and experiments do not agree perfectly: in this case, at least four calculated averages appear to be significantly different from the experimental measurements.

Step 2: Γ Minimization and θ Selection: Given the data described above we proceed by "adjusting" the simulation by assigning new weights to each frame. The new weights are such that the new computed averages are compatible with the experimental ones given as input. In practice, this is achieved by minimizing the function Γ defined in Eq. 6 with respect to the $m = 26$ Lagrange multipliers. In our implementation, we use the limited memory Broyden-Fletcher-Goldfarb-Shanno (L-BFGS) optimization algorithm implemented in the SciPy library. Since the analytic gradient can easily be calculated, the optimization is computationally inexpensive and typically takes seconds on a standard desktop computer with $m \approx 10^2$ and $n \approx 10^4$–10^5. The optimization returns m values for λ (in this case $m = 26$, the number of experimental restraints) that are used to calculate a weight for each frame via Eq. 4. By definition, the optimal weights improve the agreement with input experimental data. Different types and sources of data can be used at the same time (*see* **Note 4**). As described in the previous section, we then scan different values of the parameter θ, and consider how the fraction of effective frames N_{eff} and χ^2 varies as a function of this parameter.

In our RNA tetranucleotide example (Fig. 3b), a small value of θ corresponds to a better fit with scalar couplings (low χ^2), but

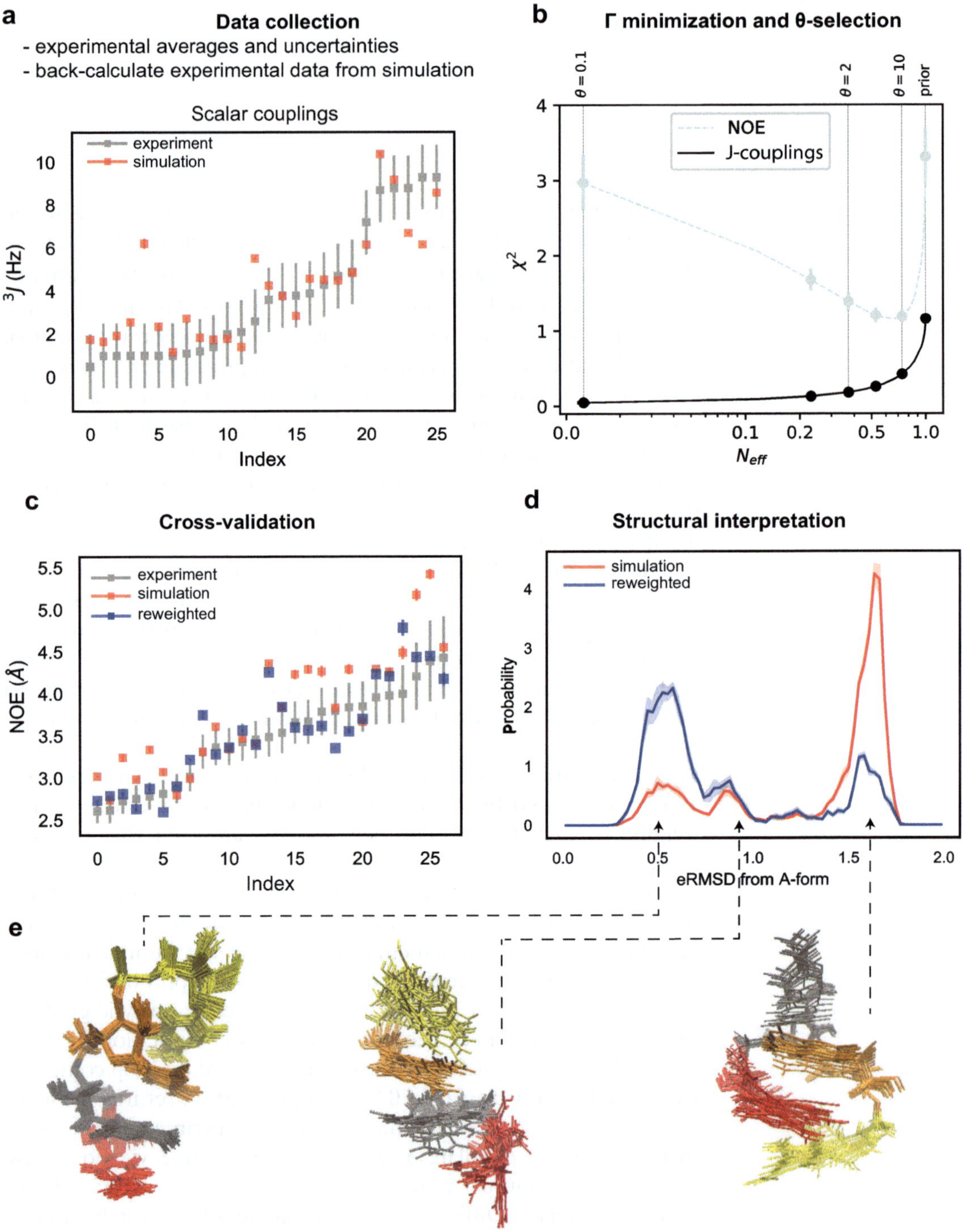

Fig. 3 Experimentally restrained simulation of an RNA tetranucleotide. (**a**) Experimental ^3J couplings (gray) compared to calculated averages using the original MD simulation (prior distribution, shown in red). Error bars indicate experimental uncertainties (gray bars) or the standard error of the mean estimated using five blocks (red bars; typically, smaller than the point). (**b**) N_{eff} versus χ^2 plot using scalar couplings restraints for different values of θ, and cross-validation using NOEs. (**c**) Experimental NOEs (gray) compared to calculated averages

obtained at the cost of a large drop in relative entropy and hence only a small effective fraction of frames used. Using a large θ, instead, we approach the χ^2 of the prior distribution. In this case one can identify useful values of θ in the range 2–10, corresponding to the *"elbow"* region in the N_{eff} versus χ^2 plot.

Step 3: Cross-Validation: When possible, it is recommended (at least initially) to split the experimental data into some used in optimization and some used for cross-validation. In our example, we monitor the agreement between calculated and experimental averages of 26 available NOE measurements that were not used as input for reweighting. By default, NOE data are treated using r^{-6} averaging: the exponent can be changed within the program (*see* **Note 5**). The agreement with NOE distances has a clear minimum around $\theta = 10$ (Fig. 3b). Note that when $\theta = 0.1$ the χ^2 relative to the NOEs is high, meaning that enforcing a tight agreement with scalar couplings is detrimental. Here, we choose $\theta = 2$ as it provides improved agreement with scalar couplings without a dramatic drop in N_{eff}. Having fixed $\theta = 2$, it is possible to use the optimal weights to calculate the NOE averages before (red) and after (blue) optimization (Fig. 3c).

Step 4: Structural Interpretation: In order to understand how the new weights affect the original MD conformational ensemble, it is useful to calculate the distribution of various structural parameters. As an example we show the histogram of the eRMSD [38, 39] from a reference A-form helix of both the original MD simulation (prior) and the restrained one (Fig. 3d). The eRMSD is a structural distance which has been designed to overcome the limitations of standard RMSD calculations in nucleic acids, and can be considered as a contact-map in which both distance and orientation between nucleobases are taken into account. The effect of the experimental restraint is to favor the presence of structures closer to A-form, as also described in our previous work [20]. From the reweighted histogram it is possible to calculate the population of the different substates shown in Fig. 3e that were generated by extracting structures from the original trajectory with the probability given by the optimized weights.

4.3 Combining SAXS Data with Simulation of Proteins

We now proceed to apply the BME method and software on a larger, more complex system with less informative experimental data. In particular, we show the results for a two-domain protein

Fig. 3 (continued) using the original MD simulation (red) and after reweighting using scalar couplings data (blue). (**d**) Histogram of the eRMSD from an A-form RNA structure calculated from the original MD simulation (red) and after reweighting (blue). The three peaks roughly correspond to three different conformations, shown in panel (**e**)

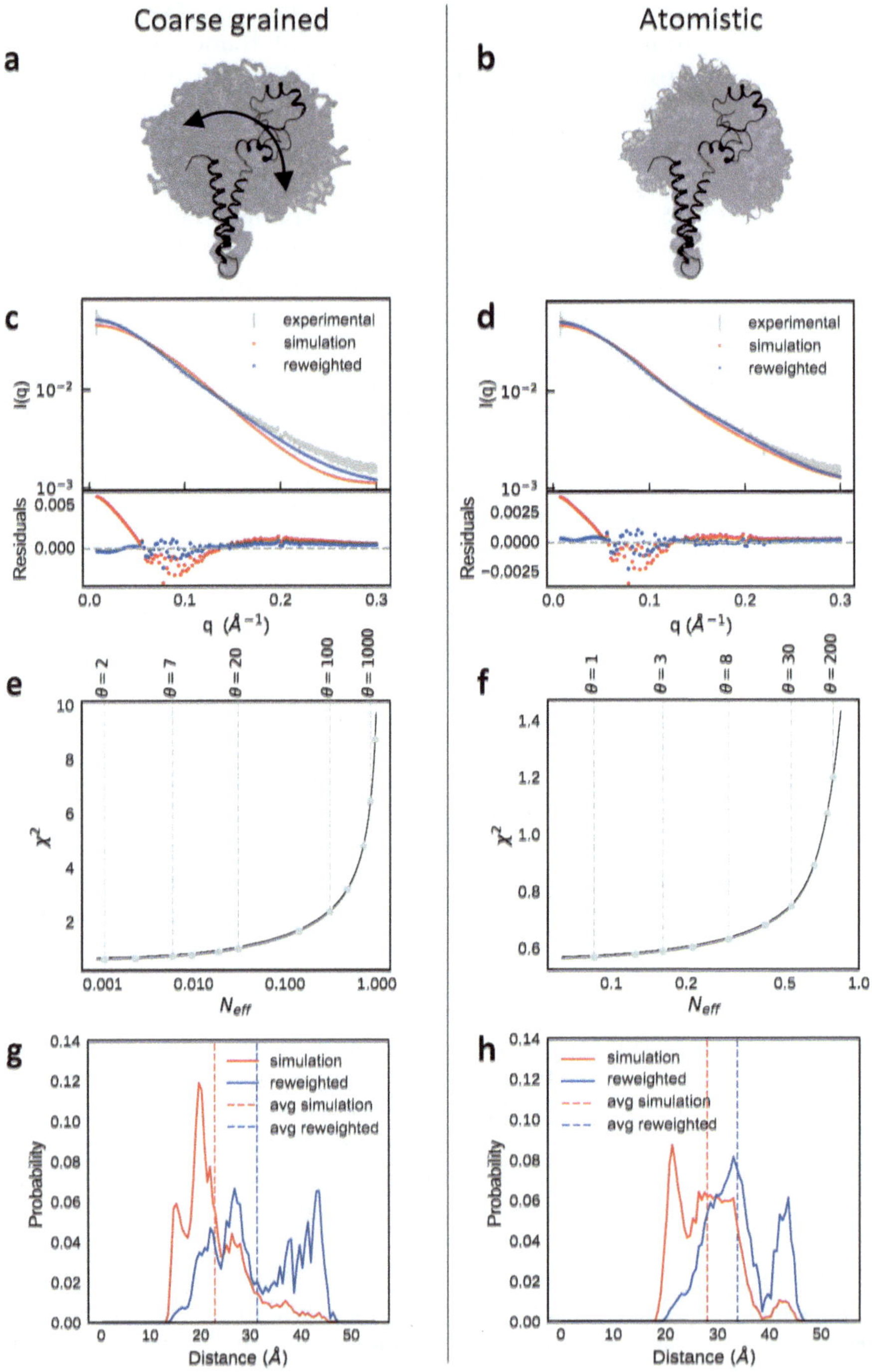

Fig. 4 SAXS refinement of coarse-grained and atomistic simulations of a flexible two-domain protein. Left (**a**, **c**, **e**, **g**), refinement of a coarse-grained MARTINI 3.0b simulation with an elastic network force constant of 500 kJ/(mol nm^2). Right (**b**, **d**, **f**, **h**), refinement of an atomistic simulation using the a99SB-*disp* force field. (**a**, **b**) The black structure is the starting configuration, and the blurred blue illustrates the sampled configurational space for each simulation by showing every fifth frame in the simulation aligned to the (bottom) NTD domain. (**c**, **d**) Calculation of SAXS data from the original MD simulations and the refined ensembles are compared to

that shows substantial structural heterogeneity of the relative orientation of the two domains, and describe how we can refine our molecular simulations by reweighting against small-angle X-ray scattering (SAXS) data. The reweighted ensemble in turn allows us to improve our description of the dynamical domain–domain motions. As object for our study we chose the protein sf3636 from the bacterium *Shigella flexneri 2a*. Previous studies have shown that the protein consists of two structural domains (NTD and CTD), with substantial interdomain motions as probed via different experiments including SAXS [40], thus providing a good example for applying BME to proteins.

Because large-scale motions in proteins might be difficult to sample with conventional simulations, they are attractive targets for coarse-grained (CG) simulations. Specifically, we here applied a recently updated parameterization of the Martini CG model [41]. In line with standard recommendations for studying protein dynamics with Martini, we applied harmonic restraints to keep each of the folded domains relatively rigid [42], whereas no restraints were applied between pairs of atoms spanning between the NTD and CTD. We performed a 4 μs long simulation using MARTINI 3.0beta with the force constant for the harmonic restraint set to the default 500 kJ/(mol nm^2) using the Gromacs software [43] and standard settings. We analyzed and reweighted an ensemble consisting of 8000 structures from the simulation by taking each 500 ps frame (Fig. 4a).

For comparison, we also performed an all-atom, explicit solvent simulation using the a99SB-*disp* force field [44]. This force field has recently been parameterized to provide an accurate balance between protein–protein and protein–water interactions, and thus should be particularly useful for looking at transient interactions between the two domains. Specifically, we performed a 2 μs long simulation using a time step of 2 fs, a temperature of 298 K and 1 bar pressure with the velocity rescaling thermostat [45] and Parrinello-Rahman barostat [46]. We analyzed and reweighted an ensemble consisting of 20,000 structures from this simulation by taking one frame every 100 ps in the simulation (Fig. 4b).

We used Pepsi-SAXS [47] to calculate the SAXS data for each of the extracted structures from the two simulations (Fig. 4c, d). In the case of the Martini simulation we first used standard approaches to reconstruct all-atom models from the CG beads [48]. As Pepsi-SAXS has several free parameters whose values may vary between proteins, we estimated these values from the ensembles. Because

Fig. 4 (continued) the experimental data. (**e**, **f**) Evaluating the effect of the global scaling parameter θ to balance the prior (force field) and the experimental data. For the atomistic simulation we chose $\theta = 30$, and for the coarse-grained simulation we chose $\theta = 100$. (**g**, **h**) Analysis of the effect of reweighting against experimental data on the interdomain distance

optimizing the values from each conformation might lead to substantial overfitting, we instead used an approach where we (for each ensemble) used the average value of such-optimized Pepsi-SAXS parameters over the entire ensemble and reran Pepsi-SAXS with these parameters fixed. We then compared the ensemble-averaged SAXS data from the atomistic and the coarse-grained ensembles with previously determined experimental values [40]. The results show that while both simulations are in reasonably good agreement with experiments, the all-atom simulations appear to provide a better description of the structure and dynamics in sf3636 (Fig. 4c, d).

Despite the good overall agreement, both simulations show systematic discrepancies with experiments, in particular at low-to-intermediate values of q. We thus use the BME procedure to refine the simulations of sf3636 against the SAXS data. Here, the experimental input file and the simulation input file with SAXS calculations for each frame in the ensemble have the following formats.

Experimental file format:

# DATA=SAXS	PRIOR=GAUSS	
q_1	$I(q_1)$	σ_1
q_2	$I(q_2)$	σ_2
⋮	⋮	⋮
q_{148}	$I(q_{148})$	σ_{148}

Simulation SAXS file format:

# label	q_1	⋯	q_{148}
frame_1	$I(q_1)_1^{CALC}$	⋯	$I(q_{148})_1^{CALC}$
frame_2	$I(q_1)_2^{CALC}$	⋯	$I(q_{148})_2^{CALC}$
⋮	⋮		⋮
frame_n	$I(q_1)_n^{CALC}$	⋯	$I(q_{148})_n^{CALC}$

We determine the χ^2 between the calculated and experimental scattering intensities over all scatter vector points, q_i, in the range 0.006–0.3 Å^{-1}. Because of difficulties in estimating or obtaining accurate errors in scattering intensities the absolute value of χ^2 may also be difficult to interpret for a SAXS experiment [49]. Thus, for each ensemble we analyzed the N_{eff}-vs-χ^2 plot to find a value of θ that reflects the compromise between the prior (simulation) and the data (Fig. 4e, f). From these we choose $\theta = 30$ for the all-atom simulation and $\theta = 100$ for the coarse-grained simulation, and the resulting calculated SAXS curves are, as expected, in much better agreement with experiment (Fig. 4c, d).

With the optimized weights it becomes possible to analyze other properties of the conformational ensembles. As an illustration, and following the previous study of this protein [40], we analyzed the distribution of the interdomain distance, quantified as the distance between the centers of mass of the NTD and CTD. The resulting histograms show that the reweighting in general has the effect of increasing the interdomain distances, suggesting that despite recent force field improvements for both coarse-grained and all-atom MD simulations they might still overestimate protein–protein interactions. Thus, the BME approach has the potential for making the resulting ensembles more robust than those from the unbiased simulations, thereby removing some of the uncertainty coming from the imperfect force fields.

5 Notes

1. When using biasing enhanced sampling techniques such as umbrella sampling or Metadynamics, the weights of the prior distribution are not uniform, i.e. $w_j^0 \neq 1/n$. The BME reweighting approach can be applied in this case by specifying the initial weights, as described in the Jupyter notebook: `Notes`, Note 1.
2. The BME software can easily be extended to include additional types of experimental data that can be calculated as the average over the weighted contribution from each frame. Indeed, for data types not explicitly supported it may be possible to use one of the current functionalities as long as the averaging is the same. Experimental data that depend also on global parameters that need to be optimized are, however, currently not supported. Data that depend on temporal correlations (e.g., kinetic data) are also not supported.
3. For NOEs, it is possible to specify upper/lower boundaries instead of average values. In such cases, the restraint is applied only if $\langle F^{calc} \rangle$ is larger/smaller than F^{EXP}. This information can be specified by flagging the experimental data file in the following fashion:

```
# DATA=NOE PRIOR=GAUSS  POWER=6
label_1     F_1^EXP   σ_1 UPPER
label_2     F_2^EXP   σ_2 LOWER
:
label_m     F_m^EXP   σ_m LOWER
```

   A practical example is given in the notebook Jupyter `Notes`, Note 3.

4. Multiple data types (NOE, couplings, chemical shifts, etc.) can be used simultaneously as restraints. A practical example is shown in the Jupyter notebook `Notes`, Note 4.
5. For most types of experiments, the ensemble-averaged value is simply a linearly weighted average over the values calculated for each frame, e.g. for scalar couplings $\langle J^{\mathrm{CALC}} \rangle = \sum_{j=1}^{n} w_j J_j^{\mathrm{CALC}}$. When using NOE data, however, the experimental data has to

be specified as a distance r^{EXP}, and the imposed restraint is proportional to the volume of the corresponding peak in a NOESY spectrum, i.e. $V/c = (r^{\mathrm{EXP}})^{-p} = \sum_{j=1}^{n} w_j (r_j^{\mathrm{CALC}})^{-p}$. The power p is by default set to 6, but it can be set to a different value (e.g., 3 [50]) using the keyword `POWER` in the header of the experimental data file.

Acknowledgements

We thank Dr. Alexander Lemak and Prof. Cheryl H. Arrowsmith for sharing the SAXS data on sf3636. We also thank Yong Wang, Mustapha Carab Ahmed, and Andreas Haahr Larsen for input and testing of BME. The research and development described here were supported by a grant from The Velux Foundations, a Hallas-Møller Stipend from the Novo Nordisk Foundation, and the Lundbeck Foundation BRAINSTRUC initiative.

References

1. Bottaro S, Lindorff-Larsen K (2018) Biophysical experiments and biomolecular simulations: a perfect match? Science 361(6400):355–360
2. Bernadó P, Mylonas E, Petoukhov MV, Blackledge M, Svergun DI (2007) Structural characterization of flexible proteins using small-angle x-ray scattering. J Am Chem Soc 129(17):5656–5664
3. Jensen MR, Communie G, Ribeiro EA, Martinez N, Desfosses A, Salmon L, Mollica L, Gabel F, Jamin M, Longhi S *et al* (2011) Intrinsic disorder in measles virus nucleocapsids. Proc Natl Acad Sci U S A 108 (24):9839–9844
4. Russel D, Lasker K, Webb B, Velázquez-Muriel J, Tjioe E, Schneidman-Duhovny D, Peterson B, Sali A (2012) Putting the pieces together: integrative modeling platform software for structure determination of macromolecular assemblies. PLoS Biol 10(1):e1001244
5. Ward AB, Sali A, Wilson IA (2013) Integrative structural biology. Science 339 (6122):913–915
6. Gaalswyk K, Muniyat MI, MacCallum JL (2018) The emerging role of physical modeling in the future of structure determination. Curr Opin Struct Biol 49:145–153
7. Boomsma W, Ferkinghoff-Borg J, Lindorff-Larsen K (2014) Combining experiments and simulations using the maximum entropy principle. PLoS Comput Biol 10(2):e1003406
8. Pitera JW, Chodera JD (2012) On the use of experimental observations to bias simulated ensembles. J Chem Theory Comput 8 (10):3445–3451
9. Ángyán AF, Gáspári Z (2013) Ensemble-based interpretations of NMR structural data to describe protein internal dynamics. Molecules 18(9):10548–10567
10. Hummer G, Köfinger J (2015) Bayesian ensemble refinement by replica simulations and reweighting. J Chem Phys 143 (24):12B634_1
11. Bonomi M, Heller GT, Camilloni C, Vendruscolo M (2017) Principles of protein structural ensemble determination. Curr Opin Struct Biol 42:106–116
12. Jaynes ET (1978) Where do we stand on maximum entropy. In: The maximum entropy formalism. MIT Press, Cambridge, pp 15–118
13. Cesari A, Gil-Ley A, Bussi G (2016) Combining simulations and solution experiments as a paradigm for RNA force field refinement. J Chem Theory Comput 12(12):6192–6200
14. Bonomi M, Camilloni C, Cavalli A, Vendruscolo M (2016) Metainference: a Bayesian inference method for heterogeneous systems. Sci Adv 2(1):e1501177

15. Dudola D, Kovács B, Gáspári Z (2017) Consensx+ webserver for the analysis of protein structural ensembles reflecting experimentally determined internal dynamics. J Chem Inf Model 57(8):1728–1734
16. Reichel K, Stelzl LS, Köfinger J, Hummer G (2018) Precision deer distances from spin-label ensemble refinement. J Phys Chem Lett 9:5748–5752
17. Köfinger J, Stelzl LS, Reuter K, Allande C, Reichel K, Hummer G (2019) Efficient ensemble refinement by reweighting. J Chem Theory Comput 15(5):3390–3401
18. Rózycki B, Kim YC, Hummer G (2011) Saxs ensemble refinement of ESCRT-III CHMP3 conformational transitions. Structure 19 (1):109–116
19. Boura E, Rózycki B, Herrick DZ, Chung HS, Vecer J, Eaton WA, Cafiso DS, Hummer G, Hurley JH (2011) Solution structure of the ESCRT-I complex by small-angle X-ray scattering, EPR, and FRET spectroscopy. Proc Natl Acad Sci U S A 108(23):9437–9442
20. Bottaro S, Bussi G, Kennedy SD, Turner DH, Lindorff-Larsen K (2018) Conformational ensembles of RNA oligonucleotides from integrating nmr and molecular simulations. Sci Adv 4(5):eaar8521
21. Graf J, Nguyen PH, Stock G, Schwalbe H (2007) Structure and dynamics of the homologous series of alanine peptides: a joint molecular dynamics/nmr study. J Am Chem Soc 129 (5):1179–1189
22. Beauchamp KA, Pande VS, Das R (2014) Bayesian energy landscape tilting: towards concordant models of molecular ensembles. Biophys J 106(6):1381–1390
23. Sanchez-Martinez M, Crehuet R (2014) Application of the maximum entropy principle to determine ensembles of intrinsically disordered proteins from residual dipolar couplings. Phys Chem Chem Phys 16(47):26030–26039
24. Salmon L, Yang S, Al-Hashimi HM (2014) Advances in the determination of nucleic acid conformational ensembles. Annu Rev Phys Chem 65:293–316
25. Leung HTA, Bignucolo O, Aregger R, Dames SA, Mazur A, Bernè che S, Grzesiek S (2015) A rigorous and efficient method to reweight very large conformational ensembles using average experimental data and to determine their relative information content. J Chem Theory Comput 12(1):383–394
26. Olsson S, Strotz D, Vögeli B, Riek R, Cavalli A (2016) The dynamic basis for signal propagation in human pin1-ww. Structure 24 (9):1464–1475
27. Brookes DH, Head-Gordon T (2016) Experimental inferential structure determination of ensembles for intrinsically disordered proteins. J Am Chem Soc 138(13):4530–4538
28. Caticha A (2004) Relative entropy and inductive inference. In: AIP conference proceedings, AIP, vol 707, pp 75–96
29. Cesari A, Reißer S, Bussi G (2018) Using the maximum entropy principle to combine simulations and solution experiments. Computation 6(1):15
30. Jaynes ET (2003) Probability theory: the logic of science. Cambridge University Press, Cambridge
31. Torrie GM, Valleau JP (1977) Nonphysical sampling distributions in Monte Carlo free-energy estimation: umbrella sampling. J Comput Phys 23(2):187–199
32. Laio A, Parrinello M (2002) Escaping free-energy minima. Proc Natl Acad Sci U S A 99 (20):12562–12566
33. Rangan R, Bonomi M, Heller GT, Cesari A, Bussi G, Vendruscolo M (2018) Determination of structural ensembles of proteins: restraining vs reweighting. J Chem Theory Comput 14(12):6632–6641
34. Oliphant T (2006) NumPy: A guide to NumPy. USA: Trelgol Publishing, http://www.numpy.org/ [Online; accessed Oct 2018]
35. Jones E, Oliphant T, Peterson P, *et al* (2001) SciPy: Open source scientific tools for Python. http://www.scipy.org/ [Online; accessed Oct 2018]
36. Tubbs JD, Condon DE, Kennedy SD, Hauser M, Bevilacqua PC, Turner DH (2013) The nuclear magnetic resonance of CCCC RNA reveals a right-handed helix, and revised parameters for AMBER force field torsions improve structural predictions from molecular dynamics. Biochemistry 52 (6):996–1010
37. Ángyán AF, Szappanos B, Perczel A, Gáspári Z (2010) Consensx: an ensemble view of protein structures and nmr-derived experimental data. BMC Struct Biol 10(1):39
38. Bottaro S, Di Palma F, Bussi G (2014) The role of nucleobase interactions in RNA structure and dynamics. Nucleic Acids Res 42 (21):13306–13314

39. Bottaro S, Bussi G, Pinamonti G, Reisser S, Boomsma W, Lindorff-Larsen K (2018) Barnaba: software for analysis of nucleic acid structures and trajectories. RNA. https://doi.org/10.1261/rna.067678.118
40. Lemak A, Wu B, Yee A, Houliston S, Lee HW, Gutmanas A, Fang X, Garcia M, Semesi A, Wang YX, Prestegard JH, Arrowsmith CH (2014) Structural characterization of a flexible two-domain protein in solution using small angle X-ray scattering and NMR data. Structure 22:1862–1874
41. MARTINI3.0 Open-beta (2018). http://www.cgmartini.nl/index.php/force-field-parameters/particle-definitions. Accessed 21 Oct 2018
42. Periole X, Cavalli M, Marrink SJ, Ceruso MA (2009) Combining an elastic network with a coarse-grained molecular force field: structure, dynamics, and intermolecular recognition. J Chem Theory Comput 5(9):1–7. https://doi.org/10.1021/ct9002114
43. Abraham MJ, Murtola T, Schulz R, Páll S, Smith JC, Hess B, Lindah E (2015) Gromacs: high performance molecular simulations through multi-level parallelism from laptops to supercomputers. SoftwareX 1–2:19–25. https://doi.org/10.1016/j.softx.2015.06.001
44. Robustelli P, Piana S, Shaw DE (2018) Developing a molecular dynamics force field for both folded and disordered protein states. Proc Natl Acad Sci U S A 115:E4758–E4766
45. Bussi G, Donadio D, Parrinello M (2007) Canonical sampling through velocity rescaling. J Chem Phys 126:014101
46. Parrinello M, Rahman A (1981) Polymorphic transitions in single crystals: a new molecular dynamics method. J Appl Phys 52 (12):7182–7190
47. Grudinin S, Garkavenko M, Kazennov A (2017) Pepsi-SAXS: an adaptive method for rapid and accurate computation of small-angle X-ray scattering profiles. Acta Crystallogr D 73:449–464
48. Wassenaar TA, Pluhackova K, Böckmann RA, Marrink SJ, Tieleman DP (2014) Going backward: a flexible geometric approach to reverse transformation from coarse grained to atomistic models. J Chem Theory Comput 10 (2):676–690. https://doi.org/10.1021/ct400617g
49. Larsen AH, Arleth L, Hansen S (2018) Analysis of small-angle scattering data using model fitting and Bayesian regularization. J Appl Crystallogr 51(4):1151–1161
50. Tropp J (1980) Dipolar relaxation and nuclear overhauser effects in nonrigid molecules: the effect of fluctuating internuclear distances. J Chem Phys 72(11):6035–6043

Chapter 16

Evaluation and Selection of Dynamic Protein Structural Ensembles with CoNSEnsX$^+$

Dániel Dudola, Bertalan Kovács, and Zoltán Gáspári

Abstract

Understanding protein function at atomistic detail is not possible without accounting for the internal dynamics of these molecules. Ensemble-based models are based on the premise that single conformers cannot account for all experimental observations on the given molecule. Rather, a suitable set of structures, representing the internal dynamics of the protein at a given timescale, are necessary to achieve correspondence to measurements. CoNSEnsX$^+$ is a service specifically designed for the investigation of such ensembles for compliance with NMR-derived parameters. In contrast to common structure evaluation tools, all parameters are treated as an average over the ensemble, if are not themselves an ensemble property like order parameters. CoNSEnsX$^+$ is also capable of selecting a sub-ensemble with increased correspondence to a set of user-defined experimental parameters. CoNSEnsX$^+$ is available as a web server at http://consensx.itk.ppke.hu, and the full Python source code is available on GitHub.

Key words Protein internal dynamics, Ensemble selection, NMR spectroscopy, Chemical shift, Order parameter, Residual dipolar coupling, Scalar coupling

1 Introduction

Description of the internal dynamics of proteins is key to understand how they function. Dynamics can be studied by theoretical methods such as molecular dynamics simulations as well as with experimental approaches like NMR spectroscopy. Today, NMR spectroscopy is the only method capable of providing information on protein dynamics at the atomic/residue level and on multiple timescales [1]. The most fruitful method of the interpretation of NMR-derived parameters is the construction of structural ensembles that are in reasonable agreement with the measurements [2, 3]. Such ensembles can be generated by restrained molecular dynamics or selection/reweighting of a structural ensemble

Electronic supplementary material: The online version of this chapter (https://doi.org/10.1007/978-1-0716-0270-6_16) contains supplementary material, which is available to authorized users.

Zoltán Gáspári (ed.), *Structural Bioinformatics: Methods and Protocols*, Methods in Molecular Biology, vol. 2112, https://doi.org/10.1007/978-1-0716-0270-6_16, © Springer Science+Business Media, LLC, part of Springer Nature 2020

generated using theoretical methods. Chapter 15 of this volume describes BME, a method for reweighting structural ensembles derived from molecular dynamics calculations [4]. This chapter describes CoNSEnsX$^+$, a program for the quick visual evaluation of the correspondence between experimental data and those back-calculated from the ensemble [5]. CoNSEnsX$^+$ is also capable of performing a simple selection of a sub-ensemble that matches the experimental parameters better than the originally submitted one. The CoNSEnsX$^+$ server is designed to work with data readily obtainable from public databases such as the Protein Data Bank and BMRB without further formatting. The server web page contains some scripts that can be used to prepare input files that come from other sources and do not necessarily match all the requirements of the formats accepted. Our source code is available on GitHub (https://github.com/PPKE-Bioinf/consensx.itk.ppke.hu) and can be used for a local install and/or further modifications as needed. To simplify local deployments, a Docker Compose configuration file is published in the source code repository, which will pull and run the CoNSEnsX$^+$ Docker images and will automatically set up the dockerized database connection. This way you can locally deploy a CoNSEnsX$^+$ setup on any host which is capable of running Docker. To parallelize longer calculations, the user is encouraged to edit the Docker Compose configuration file to start up more CoNSEnsX$^+$ containers rather than increasing the load on a single container.

In this chapter, we will use the term "back-calculation" to describe the process where parameters that can be derived from NMR measurements such as chemical shifts, scalar couplings, etc., are estimated/calculated from structural coordinates.

2 Materials

2.1 The CoNSEnsX$^+$ Web Server

The CoNSEnsX$^+$ web server is available at http://consensx.itk.ppke.hu/. Test files are available by clicking on the "Usage" tab on the right. Here, a link on a detailed description page is also available, where a set of Perl scripts that might be helpful in input data preparation are listed.

2.2 CoNSEnsX$^+$ on GitHub

The entire underlining source code to the web server can be found on GitHub under the link https://github.com/PPKE-Bioinf/consensx.itk.ppke.hu or by clicking on the top-right corner on the CoNSEnsX$^+$ web page. The GitHub description page also provides information on the required content and format of the submitted structure and experimental data files.

3 Methods

3.1 Preparation of Input Files

The CoNSEnsX$^+$ method can take three files of input. A PDB format structural ensemble file is mandatory. Please note that at present CoNSEnsX$^+$ supports only monomeric structures, i.e., multichain entries will not be processed. In addition, either an NOE distance restraint list or an NMR parameter file, both in NMR-STAR format, must be supplied to start a calculation (*see* **Note 1** on how files are read in). The input form of the server is shown in Fig. 1. Care should be taken that the PDB, NOE, and/or NMR parameter files have the same atom and residue nomenclature. We recommend that all files conform to the BMRB atom nomenclature as detailed below.

The PDB format file should contain multiple structural models flanked by the "MODEL" and "ENDMDL" keywords. A PDB file downloaded directly from the Protein Data Bank (www.rcsb.org) and representing a structural ensemble determined by solution NMR spectroscopy should usually work fine. Structural ensembles obtained from molecular dynamics simulations or other modeling approaches can also be used provided they contain the MODEL/ENDMDL keywords and have atom names matching the parameter file(s). It is also important that the PDB file should contain explicit hydrogen atoms as they are required for the back-calculation of most NMR parameters from structural coordinates. If needed, explicit hydrogen atoms can be added using a molecular modeling software. Note that the hydrogens added are not necessarily expected to match the nomenclature present in the NOE/NMR parameter files and additional adjustments are needed to ensure compatibility as described below.

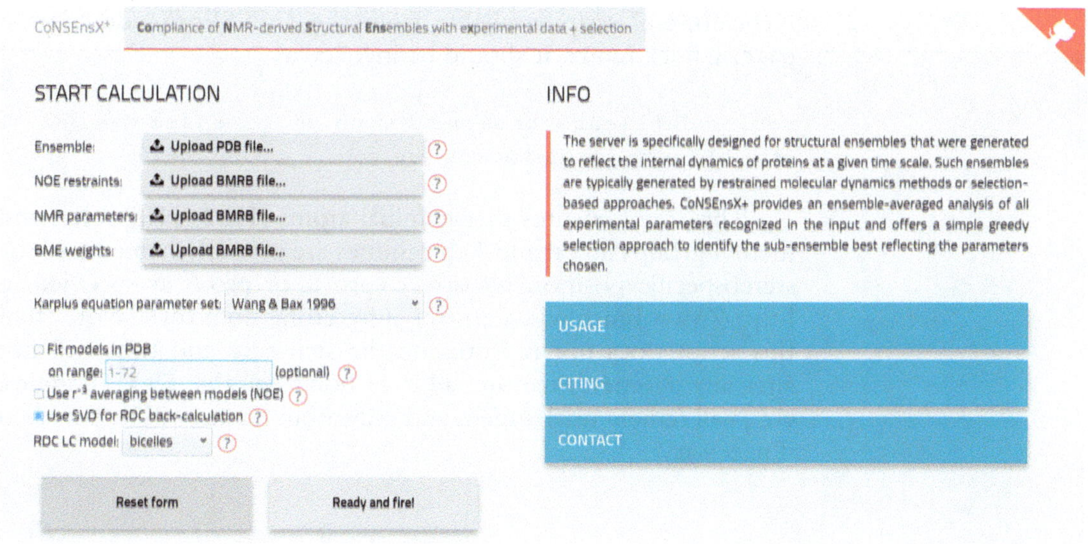

Fig. 1 Input form of the CoNSEnsX$^+$ web server

The NMR parameter file should be a standard NMR-STAR 3.1 format file used by the BMRB database. A file downloaded directly from BMRB (http://www.bmrb.wisc.edu) should be OK. In practice, such a file might not contain all experimental data required for the planned analysis, and the insertion of additional data (e.g., order parameters, scalar couplings, RDCs) might be needed. *See* Chap. 14 on more details of the generation of NMR-STAR format from formats used by common NMR data processing software.

The NOE distance restraint file should also be in NMR-STAR format. Restraint files available from the structure pages of www.rcsb.org denoted "v2 NMR restraints" can be used. Refer to Chap. 14 for tools on data conversion, if necessary.

We strongly recommend visual inspection of the files to make sure that the residue numbering and atom nomenclature of the files match. There are a number of possible atom nomenclatures in use (*see* http://www.bmrb.wisc.edu/ref_info/atom_nom.tbl), and it is not always trivial that these match between the files. Note that there might not be a 100% safe way to ensure such correspondence when using files from an external source (e.g., a database). Below, we provide a schematic protocol to obtain files that are usable with CoNSEnsX$^+$.

The first step is to obtain a PDB format structural ensemble using one of the following steps: Download a structure file derived from a solution NMR experiment from the Protein Data Bank, or generate an ensemble with an MD run or other conformational sampling technique such as Monte Carlo simulation. Save the ensemble in PDB format, making sure that the coordinates for all individual structures are separated by MODEL/ENDMDL keywords. If necessary, add hydrogen atoms with a molecular modeling software (*see* **Note 2** on adding hydrogens). To make sure you have BMRB nomenclature, you might use the pdbfile2bmrbnomenclature.pl script available from the CoNSEnsX$^+$ description page. Under Linux, it should be invoked as:

```
perl pdbfile2bmrbnomenclature.pl < ensemble.pdb >
ensemble_with_bmrb_nomenclature.pdb
```

This script ensures that geminal atoms (HB2/HB3, etc.) and methyl groups in Val and Leu residues are properly named and the stereospecific position is correct (pro-R or pro-S as specified in http://www.bmrb.wisc.edu/ref_info/atom_nom.tbl). Note that this script does not reprotonate the structure and thus will not alter any geometry feature, all X-H bond lengths, H-X-H angles, etc., will remain unchanged, and only atom names will be switched if necessary.

To use the server, an NMR parameter file and/or an NOE distance restraint file in NMR-STAR format is also required. The NOE distance restraint file used to determine the structure deposited in the PDB is usually available from the PDB web site (www.rcsb.org) by selecting the "v2 NMR restraint file" for download on the page of the corresponding structure. All other NMR parameters such as chemical shifts are usually available from the BMRB web page where a search using the corresponding PDB ID should return the corresponding NMR-STAR file(s). PDB-derived v2 NMR distance restraint files and BMRB-derived parameter files use the BMRB nomenclature, and in general they should match the numbering with the PDB-derived coordinate file. However, if any conversions are needed, the atomcoverter_bmrb.pl script can be used, available on the CoNSEnsX$^+$ description page. Care should be taken to specify the correct input format; if in doubt, it is useful to consult the formats detailed on http://www.bmrb.wisc.edu/ref_info/atom_nom.tbl or invoke the script with the "-h" switch that will display the same table extended with additional formats. The script can be used, e.g., to convert from X-PLOR nomenclature to BMRB:

```
atomconveter_bmrb.pl -f XPLOR -t BMRB < input.str > output.str
```

Please note that this script provides a simple mapping based on the nomenclature table. In contrast to PDB files, where the spatial position of hydrogen atoms makes unambiguous stereospecific assignment available, there is no such information in NMR-STAR files. Thus, the exact correspondence between PDB and these files, e.g., the proper naming of a stereospecific NOE restraint, relies on the content of the original the NMR-STAR file, usually compiled by the original depositor and the person performing the conversion.

We strongly advise to check the correspondence of the residue numbers/atom names between the files. Before uploading them to the CoNSEnsX$^+$ server, we recommend to have a look at each of the files using a text editor and to check data for the two or three N-terminal residues. They should have a matching chain ID and residue numbers and the same atom names. Importantly, atoms for which there is data in the NMR-STAR files should have a coordinate line in the PDB file (*see* **Note 3** for rare errors). This step is especially important when the PDB, NOE, and BMRB files come from different sources, e.g., the PDB file is the result of a molecular dynamics simulation, or the BMRB file is generated directly from experimental data and checked against a database-derived structure.

Table 1
NMR-derived parameters supported by CoNSEnsX⁺ and methods for the calculation from coordinates

Parameter type	Atoms/groups supported	Input file with experimental value	Calculation method
Chemical shift	Cα, Cβ, Hα, H, N	BMRB parameter file	SHIFTX [6]
Scalar coupling	$^3J_{HNH\alpha}$, $^3J_{HNC\alpha}$, $^3J_{HNC\beta}$, $^3J_{HNC}$	BMRB parameter file	Based on the Karplus equation with choosable parameters [7, 8]
S^2 order parameter	Backbone (N-H, CA-HA) and side-chain methyl groups	BMRB parameter file	Structures are superimposed using the backbone atoms of the residues specified; then, Eq. 1 in ref. 9 is used on the atoms specified
Residual dipolar coupling	Any supported by PALES	BMRB parameter file	PALES [10]
NOE distance	Hydrogen atoms and groups	NOE distance file	PRIDE-NMR [11] and individual distance calculations

3.2 Brief Overview of Calculating NMR Parameters from the Structures in the Ensemble Submitted

The CoNSEnsX⁺ web server calculates NMR parameters for each structure in the submitted PDB ensemble. Types of parameters and calculation methods are listed in Table 1. In principle, all supported parameters are calculated for all atoms and atom sets that have a corresponding experimental value in the input BMRB file. For the NOE distance list, distances for all atom/group pairs listed are calculated.

All calculated parameters are averaged for the full ensemble. For all parameters except NOEs, single arithmetic averaging is used (*see* **Note 4** on RDCs). For NOE distances, r^{-6} averaging is used across members of the structural ensemble unless the "use r^{-3} averaging" box is checked on the input page. For each structural model, r^{-6} averaging is used to calculate effective distances involving more than two atoms, e.g., those involving methyl groups.

3.3 Output of the CoNSEnsX⁺ Server for Ensemble Evaluation

The result sheet is assigned a calculation ID shown at the top of the page, and using this the results can be later accessed as a permalink for a month. The main part of the result page details the correspondence of the ensemble to the experimental data. Order parameters that can only be interpreted as an ensemble property and NOE violations are also calculated on the full ensemble (*see* below). The correspondence to order parameters, chemical shifts, and coupling constants is measured as correlation of experimental and back-calculated data as well as an RMSD value. For RDCs, a Q-factor is also provided. For these types of parameters, three kinds of plots are generated (Fig. 2):

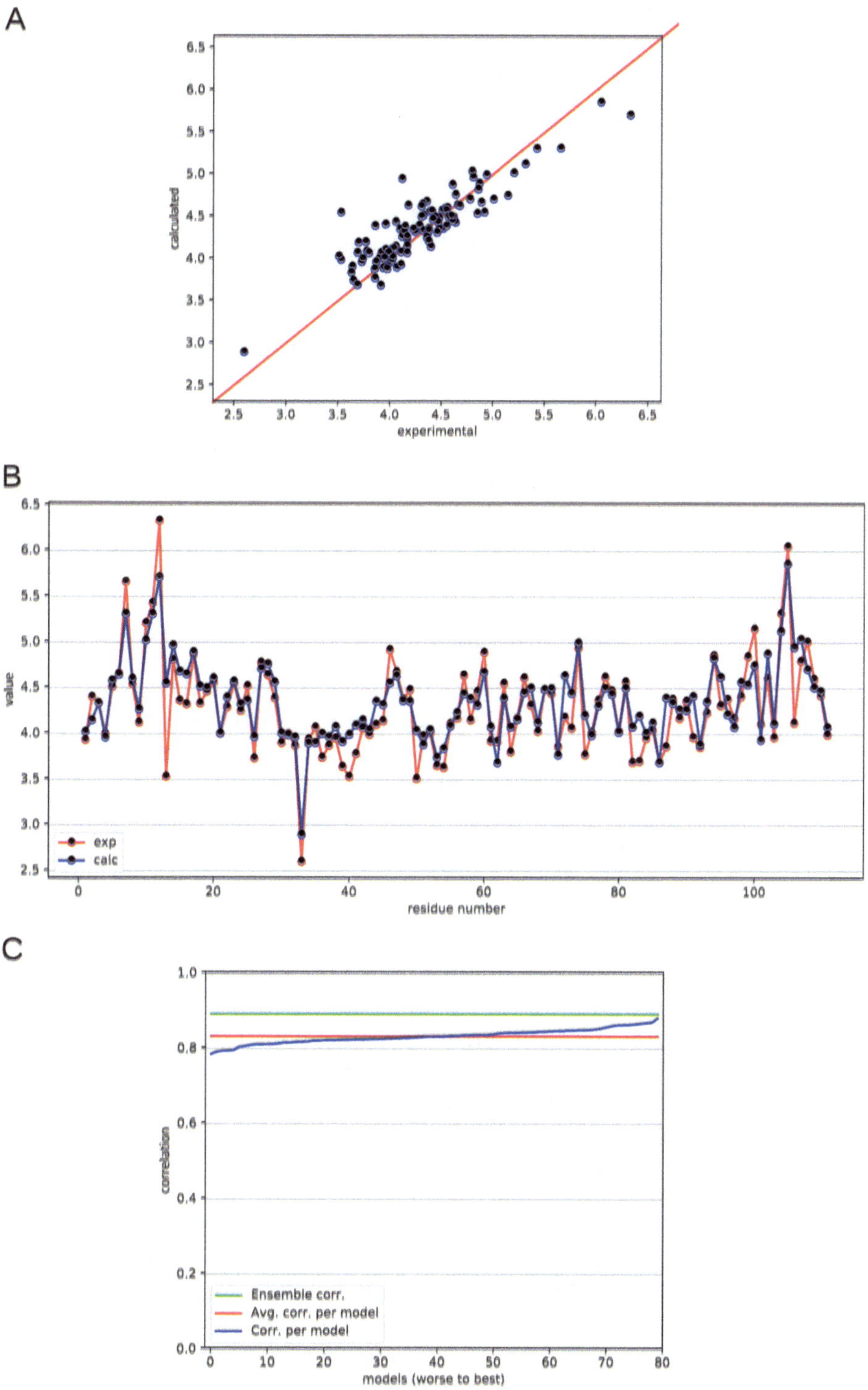

Fig. 2 Example output diagrams of the CoNSEnsX$^+$ web server, shown for Hα chemical shifts of the original ensemble of the example described in Subheading 4. (**a**) Correlation plot of experimental and back-calculated parameters. (**b**) Experimental and back-calculated parameters as a function of the sequence. The line connecting the points only serves visualization purposes. (**c**) Plot of the correlations between experimental and back-calculated data for each structure in the ensemble (blue line, ordered from worst to best) and the full ensemble (back-calculated parameters averaged over all models and correlated with the experimental data, green line). The average of per-model correlations is shown in red

1. A plot showing the experimental and the ensemble-wise back-calculated parameters for each residue. This plot helps to identify the regions where the structures do not correspond well to the experiments. It might not be expected that all factors contributing to the experimentally observed values are recaptured in a given ensemble. Outliers can be excluded from further calculations and analysis or might indicate the need to recalculate or refine the ensemble.
2. A correlation plot for the same data as in the plot above. This can further help in identifying outliers and systematic differences between experimental and recalculated data.
3. A plot showing the correlation of the experimental and back-calculated data for each structure in the ensemble. This plot also shows the average of these correlations and the correlation of the parameters obtained for the full ensemble (*see* **Note 5**). In this way, the user can judge how the ensemble representation improves the correspondence to experimental data over single structural models.

In the case of NOE restraints, the PRIDE-NMR score for each structural model in the ensemble is returned as well as a diagram on distance violations. However, the latter should be treated with caution as NOE restraint lists in the PDB can contain scaled distances depending on the structure calculation software used (*see* **Note 6**).

The CoNSEnsX$^+$ server does not return a single measure of ensemble goodness. This is because we think that it would not be straightforward to combine the correspondence to experimental data into one number, given the differences in the types, amount, and precision of the available NMR parameters for different proteins. In addition, an ensemble with a low overall score might still reflect some parameters at a satisfactory level and still be suitable to explain mechanistic features about the function of the given protein.

3.4 Using the Selection Feature of CoNSEnsX$^+$

After the first round of evaluation, CoNSEnsX$^+$ offers a sub-ensemble selection feature. This means that from the pool of the structures in the original ensemble, a smaller set can be selected that might correspond to the experimental parameters better than the original ensemble. The user can select the parameters to be included in this selection along with their relative weight on a scale of 0–10. There is an option for "bulk selection" to include parameters of the same type (e.g., chemical shifts) with a single button, which can naturally be modified for each parameter. The measure of correspondence should also be chosen here.

The selection procedure itself is a variant of a deterministic greedy approach where the best individual structure from the ensemble is selected first and additional conformers are added gradually until the overall correspondence cannot be increased further. The algorithm is capable of overcoming steps that actually decrease the correspondence provided that subsequent additions will result in a still better ensemble. This can be set with the "Overdrive" parameter on the selection input page. *See* **Note 7** for the interpretation of such selection results in more detail.

The results of the selection are summarized in a table where the values of correspondence for the user-defined set of parameters are shown for the original ensemble and the selected one. The selected sub-ensemble is available for download as a PDB format file, in which the individual structures can be uniquely mapped to those in the original ensemble as the model numbers are retained for them in the MODEL records of the PDB file. Also, the numbers of the selected models are listed at the beginning of the PDB file in a REMARK record.

A principal component analysis on the original ensemble is performed (*see* **Note 8** on details on this), and three plots showing the distribution of structures along the first four principal coordinates, 1–2, 2–3, and 3–4, are shown. On this plot, the points corresponding to the structures retained in the selected sub-ensemble are highlighted; thus, the user can visually investigate how well the sub-ensemble maps the conformational space represented by the original set of structures.

3.5 BME Support in CoNSEnsX$^+$

The current version of CoNSEnsX$^+$ provides basic support for BME applications (Chap. 15). Simple input generation for BME is performed: after a round of evaluation, input files containing chemical shifts and scalar couplings are generated that can be used in a subsequent BME calculation. In addition, the user might provide a set of BME-derived weights along with the input ensemble, and in this case the weighted averages are calculated and reported for these parameters. Note that these features do not provide full compatibility with BME but are intended to help the user evaluate different approaches to generate ensembles corresponding to experimental data.

4 Example Application

As an example, here we provide a test case derived from our previous work on parvulin-type peptidyl-proline isomerases. The protein chosen is *Staphylococcus aureus* PrsA, for which we have generated a structural ensemble [12] intended to reflect its fast (ps-ns) motions by incorporating backbone S^2 order parameters into multi-replica simulations according to the MUMO protocol [13]. For the purposes of this example, we have selected the first 80 conformers of this ensemble. It should be noted that these structures do not represent the structural differences deduced from the combined ensembles analyzed in [12]. We have omitted the first seven residues from the N-terminus, as these are the most mobile regions and dominate the motions detected by PCA obscuring functionally more relevant internal motions. We use an NMR-STAR file that is a modified version of BMRB entry 15628, simplified to contain only the chemical shifts but with the backbone S^2 restraints added. Data for the first seven residues were also deleted from this file.

For the example calculation, the user can upload the ensemble in PDB format, 2jzv_mumo_test.pdb along with the BMRB str file, 2jzv_test.str, to obtain information about the correspondence of the ensemble to the parameters. These files are available as supplementary material to the chapter.

Uploading these files to the server and pushing the "Ready and fire!" button will initiate the calculations. The results reveal that the ensemble shows an excellent correspondence to the experimental S^2 data with a correlation of 0.916 while reasonably reproducing the chemical shifts. The chemical shifts of different atom types do not depend on the structural features to a comparable extent [6], for example, the Cβ chemical shifts are largely determined by the residue type. In addition, the accuracy of shift predictions is also not uniform for all atom types. Thus, we propose to focus primarily on Hα and Cα shifts. In their case, the ensemble-wise correlation is higher than for any individual member of the ensemble, and this is more apparent for Hα shifts due to the higher deviation between individual structures (Fig. 2).

On the results page, clicking on "Toggle selection options" will display the controls for the greedy selection. For this tutorial, we shall choose "Correlation" as compliance measure and click "chemical shifts" on the bulk select line. This will select all types of chemical shifts available with a uniform weight of 10. Clicking on "Start selection" initiates the process. In this case, nine structures are selected with improved correspondence to all but the Cα and Cβ chemical shifts, for which numerically the same excellent correlation is obtained. The PCA plots below reveal that the conformational variability of the original ensemble along PC modes 1, 2, and 3 is still reasonably well captured by the selected one, despite having only nine members (Fig. 3a). These motions primarily affect the loops around the substrate-binding cleft (Fig. 3b). This is supported by our analysis performed with ProDy [14] and VMD [15] showing the overlap between the principal components obtained independently for the original and the selected ensembles (Fig. 3c). The correlation to the backbone S^2 data, which can be judged by another analysis by uploading the selected nine structures along with the original BMRB file, drops to 0.727, a value

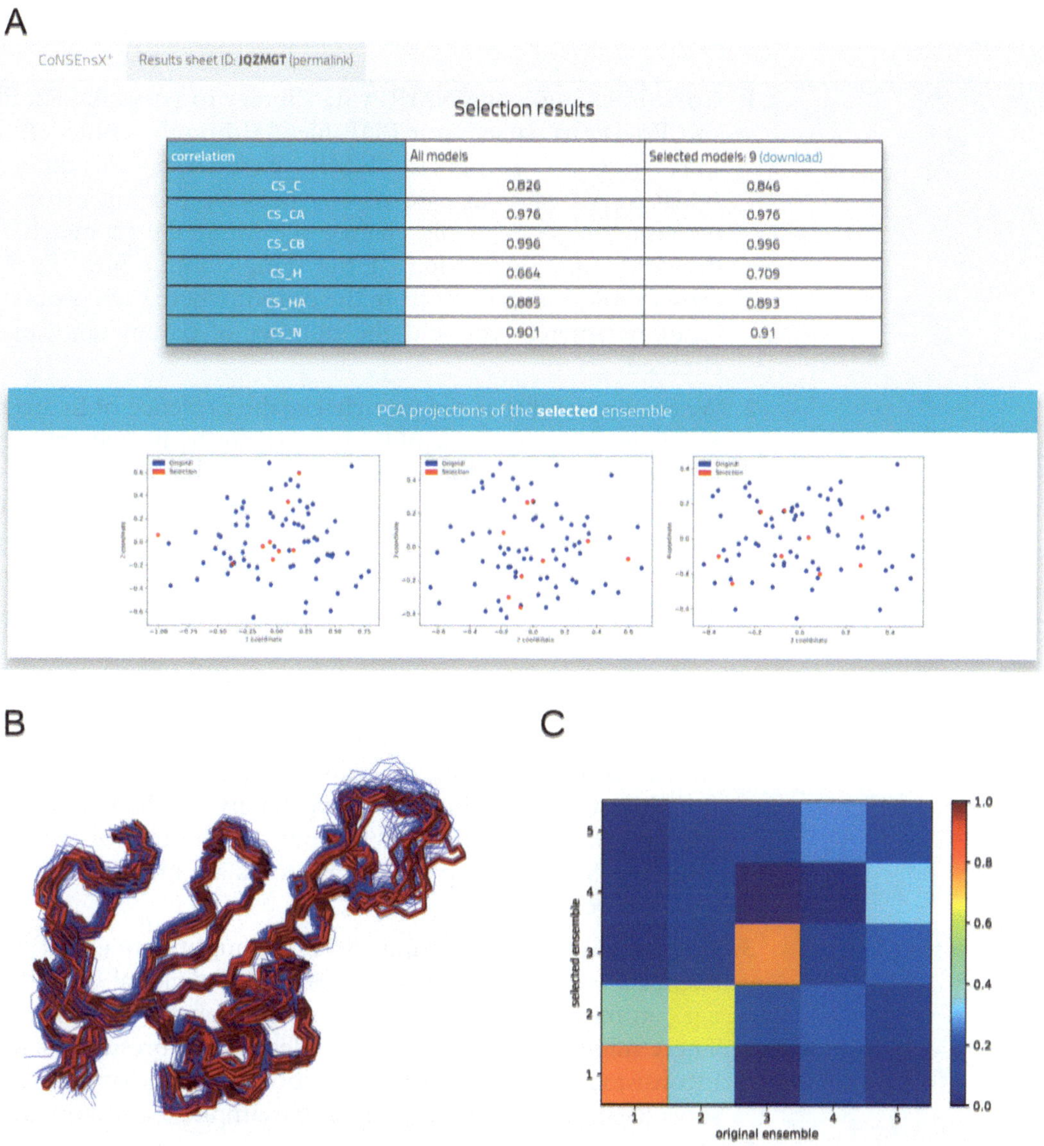

Fig. 3 Results of the ensemble selection described in Subheading 4. (**a**) Selection summary output provided by the CoNSEnsX$^+$ web server. The table shows the correspondence of the experimental parameters in the original and the selected ensemble, and the panels depict the result of a principal component analysis on the original ensemble, with the points corresponding to the selected structures highlighted in red. (**b**) Superimposed structures in the original (blue) and selected (red) ensembles. Figure prepared using MOLMOL [17]. (**c**) Overlap between principal components of the original and the selected ensemble based on an independently performed analysis. The first three modes show reasonable agreement. Figure prepared using ProDy [14]

that does not reach that of a properly restrained ensemble [16] but still indicates reasonable correspondence that is definitely higher than that expected for a common PDB-derived or unrestrained ensemble [12].

5 Notes

1. CoNSEnsX$^+$ uses the NMRPyStar library to parse BMRB files and ProDy to parse input PDB files. Although serious efforts were put to recognize NMR parameters in different NMR-STAR files, in case of any possible parsing error, we recommend to manually change the input file to match the keywords and formatting of the sample input files. Parsing errors should be evident from the output where each processed input parameter type and the number of parameters can be checked.

2. We recommend that the user checks the presence of hydrogen atoms in the input ensemble. If there are hydrogens missing, we suggest the use of a molecule modeling tool to add all hydrogen atoms. The pdb2gmx program in GROMACS might be a good choice, although by default it only works on the first model of an ensemble. Therefore, we recommend the use of our script protonate_ensemble.pl, downloadable from the CoNSEnsX$^+$ downloads page, by invoking:

```
protonate_ensemble.pl < in.pdb > out.pdb
```

   This script also invokes pdb2bmrbnomeclature.pl to ensure correct stereospecific names of the protons added.

   Note that the addition of hydrogens usually changes the atom numbering in the PDB file, but this should not cause any problems in processing the file as atoms are identified by residue number and type.

3. In some rare cases, the input PDB file might contain different chemical structures for some of the individual structures, e.g., atoms might be missing from one or more of the models. There might be no trivial way to identify the presence of such inconsistencies besides visual inspection of the file or testing the service with a smaller ensemble. For example, if a sub-ensemble containing the first two to three structures of the original ensemble works fine but the full ensemble is not processed by the server, it might well indicate that there are missing atoms in some of the models in the full original ensemble.

4. The alignment tensors are estimated separately for each model, as this might be different for them according to their overall structure and shape. PALES is invoked separately on each model. If the SVD option is checked as is the default, the alignment tensors will be optimized by PALES for best fit to the experimental RDCs. Without this option, PALES will estimate the alignment by steric considerations. Averaging of RDCs is currently done by calculating a simple arithmetic average of the values and not the alignment tensors. This is

analogous to the averaging of scalar couplings but differs from the calculation method available in GROMACS, for example.

5. Although it might not be trivial at first sight, the correlation of parameters averaged over structures is different from the average of the correlations obtained for individual structures. Usually the ensemble averaging results in a higher correlation than that obtained for the best single structure and is practically always higher than the average of structure-wise correlations.
6. Some NMR structure calculation software can use (or use by default) the "sum of r^{-6}" distance calculation method for restraints involving multiple atom pairs. This approach results in a lower effective distance than the lowest of all possible atom-atom distances, and, therefore, the corresponding distances are scaled down in the distance restraint list [16]. This primarily affects distances involving methyl groups. However, the distance restraint files usually do not contain explicit information on whether such scaling was applied or not. Using lists with scaled distances will result in incorrectly detected violations as CoNSEnsX^{+} uses the conventional r^{-6} averaging method resulting in larger effective distances. Currently, there is no error-prone way to handle this discrepancy in general. This, however, does not affect the PRIDE-NMR calculation as in that the actual distances listed in the restraint file are not explicitly used, only the presence of the restraints matters.
7. The greedy selection algorithm has the advantage of being deterministic and the disadvantage of not necessarily being able to find the optimum. However, the latter feature is shared with many other approaches. This compromise is made in order to achieve a reasonably short computation time at the expense of finding the global optimum. In practice, a sub-ensemble with better correspondence might exist and could be found with a different approach (e.g., a stochastic one). Thus, the selected sub-ensemble should be regarded as one that represents a lower limit of correspondence that can surely be achieved by some combination of the structures in the original ensemble. Another related aspect is that it returns the smallest ensemble with a good correspondence to experimental data, thereby allowing the assessment of the presence of potential overfitting, the phenomenon where the number of structures in the original ensemble is much higher than required to reflect the experimental measurements.
8. The principal component calculation in the server is done by a module of ProDy. The plots are generated for the full ensemble with the dots corresponding to the selected structures are highlighted. These should not be confused with a PC analysis performed on the selected structures alone as in those the dominant motions might be different and the correspondence

of these to the dominant motions in the original ensemble requires separate analysis, as done for the example described.

Acknowledgments

The authors acknowledge the support of the National Research, Development and Innovation Office (NKFIH) through grant no. NN124363 (to Z.G.). The research has been carried out within the project Thematic Research Cooperation Establishing Innovative Informatic and Info-communication Solutions, which has been supported by the European Union and co-financed by the European Social Fund under grant number EFOP-3.6.2-16-2017-00013.

References

1. Kovermann M, Rogne P, Wolf-Waltz M (2016) Protein dynamics and function from solution state NMR spectroscopy. Q Rev Biophys 49:e6
2. Ángyán AF, Gáspári Z (2013) Ensemble-based interpretations of NMR structural data to describe protein internal dynamics. Molecules 18:10548–10567
3. Nussinov R (2016) Introduction to protein ensembles and allostery. Chem Rev 16:6263–6266
4. Bottaro S, Bengtsen T, Lindorff-Larsen K (2019) Integrating molecular simulation and experimental data: a Bayesian/maximum entropy reweighting approach. In: Gáspári Z (ed) Structural bioinformatics, Methods in molecular biology, vol 2112, pp 219–238. Springer, New York
5. Dudola D, Kovács B, Gáspári Z (2017) CoNSEnsX+ webserver for the analysis of protein structural ensembles reflecting experimentally determined internal dynamics. J Chem Inf Model 57:1728–1734
6. Neal S, Nip AM, Zhang H, Wishart DS (2003) Rapid and accurate calculation of protein ^{1}H, ^{13}C and ^{15}N chemical shifts. J Biomol NMR 26:215–240
7. Habeck M, Rieping W, Nilges M (2005) Bayesian estimation of Karplus parameters and torsion angles from three-bond scalar couplings constants. J Magn Reson 177:160–165
8. Wang AC, Bax A (1996) Determination of the backbone dihedral angles φ in human ubiquitin from reparametrized empirical Karplus equations. J Am Chem Soc 118:2483–2494
9. Best RB, Vendruscolo M (2004) Determination of protein structures consistent with NMR order parameters. J Am Chem Soc 126:8090–8091
10. Zweckstetter M (2008) NMR: prediction of molecular alignment from structure using the PALES software. Nat Protoc 3:679–690
11. Ángyán AF, Perczel A, Pongor S, Gaspari Z (2008) Fast protein fold estimation from NMR-derived distance restraints. Bioinformatics 24:272–275
12. Czajlik A, Kovács B, Permi P, Gáspári Z (2017) Fine-tuning the extent and dynamics of binding cleft opening as a potential general regulatory mechanism in parvulin-type peptidyl prolyl isomerases. Sci Rep 7:44504
13. Richter B, Gsponer J, Varnai P, Salvatella X, Vendruscolo M (2007) The MUMO (Minimal Under-Restraining Minimal Over-Restraining) method for the determination of native state ensembles of proteins. J Biomol NMR 37:117–135
14. Bakan A, Dutta A, Mao W, Liu Y, Chennubhotla C, Lezon TR, Bahar I (2014) Evol and ProDy for bridging protein sequence evolution and structural dynamics. Bioinformatics 30:2681–2683
15. Humphrey W, Dalke A, Schulten K (1996) VMD – Visual Molecular Dynamics. J Mol Graph 14:33–38
16. Strotz D, Orts J, Chi CN, Riek R, Vögeli B (2017) eNORA2 exact NOE analysis program. J Chem Theory Comput 13:4336–4346
17. Koradi R, Billeter M, Wüthrich K (1996) MOLMOL: a program for display and analysis of macromolecular structures. J Mol Graph 14:51–55

Index

Zoltán Gáspári (ed.), *Structural Bioinformatics: Methods and Protocols*, Methods in Molecular Biology, vol. 2112, https://doi.org/10.1007/978-1-0716-0270-6, © Springer Science+Business Media, LLC, part of Springer Nature 2020

MIX
Papier aus verantwortungsvollen Quellen
Paper from responsible sources
FSC® C105338

If you have any concerns about our products,
you can contact us on
ProductSafety@springernature.com

In case Publisher is established outside the EU,
the EU authorized representative is:
Springer Nature Customer Service Center GmbH
Europaplatz 3, 69115 Heidelberg, Germany

Printed by Libri Plureos GmbH
in Hamburg, Germany